The Reticulocyte

T0174883

Author

Samuel M. Rapoport, M.D., Ph.D., Dr.h.c.
Professor Emeritus of Biochemistry
Humboldt University
Berlin, G.D.R.

CRC Press
Taylor & Francis Group
Boca Raton London New York

CRC Press is an imprint of the
Taylor & Francis Group, an **informa** business

CRC Press
Taylor & Francis Group
6000 Broken Sound Parkway NW, Suite 300
Boca Raton, FL 33487-2742

Reissued 2019 by CRC Press

A Library of Congress record exists under LC control number:

Publisher's Note
The publisher has gone to great lengths to ensure the quality of this reprint but points out that some imperfections in the original copies may be apparent.

Disclaimer
The publisher has made every effort to trace copyright holders and welcomes correspondence from those they have been unable to contact.

ISBN 13: 978-0-367-23689-2 (hbk)
ISBN 13: 978-0-367-23690-8 (pbk)
ISBN 13: 978-0-429-28117-4 (ebk)

Visit the Taylor & Francis Web site at http://www.taylorandfrancis.com and the
CRC Press Web site at http://www.crcpress.com

PREFACE

The attraction and difficulty in writing this monograph has been the unusual circumstance that its object, the reticulocyte, represents an intermediate stage of development of cell type. Even though it can be well defined by morphological and biochemical characteristics, the continuity of the differentiation process must render all delimitations unsharp and conditional. For this reason I felt the need to include an account of the differentiation stages leading up to the reticulocyte, and also to refer continuously to the erythrocyte as final stage of maturation.

The reticulocyte has attracted attention from various points of view. It has been a popular object of students of differentiation and a continuous focus of concern for hematologists. From a biochemcial point of view, it offers several advantages among which the availability in large amounts is perhaps foremost. It has been extensively used for the study of the system of protein synthesis and its components as well as the cell membrane with its receptors and transporters. The relative simplicity of the metabolic network as compared with other respiring cells has permitted deeper insight into the interrelations of metabolic pathways and largely quantitative answers. The maturation of the reticulocyte involves mainly degradation processes by which mitochondria, ribosomes, and parts of the cell membrane disappear. Similar processes must occur in all types of cells but are masked by the renewal of cell components. Thus, the reticulocyte offers the exceptional opportunity to study extent and mechanism of physiological breakdown of organelles uncontaminated by their renewal.

This is the first monograph to deal with the reticulocyte. Short and incomplete reviews are out of date, since the last one appeared more than 10 years ago. The primary literature is widely dispersed in biological, morphological, experimental, biochemical, and clinical journals of various disciplines. This monograph attempts to cover properties, molecular biology, and metabolism of the reticulocyle as well as the mechanisms of its maturation, mainly from a biological point of view. I hope that it will provide a background of information for researchers in a variety of fields and stimulate them to fill the many gaps of knowledge and understanding that still exist. Such endeavors should be particularly rewarding since much of the work on reticulocytes was performed long ago with inadequate methods and before the period of modern molecular and cell biology.

I feel impelled to express my sincere gratitude to all the people who helped me in various ways, first of all to Inge, my wife, who was my chief consultant and executant of my thoughts. I am most grateful to Drs. C. Schewe and B. Härtel, and to Mrs. G. Ryssowski for assistance in preparing the manuscript. I owe great thanks to all my co-workers, past and present, who participated with me in the adventures of research in the field and to colleagues all over the world with whom I feel connected by common interests and personal friendships.

THE AUTHOR

Samuel Rapoport, M.D., Ph.D., Dr.h.c., Dr.h.c., is Professor Emeritus of Biochemistry at Humboldt University, Division of Medicine, Berlin, German Democratic Republic. He obtained his M.D. from the University of Vienna in 1936, and his Ph.D. from the University of Cincinnati, Ohio in 1939.

Dr. Rapoport is a member of numerous scientific societies, including the American Society of Biological Chemists, and an honorary member of several societies. Dr. Rapoport is a member of the Academy of Sciences of the G.D.R. and honorary member of the Hungarian Academy of Sciences. He has received honorary degrees from Semmelweis Medical University in Budapest and the University of Leipzig. Currently he is a member of the Executive Committee of the International Union of Biochemistry.

The longstanding and current interests of Dr. Rapoport are the enzymology and the metabolism of red cells and their maturation processes. He has published several hundred research papers and has authored a textbook of medical biochemistry, which is now in its eighth edition, as well as several invited reviews. He was co-editor of the monograph *Cellular and Molecular Biology of Erythrocytes*. He has lectured widely at national and international meetings and has organized, among others, ten international symposia on the "Structure and Function of Erythroid Cells".

INTRODUCTORY REMARK: DIFFERENTIATION AND MATURATION

Differentiation is nowadays generally conceived to represent a program of gene expression, primarily determined by regulation of transcriptional processes. Maturation is a term often used to designate terminal differentiation, without discrimination of its phases. In some cell types one can distinguish a subprogram which involves practically entirely posttranscriptional events. During the differentiation of the erythron, the activity of the nucleus ceases after the stage of the basophilic erythroblast, with ensuing pycnosis of the nucleus; new formation and export of ribonucleic acids also come to an end. The transition from the basophilic to the orthochromatic erythroblast defines clearly the end of the transcriptional phase of differentiation and the beginning of the posttranscriptional phase. It seems appropriate to denominate the posttranscriptional program and I propose to reserve the term "maturation" for it. This classification will be used throughout in this monograph.

INTRODUCTORY REMARK: DIFFERENTIATION AND MATURATION

TABLE OF CONTENTS

BIOLOGY OF ERYTHROID CELLS AND RETICULOCYTES

METABOLISM OF RETICULOCYTES

Biology of Erythroid Cells and Reticulocytes

Chapter 1

SURVEY OF EARLY ERYTHROID DIFFERENTIATION

I. INTRODUCTION

Present-day research is centered on elucidation of the intrinsic and extrinsic factors instrumental in the initiation and realization of the differentiation program.

Erythroid differentiation may well be the best model for the study of general problems of differentiation, even though it has undoubtedly specific features. Among the advantages of the erythroid model are (1) the cell specialization with the predominance of one terminal product, hemoglobin, which comprises about 95% of the final complement of proteins; (2) the existence of a series of stages which can be recognized by morphological, biochemical, and immunological methods; (3) the possibility to obtain considerable numbers of differentiating cells from discrete sites; (4) the development of cell lines in which erythroid maturation has been arrested. These cells are inducible to further differentiation by a variety of chemical agents; (5) the fact that hemoglobin is the product of more than one gene and requires for its synthesis coordinate expression of several genes; also in many species there occur during ontogeny shifts in the type of populations of red cells and of hemoglobins (switching); (6) the regulation of red cell production by a defined hormonal system.

Among the problems that are being approached are (1) the nature of commitment to a particular cell type; (2) the coordination of multigene families in ontogeny and differentiation; (3) the structural arrangement of simultaneously or sequentially expressed genes; (4) the nature of gene activation and suppression; (5) the relationship between cell division, DNA synthesis, and the transitions between differentiation stages; (6) synthesis, processing, unmasking, and stability of RNA during differentiation; (7) the relation between differentiation and malignant transformation.

Several critical steps may be discerned in the differentiation programs of erythroid cells. They include: (1) proliferation of the pluripotent hematopoietic stem cells; (2) commitment of the stem cells to erythropoiesis; (3) proliferation of the committed erythroid precursors, a phase which includes several sequential stages; and (4) realization of the program characteristic of terminal differentiation.

II. STEM CELLS AND OTHER PROGENITORS

It is well known that most types of blood cells, including erythrocytes, thrombocytes, granulocytes, megakaryocytes, and monocytes, are derived from one common stem cell compartment by multistage processes. It is customary to divide the progenitor cells into several subcompartments. The most primitive pluripotential hemopoietic stem cells, which are capable both of cell replication and of generating the subsequent stages, i.e., cells committed to differentiation, are the least well defined. The compartment of committed stem cells has become accessible by the development of in vitro and in vivo clonal assays for the different cell types (see Ogawa et al[1]). In this manner cell lineages were found with more than one potential both from human and mouse sources. Of particular interest as candidates for the pluripotential stem cells were the CFU-GEMM (granulocytes, erythrocytes, macrophages, megakaryocytes).[2-4] Some observations suggest that even CFU-S and CFU-GEMM cells are still relatively mature pluripotent progenitors and that there exist truly primitive stem cells which are characterized by their extensive repopulating ability of bone marrow.[1]

Only recently success was reported in the identification of a unique type of hemopoietic colony of mice which consisted only of undifferentiated blast cells which may be the long

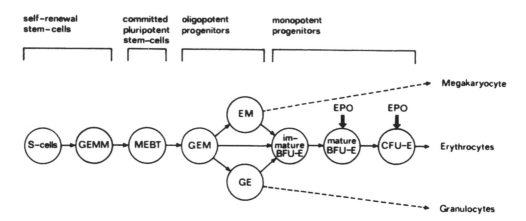

FIGURE 1.   Schematic survey of differentiation sequence of erythroid cells. S-cells, uncommitted, undifferentiated stem cells; GEMM, colony-forming unit, granulocyte, erythrocyte, macrophage megakaryocyte; MEBT, megakaryocyte, erythrocyte, B- and T-lymphocyte; GEM, granulocyte, erythrocyte, megakaryocyte; EM, erythrocyte, megakaryocyte; GE, granulocyte, erythrocyte; BFU-E, burst-forming unit, erythrocyte; CFU-E, colony-forming unit, erythrocyte; EPO, erythropoietin.

sought-after primitive stem cell.[5] These cells show no sign of terminal differentiation and appear to be the precursors of the CFU-GEMM cells. The presumptive true stem cells have both the capacity of self-renewal and of differentiation. Clonal analysis indicates that the decision of a given cell either to enter the pathway of differentiation or of self-renewal may be a random event.[6] A schematic survey of a possible sequence of the differentiation of hemopoietic stem cells based on culture assays is presented in Figure 1.

The various progenitors depend on a variety of so-called colony-stimulating factors and modulators, the discussion of which is beyond the scope of this survey. The most primitive monopotent class of erythroid progenitor cells are called BFU-E (burst forming unit erythroid) and apparently consist of two subclasses. The immature BFU-E require about 2 to 3 weeks to give rise to large colonies containing a thousand or more cells, whereas the more mature BFU-E, which are a transitory class leading to the next type of progenitor cell, the CFU-E (colony forming unit erythroid), develop more rapidly. The CFU-E cells form small colonies of no more than 60 hemoglobin-containing cells in about 7 days of culture. The immature BFU-E are not yet sensitive to erythropoietin but require one or more so-called colony-stimulating factors for growth and differentiation. Evidence for such factors was found in mitogen-stimulated spleen cells of mice, in the supernatant fluids of human bone marrow cells, in T-lymphoblasts and monocyte cell lines, activated T-cells, and various tumor cell lines.[7-9] Some of these factors have been purified to homogeneity.[10-13] Whereas one factor appears to have a selective erythroid-potentiating activity,[10] others are pluripotent and support the growth of mixed colonies.[11,12] From the available information it apppears that the factors are glycoproteins of low molecular weight. It is noteworthy that interleukin 3, originally presumed to be a growth factor for leukocytes, has turned out to be also a potent promoter of erythroid burst information and thus belongs to the group of pluripotent colony-stimulating factors. Recent work indicates that one of the effects of burst-promoting factors may be the maintenance of a high ATP concentration.[14]

The mature BFU-E are responsive both to this type of factor and to erythropoietin. However, CFU-E are far more sensitive to the latter: whereas the CFU-E respond to erythropoietin in a range of 1 to 100 mU/mℓ, the dosage required for mature BFU-E is 20- to 50-fold higher. Erythropoietin and colony-stimulating factor appear to interact with a common target cell. In rat or mouse marrow cell cultures, colony-stimulating factor suppressed erythropoietin-stimulated synthesis of hemoglobin, whereas erythropoietin suppressed the formation of colony-stimulated formation of granulocyte-macrophage colonies.[15] With the success of Nijhof et al.[16] in isolating practically pure CFU-E from mouse spleen cells it became possible to demonstrate that erythropoietin acts in the G1 phase of the cell cycle. Its primary effect is to trigger the erythroid progenitor cell to undergo proliferation followed by initiation of further steps of differentiation, including synthesis of globin.

III. DIFFERENTIATION MODELS

The early work was mainly carried out on bone marrow of man, mice, rats, and rabbits or on fetal liver. Recent work has concentrated on transformed erythroid cells in view of the limitations of physiological systems, such as their complexity, the variety of regulatory factors, and the difficulties in maintaining the cultures. The first transformed line to be used was the mouse erythroleukemia cell (Friend cell).[17] It can easily be maintained in continuous culture, exhibits a low level of spontaneous differentiation, and may be induced to differentiate by a variety of compounds such as dimethyl sulfoxide,[17] butyric acid,[18] hemin, and others (reviewed by Marks and Rifkind[19]). Upon exposure to the inducers the apparent block of differentiation is overcome and a variety of erythroid markers are expressed in a coordinate manner. One of these (besides globin mRNA) is spectrin, which is a major component of the skeleton of the erythrocyte membrane,[20,21] 5-ALA synthase,[19,20,22-26] and other early enzymes of heme synthesis; the p 53 protein also increases. With the induction event there is a prolongation of the G1 phase and a decrease in the transcription rate of ribosomal DNA sequences[26-32] but also of those encoding vimentin.[33]

What stage of the differentiation program is represented by the Friend erythroleukemic cell? The weight of evidence indicates that this cell corresponds to the CFU-E, already fully committed to erythroid development even before induction. This conclusion is based on the presence of low levels of globin mRNA,[34,35] spectrin,[20] high levels of carboanhydrase, acetylcholinesterase,[36,37] and of 2,3 bisphosphoglycerate.[38] Thus, the so-called induction does not represent initial commitment but rather the release of a block of committed cells to progress in their differentiation.

The murine leukemias induced by Friend virus differ in their characteristics among each other according to the strains of the virus. One strain causes polycythemia with an increase of the splenic pool of early and late erythroid progenitors.[39-41] The development of clones from CFU-E of mice infected with this strain is independent of added erythropoietin in contrast to normal CFU-E. On the other hand these animals appear to have greatly increased levels of burst-promoting factor. Later in the infection, the splenic BFU-E become independent of the presence of both burst-promoting factor and erythropoietin.

Hemin is an exceptional inducer. It exerts a selective enhancing influence on the colony formation of primitive erythroid progenitors and can induce the entire erythroid differentiation program,[42,43] but not terminal division in the cultures.[44]

Recently, a human cell line K 562 has become an object of interest. It was originally isolated from a patient with chronic myelogenous leukemia but found to have surface characteristics similar to those of normal human erythrocytes and to be inducible to accumulate human embryonic and fetal hemoglobins.[45-47]

Many of the inducers are common for K 562 and Friend murine erythroleukemia cells. Hemin was found to stimulate the synthesis of globin with a strong preference for the stimulation of α-globin synthesis.[48]

In other experiments different patterns for the synthesis of globin chains with the prevalence of embryonic or γ-chains were observed.[49-51] The K 562 line exhibits distinct heterogeneity of its populations and thus permits the isolation of spontaneous variant sublines. None of the clones are completely negative to benzidine without inducing agents but also none 100% positive after culture with inducers. This circumstance indicates that it is most unlikely that the heterogeneity of benzidine staining results from a mixture of fully inducible and totally resistant cell lines. Also, the clones differed in their response to various inducers.[52]

These observations are in line with many other indications for stochastic features in the differentiation program. As mentioned before, even the first alternative decision — self-renewal or differentiation — appears to be stochastic. The subsequent decisions — which type of cell erythroid or other is to be formed — appear to be governed by progressive and stochastic restrictions in the differentiation potential of the progenitors. This assumption is supported by the existence of the oligopotent hemopoietic cells which undergo terminal differentiation in two or three cell lineages in varying combinations. Also, there may be skipping of differentiation stages (see Chapter 7). The data of Nijhof and Wieranga[53] suggest that CFU-E cells may give origin to proerythroblasts or basophilic erythroblasts directly. According to their view, reports describing an intermediate cell between the CFU-E and the proerythroblast[54] may belong to the proerythroblast compartment.

These considerations may also apply to the expression of globin genes and their switching, both in embryonic development and during further stages, as well as to the reappearance of fetal hemoglobin in adult life. Stamatoyannopoulous et al.[55] described the stochastic expression of fetal hemoglobin in adult erythroid cells. Similar observations have been made repeatedly during the development of chick embryos.[56] They form early a primitive erythroid precursor cell which is succeeded by a definite erythroid cell. The same cohort of primitive cells was found to produce either embryo-specific or adult-type globins. These results are interpreted by the hypothesis that the expression of the globin genes is controlled by programs which are encoded in progenitor cells that undergo changes during the process of their differentiation, so that they lose their ability to express the program of earlier stages. It is assumed that the cells may differ among each other in their vigor to divide, with those most active starting differentiation while the primitive program of globin synthesis is still effective. On the other hand, cells that divide more slowly would enter differentiation with the later adult program.[57] This hypothesis provides a plausible explanation of acquired elevations of Hb F which are attributed to a distortion of the normal differentiation process of erythroid cells in such a way that the early progenitors undergo premature commitment.[58-60]

These conceptions may also apply to the response of experimental mammals to severe anemia. Under these conditions there appear in the peripheral blood reticulocytes of double volume. In all likelihood they originate from a differentiation program in which one cell division is skipped before enucleation occurs. Hand in hand with this dimensional change there occur several other deviations in the composition of the reticulocytes, which are discussed later.

Some contradictions in the estimates of cell cycle time may be related both to the heterogeneity in the response of a cohort and to extracellular factors including hormones, particularly erythropoietin. It is generally assumed that the normal time for erythroid cells to reach their mature stage in humans is of the order of 1 week, of which about 4 days are spent in cell division. The data on cell kinetics show wide variations among researchers who studied mostly murine systems. Monette et al.[61] found a generation time of 16 hr, Udupa and Reissmann[62] one of 10 hr, Vassort et al.[63] and Mary et al.[64] measured 8 hr, and Nijhof et al.[65] one of 6 to 7 hr.

What is the primary site of action of chemical inducers of differentiation? Several authors have suggested the plasma membrane to be the site (see Marks and Rifkind[19]). However, it should be kept in mind that many of the inducing compounds such as dimethyl sulfoxide, hexamethylene bisacetamide, butyric acid, and others permeate easily and could thus also immediately reach intracellular sites; also, some inducers are known to cause structural change of the DNA.[19] In studies on membrane dynamics by means of fluorescence polarization of diphenylhexatriene-labeled Friend leukemia cells, changes were observed after addition of inducing agents. The fluorescence of diphenylhexatriene, which is widely used as a probe of the physical state of lipids, reflects the dynamics of membrane components. It was shown that the effects of the inducers were not only exerted on the plasma membrane but also on microsomal, lysosomal, and mitochondrial membranes.[66] Furthermore, the effects on the plasma membrane were also observed in a noninducible cell variant, whereas the membranes of the intracellular organelles did not show any response. These data cast some doubt on the assumption that the plasma membrane is the site of the primary effect of inducers, and suggest that this role is played by the intracellular membranes. However, another point of attack of inducers such as the nucleus is not excluded.

A most interesting finding is the inhibitory effect of diacylglycerols and phospholipase C, which releases endogenous diacylglycerol from membrane phospholipids on the induction of differentiation of erythroleukemic cells by dimethyl sulfoxide. This effect indicates both the importance of the cell membrane and the possibility that diacylglycerols are involved in differentiation, perhaps via their action on protein kinase C, which appears to be instrumental in the regulation of transcription.[67]

There are increasing indications that changes in ion transport may play an important role in the initiation of differentiation of erythroleukemic cells. An early decrease of Na^+K^+ ATPase with enhanced influx of Na^+ was found which may facilitate Ca^{2+} influx by the Na^+/Ca^{2+} antiport.[67,68] In agreement with this hypothesis, ionophores, which facilitate the Ca^{2+} entry, eliminate the lag period of cell differentiation, whereas inhibitors of Ca^{2+} influx like EGTA and ameloride block the inducing effect of dimethyl sulfoxide and butyric acid on differentiation.[69,70] Definite evidence for the function of the Na^4/Ca^{2+} antiport was obtained, and the Na^+K^+ ATPase as well as the Ca^{2+}-stimulated Mg^{2+}-dependent ATPase were isolated and characterized from the plasma membranes of murine erythroleukemic cells.[70-72] The Na^+K^+ ATPase was found in comparatively large amounts, accounting for 0.4% of the membrane proteins. It was suggested that the enzyme may be regulated by phosphorylation. On the other hand, evidence was obtained that a decreased intracellular Na^+ concentration is an early event in the differentiation of murine erythroleukemic cells, a finding which implies a less straightforward relation between the changes of Na^+,K^+-ATPase and the actual variations of the Na^+ concentration.[72]

IV. REGULATION OF GENE EXPRESSION

A. Activation of Genes

The problems connected with activation and suppression of genes by which the final profile of the erythrocyte is determined are as yet unresolved. Extensive discussion of these would transgress the limits of this presentation. Globin accounts for less than 0.1% of the cytoplasmic protein synthesis in the uninduced mouse erythroleukemic cell, rises to 15 to 20% of the newly synthesized proteins after induction, and reaches a value of 85% on further differentiation. Thus, the proportion of globin to other proteins increases by three orders of magnitude during differentiation.[73] At the same time the synthesis of the nonglobin proteins decreases significantly. Accordingly the percentage of globin mRNA also changes, amounting to 10% of the total mRNA population in erythroblasts and reaching 90% in reticulocytes.

During differentiation there is also a great reduction of the number of genes expressed. Compared with the 10^5 to 10^6 gene sequences in the nucleus, the repertoire of erythroid cells is much more limited. Fewer than 4000 species of Poly(A)-containing nuclear RNAs and fewer than 100 of polysomal RNAs are found in immature erythroid cells from anemic chickens.[74] Similarly, the complexity of the mRNA population decreases during the differentiation of Friend leukemic cells and mouse spleen erythroblasts.[75,76] The rates of transcription for individual mRNAs were shown to differ by factors of up to 200 (Tobin et al.[77]), indicating that most of the divergences in the accumulated levels of mRNAs are caused by this circumstance. In a comparison of the in vitro translation of the reticulocyte mRNAs, only few translation products in addition to globin were found by Lodish et al.[78]

However, it was estimated that approximately 250 messages — most of them rare and a few of them of middle abundance — exist in the mouse reticulocyte, whereas in uninduced Friend leukemia cells the abundant classes predominated. Spleen erythroblasts showed an intermediate pattern.[76] Most of the mRNAs were not erythroid-specific. To some extent the reduction of the number of mRNAs during erythroid differentiation may be apparent only. The existence of translationally inactive mRNAs in form of cytoplasmic mRNPs should be considered. It was estimated that in duck erythroblasts, in which 200 species of mRNA are found in polyribosomes (a number similar to that observed in mouse reticulocytes), more than 1200 other types of mRNA were detectable in the inactive mRNP.[79]

In addition to the differences in transcription, those of stability appear to account for the progressive change in the abundance of globin mRNA during differentiation. The work of Aviv et al.[80] indicated a half-life of about 60 hr for globin mRNA in early erythroblasts, whereas the nonglobin RNAs fell into two classes with half-lives of 3 and 30 hr. They also found a decrease of the half-life of globin mRNA in terminally differentiated erythroblasts and in reticulocytes, as was also reported by Lowenhaupt.[81,82] However, Volloch and Housman[83] demonstrated that by addition of ficoll or albumin, differentiated Friend cells and reticulocytes could be stabilized and showed half-lives for globin mRNA of 60 hr, which is the value in early erythroblasts. A possible explanation for the decreased stability of nonglobin mRNAs is given by the observation that the activity of poly-A-polymerase decreases during differentiation of mouse erythroleukemic cells;[84] by loss of poly-A the half-life of the mRNAs is drastically reduced.

Another important factor regulating the abundance of mRNAs in the cytoplasm appears to be their export from the nucleus. Evidence has been adduced that nuclear transport is a significant control step in the final differentiation of murine erythroleukemia cells.[85]

The question arises to what extent the nonglobin mRNAs making their appearance with the onset of differentiation represent a set of genes coding for characteristic erythroid proteins. This question is not easy to answer. Membrane proteins, such as spectrin, the enzymes of the heme pathway, and carbonic anhydrase, which appear early in differentiation, are not exclusive for red cell lineage; they also occur in other cell types. Nevertheless, their appearance and that of one kind of lipoxygenase which is strictly erythroid-specific in coordination with that of globin mRNAs indicates a specific reaction which in its pattern is characteristic for the erythroid cell. From the pattern of expression, the mRNA encoding a 19-kD polypeptide, ep 19, of unknown function also belongs in this group.[86,87]

Recently, carbonic anhydrase I was reported to be a novel erythrocyte-specific marker useful for studying differentiation.[87]

Indication for the specificity of the coordinate expression of globin and nonglobin genes was obtained in a recent study on hybrids between erythroid and other cells in which erythroleukemic cells were fused with T-lymphoma or neuroblastoma cells of mice.[88] The expression of β-globin, lipoxygenase, and ep 19 genes was tested. It was found that all three mRNAs were either expressed or repressed together. Also it was found that the sensitivity to DNase I, characteristic for the genes in the erythroid parent, is lost in the hybrids in which their expression was suppressed.

B. Changes of Chromatin Associated with Activation of Globin Genes

Studies of the organization of chromatin by means of digestion with nucleases have indicated that the domains of chromatin sensitive to DNase I include the α- and β-globin gene clusters.[89] Within these domains, special regions, the hyper-sensitive sites, are found to be particularly susceptible to DNase I action. They arise after induction of the murine erythroleukemic cells. The S1 nuclease, which is specific for single strands, also cleaves chromatin at positions close, but not identical to, the sites hypersensitive to DNase I. The changes in the susceptible sites during differentiation are similar for both enzymes.[90-92]

However, the formation of such sensitive sites which are often upstream (5′) to the globin genes[93] seems to be a necessary but not sufficient condition for their transcription, as suggested by the fact that in chicken and in human embryonic erythroid cells both the embryonic-fetal and the adult globin genes are sensitive to DNase although the adult genes are not transcribed. On the other hand, the embryonic globin genes do become resistant to DNase I in adult erythroid cells in parallel with cessation of their transcription. A similar case is presented by the HEL human erythroid cell line;[94] here one finds hypersensitive sites in the 5′ regions of the α- and β-globin genes which are not transcribed.

In view of the recent work connecting ubiquitin activation to the cell cycle, it appears of interest that the content of protein A24, an adduct of histone 2A and ubiquitin, is six times higher in the transcriptionally active nuclei of chicken reticulocytes than in those of mature erythrocytes.[96] The decrease of protein A24 was compensated by the increase of histone 2A.

C. The Possible Role of DNA Methylation in the Regulation of Gene Expression

The signals for activation of genes are an object of ongoing research. There are several reports in the literature which indicate that there exists a negative correlation between the methylation of cytosin in C_pG dinucleotide sequences of DNA and the level of gene activity (see Doerfler for review[97]),[98-101] with undermethylation being associated with transcription of the gene. Specific unmethylated sites surrounding the globin genes were found both in human and rabbit erythroid cells.[99,100] Recent experiments[102] would indicate that it is not so much the methylation of C_pG sequences in the structural gene, but that of upstream sequences which prevents transcription.

A strong correlation was found in human embryonic, fetal, and adult erythroblasts producing ε, γ, and β chains, respectively, between hypomethylation in close flanking sequences of the globin genes, suggesting a causal relationship to switching of the genes.[95] On the other hand, no correlation between the variable degree of methylation and differentiation was found in induced murine erythroleukemic cells.[92]

Further evidence for the role of methylation was obtained by employment of an inhibitor of methylation, 5-azacytidine. Haigh et al.[103] studied the presence or absence of methylation of cytosin at specific sites of the chick α-globin gene cluster in DNA from embryonic and adult erythroid cells in comparison with DNA from brain and sperm cells by means of restriction enzymes. They found the sites to be totally methylated in sperm DNA and with some exceptions in brain DNA. In agreement with the basic hypothesis the sites in or near the globin genes that are expressed in the embryonic or adult cells, respectively, were either completely unmethylated or undermethylated. This correlation is of particular interest since methylated C_pG sequences will allow the stretch of DNA in which they occur to assume the left-handed Z-conformation which may be of significance for its function.[104] Indirect evidence was obtained by De Simone et al.[105] They administered 5-azacytidine to baboons made anemic by bleeding and found a strong stimulation of the synthesis of HbF.

Again it would seem that loss of methylation is not a sufficient condition for the transcription of globin genes. An example to support this conclusion are observations that one clone of chick erythroblasts transformed by a temperature-sensitive viral strain exhibits unmethylated C_pG sites near the α-globin gene even though the gene was not transcribed.[106] Similar findings were obtained with respect to several other genes.

D. The Possible Role of Translationally Inactive RNA

The existence of mRNA-protein particles in which the mRNA is masked and thus prevented from being translated has been known for a long time and has received extensive study. Initially the work was concentrated on the early stages of embryogenesis. Oocytes contain large amounts of mRNA in an untranslated form which are utilized after fertilization (for a review see Davidson[107]). Spirin[108] proposed as early as 1964 the general existence of masked mRNA in form of mRNPs and coined the designation "informosomes". Scherrer and his group devoted several investigations to the characterization of RNPs in the cytoplasm of duck erythroblasts (for a summary see Vincent[109]). They found a 20S globin-mRNA protein-complex in the postpolyribosomal supernatant lysate which was translationally inactive, whereas the deproteinized mRNA efficiently directed the synthesis of both α- and β-globin.[110] The mRNA-protein particles were heterogeneous in size and contained a wide spectrum of polypeptides. Whether such particles serve a regulatory role with respect to the synthesis of globin is unclear since only about 10% of the globin RNA are bound to them. In later work two types of mRNPs differing in size were separated. The smaller 20S mRNPs contained mainly globin mRNA, whereas the larger 35S mRNPs possessed a heterogeneous nonglobin mRNA population. In both particles a large number of polypeptides were found, ranging in molecular mass from 20 to 120 kD, which differed to some extent.[111] In one case the repressed mRNA could be identified as that for the poly(A)-binding protein with a molecular mass of 73 kD.[112,113]

In a study of mouse erythroleukemic cells, several abundant mRNA species were found, a large proportion of which were nontranslated.[114] Exposure of the cells to dimethyl sulfoxide led to an early decrease of the total polysomal mRNA, which was to some extent selective. For two of the species the proportion of molecules engaged in translation decreased rapidly, leading to an accumulation of untranslated species. These changes precede the increase of globin mRNA.

A great difficulty in all these studies is the biological heterogeneity of the cytosolic mRNPs, which is apparent in the hypothetical scheme of Scherrer and co-workers.[109] For instance, the 9S mRNA isolated from cytosolic mRNPs of rabbit reticulocytes were found to have shorter poly(A) chains and it was concluded that the mRNPs are "old" and in a posttranslational phase.[115] An outstanding example of posttranscriptional regulation is lipoxygenase (see Chapter 12). Lipoxygenase mRNA is present in immature reticulocytes and erythroblasts in the form of translationally inactive mRNA-protein particles, and are unmasked at a late stage of differentiation. A second instance may be ferritin, which is formed in large amounts in frog erythroid cells on stimulation of erythropoiesis. Here a large portion of the respective mRNA is found in translationally inactive form.[116,117] The signals by which these mRNAs become unmasked have so far not been defined precisely.

The discovery of small RNAs which have the property to inhibit mRNA translation in vitro open a further possibility of regulation.[118,119] Such RNAs occur as ribonucleoprotein particles and it has been claimed that these constitute a ubiquitous, morphologically distinct structure with a characteristic set of proteins.[120] The name "prosome" was proposed for them. However, it has not been demonstrated so far that the inhibition exerted is selective; thus, prosome significance is as yet unclear.

E. The Differential Expression of the Globin Genes

The genes which encode for human hemoglobin are arranged in two multigene clusters.[121] The α-like genes are found on chromosome 16. They include a single functional ζ-gene which is expressed in early embryonic development and two α-genes which take its place and continue to be expressed throughout the entire life. The β-like genes are located on Chromosome 11. They include a single functional ε-gene expressed in the early embryonic period followed by linked duplicated γ-genes which are characteristic for the fetal period and single β- and ϑ-genes which become activated at the end of gestation in a ratio of about 30:1.

The further discussion will limit itself to the regulation of α-, β-, and ϑ-like genes. For such studies, transfection of globin genes into cells generally is employed using various expression vectors. It has been established that two conserved regions, the "ATA" box at about 30 base pairs and the "CCAAT" box at about 80 base pairs, upstream from the site of transcription initiation are important for the function of many cellular genes. These two regions may well be part of the basic recognition system for initiation of transcription rather than regulatory elements specific for globin genes. Sequences further upstream may be of importance. There appear to be considerable differences in the regulation between the α- and β-genes. For various β-globin genes a common CACCC sequence has been found 100 to 150 base pairs upstream from the initiation site which appears to be significant for optimal transcription. No corresponding element has been found for the α-globin gene.[122,123] The in vivo transcription of β-globin genes cloned in an expression vector (SV 40) is fully dependent on the presence of an "enhancer sequence" which increases the transcription of the vector, whereas the α-globin gene is independent of the enhancer element.[122-125]

The low level of expression of the ϑ-globin gene has been related to the instability of the ϑ-mRNA since it is found in bone marrow cells but not in peripheral reticulocytes.[126] The promoter function of the ϑ-globin gene in transfection experiments was found to be dependent on the enhancer element for SV 40 like that for the β-globin gene. However, 30 to 50 times more correctly initiated β-globin mRNA molecules were found.[123] These data suggest that poor function of the ϑ-globin gene promoter may be directly responsible for the low level of expression of this gene.

F. Processing of Globin Transcripts

Both experiments on erythroblasts and direct measurements of transcription with isolated nuclei have demonstrated that the transcripts of globin gene have 5'- and 3'-termini that correspond to those of mature globin mRNA but are considerably larger (15S rather than 9S).[127,128] The large transcripts are processed by sequential removal of introns with the intermediate formation of various spliced transcripts until finally the mature globin mRNA results.[127] A minority of the transcripts are much larger, owing to the circumstance that their transcription either starts in the 5'-flanking region of the gene or extends downstream past the polyadenylation signal sequence.[128]

V. ERYTHROPOIETIN AND OTHER EXTRINSIC FACTORS

A. Erythropoietin and Other Hormones

The rate of erythropoiesis in a mammalian organism is regulated by the level of a specific hormone, erythropoietin. The hormone is made to about 90% in the juxtaglomerular cells of the kidney, the remainder being presumably produced by the liver. Erythropoietin is an acid glycoprotein with a molecular mass of about 40 kD (Goldwasser,[129] Miyake et al.,[130] and Sasaki et al.[131]). Recently the cDNA for erythropoietin has been cloned and sequenced, and so the sequence of the amino acids has been determined, including the location of disulfide bridges and the binding sites of the sialic acids.[131,132] There appear to be no qualitative dissimilarities in the action among erythropoietins of different species, but there may be quantitative differences since pure human erythropoietin was found to be eight times more active than the corresponding sheep preparation in tests in vivo. The acid character of the erythropoietin is mainly derived from the presence of several sialic acid residues. These are not essential for cellular action; erythropoietin which has lost all its terminal sialic acids was found to be at least as active as sialo-erythropoietin in vitro but inactive under conditions in vivo. The sialic acid component is apparently necessary to prevent a rapid clearance of the hormone from the blood plasma,[129] which is effected by hepatic cells. The new terminal galactose residues exposed by desialation are the signal for the uptake by the liver cells. By oxidation of the galactose groups with galactose oxidase the affinity of the hepatic cells for them is abolished and the in vivo biological activity of the desialated hormone is restored.

The concentration of erythropoietin in the blood plasma is normally about $5 \times 10^{-11} \ M$

but may increase up to 100-fold under hypoxic stress.[132] The rate of synthesis of erythropoietin in the kidney is strictly dependent on the oxygen tension at the site of its formation.

Despite various efforts over many years there is as yet no detailed insight into the molecular details of the action of erythropoietin, beginning with its interaction with the target cells and the type of second message within the cell, up to events which lead to the reprogramming of transcription. Erythropoietin stimulates both proliferation and differentiation of erythroid cells.[133-138]

The effect of erythropoietin presupposes an interaction with a receptor in the membrane of a hormone-sensitive cell for which there is some evidence.[139,140]

By immunocytochemical methods it was found that early polychromatic erythroblasts exhibit the largest number of binding sites, followed by the basophilic erythroblasts, pro-erythroblasts, and late polychromatophilic erythroblasts, in that order, whereas normoblasts and reticulocytes are only minimally labeled if at all.[140] It is very likely that the development of erythropoietin sensitivity in the progenitors is determined by the appearance of erythro-poietin receptors.

Mounting evidence suggests that the interaction of erythropoietin with the cell membrane is transmitted to the interior of the cell via adenylate cyclases. Early increases of cAMP concentrations were found in the presence of erythropoietin in rabbit bone marrow cells and in fetal liver cells;[141-145] the same holds for cGMP.[146,147] Addition of cAMP to cultures of murine bone marrow or of fetal liver activated the CFU-E.[148-150] β_2-Adrenergic agonists in combination with erythropoietin increase the number of CFU-E of rabbit bone marrow cultures significantly, an effect which was blocked by β_2-adrenergic antagonists.[151] The increases of cAMP and cGMP may be related to the cell cycle. In proliferating cells without erythropoietin the concentrations of the two nucleotides changed in a pattern related to the stages of the cell cycle, with increases of cAMP during the G1 and G2 phases and of cGMP during mitosis and the period of DNA synthesis. Erythropoietin decreased the duration of the G1 phase to such an extent that the total cycle time was reduced by about 40% and appeared to increase the nucleotides without disturbing the pattern of their changes.[143,144] In an interesting study it was reported that the cyclic nucleotides and even stronger erythropoietin inhibited the methylation of DNA.[152]

By analogy with other reports one may surmise that activation of some cAMP-dependent protein kinases occurs which could have diverse effects including those on the machinery of transcription. It is conceivable that the independence of tumor erythroid cells of eryth-ropoietin might be due to viral-gene-directed activation of protein kinases. At any rate there is old and still valid evidence that the earliest action of erythropoietin is effected on tran-scription. Within 5 min of exposure to erythropoietin, adult rat marrow cells were found to synthesize a small amount of very large RNA. After 15 min new ribosomal RNA was observed in the form of 45S RNA precursor and an increase of total transcription (see Chang et al.,[139] Gross and Goldwasser,[153,154] Miniatis et al.,[155] and Ramirez et al.[156]).

Only later is there an increase in globin mRNA which continues to accumulate for many hours.[155] Evidence was obtained that the reprogramming of the cells to initiate synthesis of globin mRNA requires DNA synthesis and is thus a cell cycle-dependent event.[158,159] How-ever, from other equally clear-cut experiments the opposite conclusion was reached, so that the question appears to be open.[160,161]

Early sequential induction of the enzymes in the pathway of heme synthesis prior to the formation of hemoglobin[26] during erythropoietin-induced differentiation was reported in the spleen of polycythemic mice and in fetal mouse liver and in normal human bone marrow cultures.[22-24,26] In a recent study on the action of erythropoietin on an anemia strain of mouse erythroleukemic cells in vitro, the following effects of the hormones were found: induction of uroporphyrinogen I synthase; increased iron incorporation in heme; a massive increase of mRNAs for α and β chains of globin, as well as changes leading to condensation and finally extrusion of the nucleus.[162]

How the molecular changes are related to the effects of erythropoietin on the vitality of erythroid cell cultures is as yet unclear. In the absence of erythropoietin proerythroblast populations die rapidly whereas they are able to divide and can be maintained in culture for more than 10 days in its presence (for review see Nigon[163]).

The observations that erythropoietin stimulates iron transport before affecting hemoglobin synthesis in cultures of rat marrow cells could be interpreted to indicate that the induction of transferrin receptors in the cell membrane precedes that of globin.

What about other hormonal influences on erythropoiesis? Testosterone triggers CFU-E into the S phase of the cell cycle. This effect appears to be independent of erythropoietin and not to be mediated via a cyclase system.[164,165] It was suggested that the action is exerted upon the ion conductance of the cell membrane. In line with this assumption are the results on the importance of Ca^{2+} influx for the initiation of differentiation.[69,70] A further direct cellular effect has been demonstrated with respect to glucocorticoids.[166] The previously described inhibitory effect of steroids on dimethyl sulfoxide-induced differentiation could be related to the existence of a large number of glucocorticoid receptors with a high affinity (K_d $5 \cdot 10^{-9}$ M). The cells induced by dimethyl sulfoxide showed a decrease in the number of binding sites by one order of magnitude. The glucocorticoids also decreased the cloning efficiency of the leukemic cells in apparent contradiction to their stimulatory effect on normal murine and human erythroid progenitor cells.[167] In a recent study it was reported that insulin stimulated erythroid colony formation of CFU-E cells from fetal mouse liver and adult bone marrow independently of erythropoietin, however, only in superphysiological concentrations.

Thyroid and growth hormone and in some studies prolactin and placental lactogen have been reported to stimulate either colony formation or erythropoiesis. All of the factors mentioned appear to act as modulators of erythroid differentiation rather than as inducers.[168,169] An inhibitory effect of prostaglandin $PGF_{2\alpha}$ on Friend virus induced erythropoiesis was reported.[170] Studies performed on whole animals are of course difficult to interpret.

B. The Effect of Iron

The synthesis of hemoglobin requires, of course, the import of iron. This process (see Chapter 10) occurs mainly by way of an interaction between the Fe-transferrin complex with specific receptors which are then internalized. Studies on the transferrin receptors have indicated a correlation of their number with the stage of differentiation and the intensity of hemoglobin synthesis. In human and murine normal bone marrow cells it was found that the early progenitor cells (CFU-E) exhibited little or no transferrin receptors as judged by binding of fluorescence-labeled ferritin or labeling with antitransferrin-receptor monoclonal antibodies.[171,172] Quantitative data which are in basic agreement were obtained on cultures of mouse erythroleukemia cells. The noninduced cells did not take up transferrin; the number of receptors increased during differentiation to reach a maximum in the intermediate (polychromatic) stage. Lack of iron suppresses the actions of erythropoietin[138,173] and possibly delays the inactivation of the nucleus. Cobalt as well as folic acid were reported to stimulate differentiation of erythroid cells in cultures of human bone marrow as judged by differentiated cell and mitosis counting.[138]

C. The Effects of Hemin

As mentioned previously, hemin can induce the expression of the complete erythroid differentiation program. On the other hand, deficient heme synthesis may be the cause for the noninducibility of some erythroleukemic cell lines of mice.[174] In an interesting study of a clone of mouse erythroleukemia cells which cannot complete erythroid differentiation and do not accumulate heme in response to dimethyl sulfoxide, it was shown that this type of cell displays both globin synthesis and formation of the enzymes of heme synthesis. Provision of hemin in addition to dimethyl sulfoxide led to partial correction of the defect.[175] In later studies it was shown that such variants fail to increase transferrin binding, indicating a lack of induction of transferrin receptors[176] and that hemin inhibits their induction.[177] Hemin exhibits rapid inducing effects in both mouse and human erythroleukemic cells. One may observe both early syntheses and terminal differentiation.[178-184] Hemin increases the utilization of globin mRNA and stimulates the enzymes of the heme biosynthetic pathway.[182]

In a recent report it was demonstrated that hemin increased transcription, particularly of poly-A-rich mRNAs, but also exerted a stabilizing effect on them.[185] The differentiation of immature erythroblasts isolated from anemic rabbit bone marrow was accelerated by addition of hemin. Both the proportion of benzidine-positive cells and the synthesis of heme — relative to total protein — were increased, whereas cell growth and DNA synthesis were decreased.[186] The authors suggested that the concentration of heme in plasma, which normally is about 1 μM and is increased 5- to 10-fold in hemolytic anemias, may possibly exert a stimulatory compensatory effect on erythropoiesis.

There is generally a close coordination between the amounts of heme and globin synthesized, so that there is little or no detectable free heme accumulated. Closer analysis indicates that heme synthesis may precede the appearance of hemoglobin.[187] This observation is in line with recent work in which it was argued that heme is needed for the synthesis of the cytochrome during new formation of mitochondria.[53]

A dissociation between heme biosynthesis and commitment to differentiation of murine erythroleukemic cells was demonstrated in an investigation in which imidazole was used as an inhibitor of iron uptake and heme synthesis.[188] On the other hand, it was shown[189] that succinyl acetone, a specific inhibitor of 5-ALA-dehydratase, decreased iron uptake and the synthesis of heme and globin in a coordinated manner. The stimulating effect of erythropoietin on globin synthesis was abolished and the conclusion was drawn — which certainly needs further confirmation — that the effect of the hormone requires heme synthesis.

VI. CONCLUSIONS

All the available information would indicate that it is unlikely that the specific commitment of erythroid cells is connected with the activation of a single "master" gene serving a coordinating function; it is rather evident that several unlinked genes are activated in the earliest period of differentiation which include, among others, erythropoietin and transferrin receptors, 5-ALA-synthase, spectrin, lipoxygenase, and ep 19. The mechanisms instrumental in the coordination of the complex pattern of gene activation are as yet unclear.

The early stages of differentiation are a multistep process which in some unknown manner are determined by an "internal clock" before final commitment occurs. The definition of commitment at the molecular level remains to be clarified. It should be kept in mind that most of the recent work has been carried out on erythroid tumor cells. One may expect that the patterns of activation of genes may differ to some extent from those of normal erythroid progenitors.

REFERENCES

1. **Ogawa, M., Porter, P. N., and Nakahata, T.,** Renewal and commitment to differentiation of hemopoietic stem cells, an interpretive review, *Blood,* 61, 823, 1983.
2. **Johnson, G. R. and Metcalf, D.,** Pure and mixed erythroid colony formation in vitro stimulated by spleen conditioned medium with no detectable erythropoietin, *Proc. Natl. Acad. Sci. U.S.A.,* 74, 3879, 1977.
3. **Hara, H. and Ogawa, M.,** Murine hemopoietic colonies in culture containing normoblasts, macrophages and megakaryocytes, *Am. J. Hematol.,* 4, 23, 1978.
4. **Fauser, A. A. and Messner, H. A.,** Granulo-erythropoietic colonies in human marrow, peripheral blood, and cord blood, *Blood,* 52, 1243, 1978.

5. **Nakahata, T. and Ogawa, M.,** Identification in culture of a class of hemopoietic colony-forming units with extensive capability to self-renew and generate multipotential hemopoietic colonies, *Proc. Natl. Acad. Sci. U.S.A.,* 79, 3843, 1982.

6. **Humphries, R. K., Eaves, A. C., and Eaves, C. J.,** Self-renewal of hemopoietic stem cells during mixed colony formation in vitro, *Proc. Natl. Acad. Sci. U.S.A.,* 78, 3629, 1981.

7. **Wagemaker, G.,** Hemopoietic factors required for differentiation of multipotential cells in vitro, in *Hemoglobins in Development of Differentiation,* Alan R. Liss, New York, 1981, 85.

8. **Porter, P. N., Ogawa, M., and Leary, A. G.,** Enhancement of the growth of human early erythroid progenitors by bone marrow conditioned media, *Exp. Hematol.,* 8, 83, 1980.

9. **Hamburger, A. W.,** Enhancement of human erythroid progenitor cell growth by media conditioned by a human T-lymphocyte line, *Blood,* 56, 633, 1980.

10. **Westbrook, C. A., Gasson, J. C., Gerber, S. E., Selsted, M. E., and Golde, D. W.,** Purification and characterization of human T-lymphocyte-derived erythroid-potentiating activity, *J. Biol. Chem.,* 259, 9992, 1984.

11. **Welte, K., Platzer, E., Ju, L., Gabrilove, J. L., Levi, E., Mertelsmann, R., and Moore, M. A. S.,** Purification and biochemical characterization of human pluripotent haematopoietic colony-stimulating factor, *Proc. Natl. Acad. Sci. U.S.A.,* 82, 1526, 1985.

12. **Jubinsky, P. T. and Stanley, E. R.,** Purification of hemopoietin 1:A multiline-age hemopoietic growth factor, *Proc. Natl. Acad. Sci. U.S.A.,* 82, 2764, 1985.

13. **Goodman, J. W., Hall, E. A., Miller, K. L., and Shinpock, S. G.,** Interleukin 3 promotes erythroid burst formation in "serum-free" cultures without detectable erythropoietin, *Proc. Natl. Acad. Sci. U.S.A.,* 82, 3291, 1985.

14. **Whetton, A. D. and Dexter, T. M.,** Effect of haematopoietic cell growth factor on intracellular ATP levels, *Nature,* 303, 629, 1983.

15. **Van Zant, G. and Goldwasser, E.,** Simultaneous effects of erythropoietin and colony-stimulating factor on bone marrow cells, *Science,* 198, 733, 1977.

16. **Nijhof, W. and Wieranga, P. K.,** Isolation and characterization of the erythroid progenitor cell, *J. Cell Biol.,* 96, 386, 1983.

17. **Friend, C., Scher, W., Holland, J. G., and Sato, T.,** Hemoglobin synthesis in murine virus-induced leukemia cells in vitro: stimulation of erythroid differentiation by dimethyl sulfoxide, *Proc. Natl. Acad. Sci. U.S.A.,* 68, 378, 1971.

18. **Leder, A. and Leder, P.,** Butyric acid, a potent inducer of erythroid differentiation in cultured erythroleukemia cells, *Cell,* 5, 319, 1975.

19. **Marks, P. A. and Rifkind, R. A.,** Erythroleukemic differentiation, *Annu. Rev. Biochem.,* 47, 419, 1978.

20. **Eisen, H., Bach, R., and Emerg, R.,** Induction of spectrin in erythroleukemic cells transformed by Friend virus, *Proc. Natl. Acad. Sci. U.S.A.,* 74, 3898, 1977.

21. **Maniatis, G. M.,** Erythropoiesis: a model for differentiation, in *Cell Function and Differentiation,* (Part A), Akoyunoglou, G., Evangelopoulos, A. E., Georgatsos, J., et al., Eds., Alan R. Liss, New York, 1982, 13.

22. **Sassa, S.,** Sequential induction of heme pathway enzymes during erythroid differentiation of mouse Friend leukemia virus-infected cells, *J. Exp. Med.,* 143, 305, 1976.

23. **Sassa, S.,** Heme biosynthesis in erythroid cells: distinctive aspects of the regulatory mechanism, in *Regulation of Hemoglobin Biosynthesis,* Goldwasser, E., Ed., Elsevier, New York, 1983, 359.

24. **Sassa, S. and Urabe, A.,** Uroporphyrinogen-I synthase induction in normal human bone marrow cultures: an early and quantitative response of erythroid differentiation, *Proc. Natl. Acad. Sci. U.S.A.,* 76, 5321, 1979.

25. **Rutherford, T., Thompson, G. G., and Moore, M. R.,** Heme biosynthesis in Friend erythroleukemia cells: control by ferrochelatase, *Proc. Natl. Acad. Sci. U.S.A.,* 76, 833, 1979.

26. **Freshney, R. I. and Paul, J.,** The activities of three enzymes of heme synthesis during hepatic erythropoiesis in the mouse embryo, *J. Embryol. Exp. Morphol.,* 26, 313, 1971.

27. **Shen, D.-W., Real, F. X., DeLeo, A. B., Old, L. J., Marks, P. A., and Rifkind, R. A.,** Protein p 53 and inducer-mediated erythroleukemia cell commitment to terminal cell division, *Proc. Natl. Acad. Sci. U.S.A.,* 80, 5919, 1983.

28. **Terada, M., Fried, J., Nudel, U., Rifkind, R. A., and Marks, P. A.,** Transient inhibition of initiation of S-phase associated with dimethyl sulfoxide induction of murine erythroleukemia cells to erythroid differentiation, *Proc. Natl. Acad. Sci. U.S.A.,* 74, 248, 1977.

29. **Gambari, R., Terada, M., Bank, A., Rifkind, R. A., and Marks, P. A.,** Synthesis of globin mRNA in relation to the cell cycle during induced murine erythroleukemia differentiation, *Proc. Natl. Acad. Sci. U.S.A.,* 75, 3801, 1978.

30. **Gambari, R., Marks, P. A., and Rifkind, R. A.,** Murine erythroleukemia cell differentiation: relationship of globin gene expression and of prolongation of G_1 to inducer effects during G_1/early S, *Proc. Natl. Acad. Sci. U.S.A.,* 76, 4511, 1979.

31. **Tsiftsoglou, A. S., Wong, W., Volloch, V., Gusella, J., and Housman, D.,** Commitment of murine erythroleukemia (MEL) cells to terminal differentiation is associated with coordinated expression of globin and ribosomal genes, in *Cell Function and Differentiation,* (Part A), Akoyunoglou, G., Evangelopoulos, A. E., Georgatos, J., et al., Eds., Alan R. Liss, New York, 1982, 69.

32. **Reuben, R. C., Rifkind, R. A., and Marks, P. A.,** Chemically induced murine erythroleukemic differentiation, *Biochim. Biophys. Acta,* 605, 325, 1980.

33. **Conkie, D., Affara, N., Harrison, P. R., Paul, J., and Jones, K.,** *In situ* localization of globin messenger RNA formation. II. After treatment of Friend virus-transformed mouse cells with dimethyl sulfoxide, *J. Cell Biol.,* 63, 414, 1974.

34. **Ngai, J., Capetanaki, Y. G., and Lazarides, G.,** Differentiation of murine erythroleukemia cells results in the rapid repression of vimentin gene expression, *J. Cell Biol.,* 99, 306, 1984.

35. **Opitz, V., Seidel, H. J., and Bertoncello, I.,** Erythroid stem cell in Friend virus-infected mice. *J. Cell. Physiol.,* 96, 95, 1978.

36. **Conscience, J.-F., Miller, R. A., Henry, J., and Ruddle, F. H.,** Acetylcholinesterase, carbonic anhydrase and catalase activity in Friend erythroleukemic cells, non-erythroid mouse cell lines and their somatic hybrids, *Exp. Cell Res.,* 105, 401, 1977.

37. **Conscience, J.-F. and Meier, W.,** Coordinate expression of erythroid marker enzymes during dimethyl-sulfoxide-induced differentiation of Friend erythroleukemic cells, *Exp. Cell Res.,* 125, 111, 1980.

38. **Yeoh, G. C. T.,** Levels of 2,3-diphosphoglycerate in Friend leukemic cells, *Nature (London),* 285, 108, 1980.

39. **Peschle, C., Migliaccio, G., Lettieri, F., Migliaccio, A. R., Ceccarelli, R., Barba, P., Titti, F. L., and Rossi, G. B.,** Kinetics of erythroid precursors in mice infected with the anemic or the polycythemic strain of Friend leukemia virus, *Proc. Natl. Acad. Sci. U.S.A.,* 77, 2054, 1980.

40. **Peschle, C.,** Erythropoiesis, *Annu. Rev. Med.,* 31, 303, 1980.

41. **Peschle, C., Rossi, G. B., Migliaccio, G., Covelli, A., Gabbianelli, M., and Mastroberardino, G.,** The early stage of Friend virus erythroleukemias: a tentative model, in *Cell Function and Differentiation,* (Part A), Akoyunoglou, G., Evangelopoulos, A. E., Georgatsos, J., et al., Eds., Alan R. Liss, New York, 1982, 59.

42. **Monette, F. C. and Holden, S. A.,** Hemin enhancement of primitive erythroid progenitors in vitro: relationship to burst-promoting activity (BPA), *Exp. Hematol.,* 10, 281, 1982.

43. **Lu, L. and Broxmeyer, H. E.,** The selective enhancing influence of hemin and products of human erythrocytes on colony formation by human multipotential (CFU-GEMM) and erythroid (BFU-E) progenitor cells in vitro, *Exp. Hematol.,* 11, 721, 1983.

44. **Hofer, E., Hofer-Warbinek, R., and Darnell, J. E., Jr.,** Globin RNA transcription: a possible termination site and demonstration of transcriptional control correlated with altered chromatin structure, *Cell,* 29, 887, 1982.

45. **Gahmberg, C. G. and Anderson, L. C.,** K 562 — a human leukemic cell line with erythroid features, *Semin. Hematol.,* 18, 72, 1981.

46. **Anderson, L. C., Jokinen, M., and Gahmberg, C. G.,** Induction of erythroid differentiation in the human leukemia cell line, *Nature (London),* 278, 364, 1979.

47. **Rutherford, T. R., Clegg, J. B., and Weatherall, D. J.,** K 562 human leukemic cells synthesize embryonic hemoglobin in response to hemin, *Nature (London),* 280, 164, 1979.

48. **Rowley, P. T., Ohlsson-Wilhelm, B. M., Hicks, D. G., Rudolph, N. S., Farley, B., Kosciolek, B., and La Bella, S.,** Regulation of hemoglobin synthesis in K 562 human erythroleukemia cells, in *Regulation of Hemoglobin Biosynthesis,* Goldwasser, E., Ed., Elsevier, New York, 1983, 333.

49. **Alter, B. P. and Goff, S. C.,** Electrophoretic separation of human embryonic globin demonstrates "α-thalassemia" in human leukemia cell line K 562, *Biochem. Biophys. Res. Commun.,* 94, 843, 1980.

50. **Benz, E. J., Jr., Murnane, M. J., Tonkonow, B. L., Berman, B. W., Mazur, E. M., Cavallesco, C., Jenko, T., Snyder, E. L., Forget, B. G., and Hoffman, R.,** Embryonic-fetal erythroid characteristic of human leukemic cell line, *Proc. Natl. Acad. Sci. U.S.A.,* 77, 3509, 1980.

51. **Cioe, L., McNab, A., Hubbel, H. R., Meo, P., Curtis, P., and Rovera, G.,** Differential expression of the globin genes in human leukemia K 562(S) cells induced to differentiate by hemin or butyric acid, *Cancer Res.,* 41, 237, 1981.

52. **Rowley, P. T., Ohlsson-Wilhelm, B. M., Farley, B. A., and La Bella, S.,** Inducer of erythroid differentiation in K 562 human leukemia cells, *Exp. Hematol.,* 9, 32, 1981.

53. **Nijhof, W. and Wierenga, P. K.,** A new system for the study of erythroid cell differentiation, *Exp. Hematol.,* 12, 115, 1984.

54. **Monette, F. C., Weiner, E. J., and Faletra, P. P.,** The state of differentiation of erythroid cells forming clusters in vitro, *Exp. Hematol.,* 9, 711, 1981.

55. **Stamatoyannopoulos, G., Kurnit, D. M., and Papayannopoulou, T.,** Stochastic expression of fetal hemoglobin in adult erythroid cells, *Proc. Natl. Acad. Sci. U.S.A.,* 78, 7005, 1981.

56. **Schalekamp, M., de Jonge, P., and van Goor, D.,** Is erythroid cell differentiation a matter of all-or-none transcription only?, in *Cell Function and Differentiation,* (Part A), Akoyunoglou, G., Evangelopoulos, A. E., Georgatsos, J., et al., Eds., Alan R. Liss, New York, 1982, 25.

57. **Stammatoyannopoulos, G., Papayannopoulou, T., and Martin, P.,** Cell biology of hemoglobin switching. III. Studies of cell lines and mutants, in *Regulation of Hemoglobin Synthesis,* Goldwasser, E., Ed., Elsevier, New York, 1983, 401.

58. **Papayannopoulou, T., Kalmantis, T., and Stammatoyannopoulos, G.,** Cellular regulation of hemoglobin switching: evidence for inverse relationship between fetal hemoglobin synthesis and degree of maturity of human erythroid cells, *Proc. Natl. Acad. Sci. U.S.A.,* 76, 6420, 1979.

59. **Papayannopoulou, T., Vichinsky, E., and Stammatoyannopoulos, G.,** Foetal Hb production during acute erythroid expansion. I. Observations in patients with transient erythroblastopenia and post-phlebotomy, *Br. J. Haematol.,* 44, 535, 1980.

60. **Stammatoyannopoulos, G., Papayannopoulou, T., Brice, M., Kurachi, S., Nakamoto, B., Lim, G., and Farguhar, M.,** Cell biology of hemoglobin switching. I. The switch from fetal to adult hemoglobin formation during ontogeny, in *Hemoglobins in Development and Differentiation,* Stammatoyannopoulos, G. and Nienhuis, A. W., Eds., Alan R. Liss, New York, 1981, 287.

61. **Monette, F. C., Kent, R. B., Weiner, E. J., Jarris, R. F., Quellette, P. L., Thorson, J. A., and Zelick, R. D.,** Cell cycle properties and proliferation kinetics of late erythroid progenitors in murine bone marrow, *Exp. Hematol.,* 8, 484, 1980.

62. **Udupa, K. D. and Reissman, K. R.,** Cell kinetics of erythroid colony forming cells (CFU-E) studied by hydroxyurea injections and sedimentation velocity profile, *Exp. Hematol.,* 6, 398, 1978.

63. **Vassort, F., Winterholer, M., Frindel, E., and Tubiana, M.,** Kinetic parameters of bone marrow stem cells using in vivo suicide by tritiated thymidine or by hydroxyurea, *Blood,* 41, 789, 1973.

64. **Mary, J. Y., Valleron, A. J., Croizat, H., and Frindel, E.,** Mathematical analysis of bone marrow erythropoiesis: application to C_3H data, *Blood Cells,* 6, 241, 1980.

65. **Nijhof, P. K., Wierenga, J. P., and Bloem, R.,** Cell kinetic behaviour of a synchronized population of erythroid precursor cells in vitro, *Cell Tissue Kinet.,* 17, 629, 1984.

66. **Billard, C., Billard, M., and Mishal, Z.,** Effects of differentiation inducers on diphenylhexatriene fluorescence polarization in intracytoplasmic and plasma membranes from Friend erythroleukemia cells, *Biochem. Biophys. Res. Commun.,* 117, 294, 1983.

67. **Pincus, S. M., Beckmann, B. S., and George, W. J.,** Inhibition of dimethylsulfoxide-induced differentiation in Friend erythroleukemic cells by diacylglycerols and phospholipase C, *Biochem. Biophys. Res. Commun.,* 125, 491, 1985.

68. **Mager, D. and Bernstein, A.,** Early transport changes during erythroid differentiation of Friend leukemic cells, *J. Cell. Physiol.,* 94, 275, 1978.

69. **Levenson, R., Macara, J. G., Smith, R. L., Cantley, L., and Housman, D.,** Role of mitochondrial membrane potential in the regulation of murine erythroleukemia cell differentiation, *Cell,* 28, 855, 1982.

70. **Smith, R. L., Macara, J. G., Levenson, R., Housman, D., and Cantley, L.,** Evidence that a Na^+/Ca^{2+} antiport system regulates murine erythroleukemia cell differentiation, *J. Biol. Chem.,* 257, 773, 1982.

71. **Yeh, L.-A., Ling, L., English, L., and Cantley, L.,** Phosphorylation of the (Na,K)-ATPase by a plasma membrane-bound protein kinase in Friend erythroleukemia cells, *J. Biol. Chem.,* 258, 6567, 1983.

72. **Lannigan, D. A., and Knauf, P. A.** Decreased intracellular Na^+ concentration is an early event in murine erythroleukemic cell differentiation, *J. Biol. Chem.,* 260, 7322, 1985.

73. **Housman, D., Volloch, V., Tsiftsoglou, A. S., and Levenson, R.,** The use of the MEL cell system as a model for the control of terminal erythroid differentiation, in *Regulation of Hemoglobin Biosynthesis,* Goldwasser, E., Ed., Elsevier, New York, 1983, 311.

74. **Lasky, L. A., Nowck, N., and Tobin, A. J.,** Few transcribed sequences are translated in avian erythroid cells, *Dev. Biol.,* 67, 23, 1978.

75. **Obinata, M. and Ikawa, Y.,** Change in message sequences during erythrodifferentiation, *Nucleic Acids Res.,* 8, 4271, 1980.

76. **Mishina, Y., Natori, S., and Obinata, M.,** Cloning and characterization of genes expressed in the mouse reticulocytes, *Dev. Growth Differ.,* 26, 311, 1984.

77. **Tobin, A. J., Hansen, D. A., Seftor, E. A., McCabe, J. B., and De Kloe, J.,** Transcription and processing of mRNAs during erythroid development on chickens, in *Regulation of Hemoglobin Biosynthesis,* Goldwasser, E., Ed., Elsevier, New York, 1983, 271.

78. **Lodish, H. F. and Desalu, O.,** Regulation of synthesis of nonglobin proteins in cell-free extracts of rabbit reticulocytes, *J. Biol. Chem.,* 248, 3520, 1973.

79. **Imaizumi-Scherrer, M.-T., Maundrell, K., Civelli, O., and Scherrer, K.,** Transcriptional and post-transcriptional regulation in chick erythroblasts, *Dev. Biol.,* 93, 126, 1982.

80. **Aviv, H., Voloch, Z., Bastos, R., and Levy, S.,** Biosynthesis and stability of globin mRNA in cultured erythroleukemic Friend cells, *Cell,* 8, 495, 1976.

81. **Lowenhaupt, K. and Lingrel, J. B.,** A change in the stability of globin mRNA during the induction of murine erythroleukemia cells, *Cell,* 14, 337, 1978.

82. **Lowenhaupt, K. and Lingrel, J. B.,** Synthesis and turnover of globin mRNA in murine erythroleukemia cells induced with hemin, *Proc. Natl. Acad. Sci. U.S.A.,* 76, 5173, 1979.

83. **Volloch, V. and Housman, D.,** Stability of globin mRNA in terminally differentiating murine erythroleukemia cells, *Cell,* 23, 509, 1981.

84. **Adolf, G. R. and Swetly, P.,** Poly(A)polymerase activity during cell cycle and erythropoietic differentiation in erythroleukemic mouse spleen cells, *Biochim. Biophys. Acta,* 518, 334, 1978.

85. **Shaul, Y., Ginzburg, J., and Aviv, H.,** Preferential transcription and nuclear transport of globin gene sequences, as control steps leading to final differentiation of murine erythroleukemic cells, *Eur. J. Biochem.,* 128, 637, 1982.

86. **Goldfarb, P. S., O'Prey, J., Affara, N., Yang, O. S., and Harrison, P. R.,** Isolation of non-globin genes expressed preferentially in mouse erythroid cells, *Nucleic Acids Res.,* 11, 3517, 1983.

87. **Konialis, Ch. P., Barlow, J. H., and Butterworth, P. H. W.,** Cloned cDNA for rabbit erythrocyte carbonic anhydrase I: A novel erythrocyte-specific probe to study development in erythroid tissues, *Proc. Natl. Acad. Sci. U.S.A.,* 82, 663, 1985.

88. **Affara, N., Fleming, J., Goldfarb, P. S., Black, E., Thiele, B., and Harrison, P. R.,** Analysis of chromatin changes associated with the expression of globin and non-globin genes in cell hybrids between erythroid and other cells, *Nucleic Acids Res.,* 13, 5629, 1985.

89. **Weintraub, H., Larsen, A., and Grondine, M.,** α-Globin-gene switching during the development of chick embryos: expression of chromosome structure, *Cell,* 24, 333, 1981.

90. **Larsen, A. and Weintraub, H.,** An altered DNA conformation detected by S1 nuclease occurs at specific regions in active chick globin chromatin, *Cell,* 29, 609, 1982.

91. **Balcarek, J. M. and McMorris, F. A.,** DNase I hypersensitive sites of globin genes of uninduced Friend erythroleukemia cells and changes during induction with dimethyl sulfoxide, *J. Biol. Chem.,* 258, 10622, 1983.

92. **Sheffery, M., Rifkind, R. A., and Marks, P. A.,** Murine erythroleukemia cell differentiation: DNase I hypersensitivity and DNA methylation near the globin genes, *Proc. Natl. Acad. Sci. U.S.A.,* 79, 1180, 1982.

93. **Stalder, J., Larsen, A., Engel, J. D., Dolan, M., Groudine, M., and Weintraub, H.,** Tissue-specific DNA cleavages in the globin chromatin domain introduced by DNase I, *Cell,* 20, 451, 1980.

94. **Groudine, M., Kohwi-Shigematsu, T., Gelinas, R., Stamatoyannopoulos, G., and Papayannopoulou, T.,** Human fetal to adult hemoglobin switching: Changes in chromatin structure of the β-globin gene locus, *Proc. Natl. Acad. Sci. U.S.A.,* 80, 7551, 1983.

95. **Mavilio, F., Giampaolo, A., Care, A., Migliaccio, G., Calandrini, M., Russo, G., Pagliardi, G. L., Mastroberardino, G., Marinucci, M., and Peschle, C.,** Molecular mechanisms of human hemoglobin switching: selective undermethylation and expression of globin genes in embryonic, fetal, and adult erythroblasts, *Proc. Natl. Acad. Sci. U.S.A.,* 80, 6907, 1983.

96. **Goldknopf, I. L., Wilson, G., Ballal, N. R., and Busch, H.,** Chromatin conjugate protein A 24 is cleaved and ubiquitin is lost during chicken erythropoiesis, *J. Biol. Chem.,* 255, 10555, 1980.

97. **Doerfler, W.,** DNA methylation and gene activity, *Annu. Rev. Biochem.,* 52, 93, 1983.

98. **Razin, A. and Riggs, A. D.,** DNA methylation and gene function, *Science,* 210, 604, 1980.

99. **Shen, C. K. and Maniatis, T.,** Tissue specific DNA methylation in a cluster of rabbit β-like globin genes, *Proc. Natl. Acad. Sci. U.S.A.,* 77, 6634, 1980.

100. **van der Ploeg, L. H. T. and Flavell, R. A.,** DNA methylation in the human $\gamma\delta\beta$-globin locus in erythroid and nonerythroid tissues, *Cell,* 19, 947, 1980.

101. **Groudine, M., Eisenman, R., and Weintraub, A.,** Chromatin structure of endogenous retroviral genes and activation by an inhibitor of DNA methylation, *Nature (London),* 292, 311, 1981.

102. **Busslinger, M., Hurst, J., and Flavell, R. A.,** DNA methylation and the regulation of globin gene expression, *Cell,* 34, 197, 1983.

103. **Haigh, L. S., Hellewell, S., Roninson, I. B., Owens, B. B., and Ingram, V. M.,** Control of hemoglobin expression in chick embryonic development, in *Cell Function and Differentiation,* (Part A), Akoyunoglou, G., Evangelopoulos, A. E., Georgatsos, J., et al., Eds., Alan R. Liss, New York, 1982, 35.

104. **Rich, A., Nordheim, A., and Wang, A. H. J.,** The chemistry and biology of left-handed Z-DNA, *Annu. Rev. Biochem.,* 53, 791, 1984.

105. **De Simone, J., Heller, P., Hall, L., and Zwiers, D.,** 5-Azacytidine stimulates fetal hemoglobin (HbF) synthesis in anemic baboons, in *Regulation of Hemoglobin Biosynthesis,* Goldwasser, E., Ed., Elsevier, New York, 1983, 351.

106. **Weintraub, H., Beug, H., Groudine, M., and Graf, T.,** Temperature-sensitive changes in the structure of globin chromatin in lines of red cell precursors transformed by ts-AEV, *Cell,* 28, 931, 1982.

107. **Davidson, E. H.,** *Gene Activity in Early Development,* Academic Press, New York, 1976.

108. **Spirin, A. S.,** Informosomes, *Eur. J. Biochem.,* 10, 20, 1969.

109. **Vincent, A., Goldenberg, S., Standart, N., Civelli, O., Imaizumi-Scherrer, T., Maundrell, K., and Scherrer, K.,** Potential role of mRNP proteins in cytoplasmic control of gene expression in duck erythroblasts, *Mol. Biol. Rep.,* 7, 71, 1981.

110. **Civelli, O., Vincent, A., Maundrell, K., Buri, J.-F., and Scherrer, K.,** The translational repression of globin mRNA in free cytoplasmic ribonucleoprotein complexes, *Eur. J. Biochem.,* 107, 577, 1980.

111. **Vincent, A., Akhayat, O., Goldenberg, S., and Scherrer, K.,** Differential repression of specific mRNA in erythroblast cytoplasm: a possible role for free mRNP proteins, *EMBO J.,* 2, 1869, 1983.

112. **Maundrell, K., Imaizumi-Scherrer, M. T., Maxwell, E. S., Civelli, O., and Scherrer, K.,** Messenger RNA for the 73,000-dalton poly(A)-binding protein occurs as translationally repressed mRNP in duck reticulocytes, *J. Biol. Chem.,* 258, 1387, 1983.

113. **Akhayat, O., Vincent, A., Goldenberg, S., Person, A., and Scherrer, K.,** The translation of the messenger for the poly(A)-binding protein-associated with translated mRNA is suppressed. A case of cytoplasmic repression in duck erythroblasts, *FEBS Lett.,* 162, 25, 1983.

114. **Yenofsky, R., Cereghini, S., Krowczynska, A., and Brawerman, G.,** Regulation of mRNA utilization in mouse erythroleukemic cells induced to differentiate by exposure to dimethyl sulfoxide, *Mol. Cell. Biol.,* 3, 1197, 1983.

115. **Princen, H. M. G., van Eekelen, Ch. A. G., Asselberg, F. A. M., and van Venrooij, W. J.,** Free cytoplasmic messenger ribonucleoprotein complexes from rabbit reticulocytes, *Mol. Biol. Rep.,* 6, 59, 1979.

116. **Shull, G. E. and Theil, E. C.,** Translational control of ferritin synthesis by iron in embryonic reticulocytes of the bullfrog, *J. Biol. Chem.,* 257, 14187, 1982.

117. **Shull, G. E. and Theil, E. C.,** Regulation of ferritin mRNA: a possible gene-sparing phenomenon, *J. Biol. Chem.,* 258, 7921, 1983.

118. **Kühn, B., Villringer, A., Falk, H., and Heinrich, P. C.,** Inhibition of cell-free protein synthesis by low-molecular weight RNAs from free cytoplasmic ribonucleoprotein particles, *Eur. J. Biochem.,* 126, 181, 1982.

119. **McCarthy, T. L., Siegel, E., Mroczkowski, B., and Heywood, S. M.,** Characterization of translational-control ribonucleic acid isolated from embryonic chick muscle, *Biochemistry,* 22, 935, 1983.

120. **Schmid, H. P., Akhayat, O., De Sa, C. M., Puvion, F., Koehler, K., and Scherrer, K.,** The prosome: an ubiquitous morphologically distinct RNP particle associated with repressed mRNPs and containing specific Sc RNA and a characteristic set of proteins, *EMBO J.,* 3, 29, 1984.

121. **Stammatoyannopoulos, G. and Nienhuis, A. W., Eds.,** *Organization and Expression of Globin Genes,* Alan R. Liss, New York, 1981.

122. **Mellon, P., Gluzman, Y., Parker, R., and Maniatis, T.,** Identification of DNA sequences required for transcription of the human α-globin gene using a new SV 40 host vector system, *Cell,* 27, 279, 1981.

123. **Dierks, P., van Ooyen, A., Cochran, M. D., Dobkin, C., Reiser, J., and Weisemann, C.,** Three regions upstream from the cap site are required for efficient and accurate transcription of the rabbit β-globin gene in mouse 3T6 cells, *Cell,* 32, 695, 1983.

124. **Humphries, R. K., Ley, T., Turner, P., Moulton, A. D., and Nienhuis, A. W.,** Differences in human alpha, beta and delta globin gene expression in monkey kidney cells, *Cell,* 30, 173, 1982.

125. **Treisman, R., Green, M. R., and Maniatis, T.,** Cis and trans activation of globin gene transcription in transient assays, *Proc. Natl. Acad. Sci. U.S.A.,* 80, 7428, 1983.

126. **Wood, W. G., Old, J. M., Roberts, A. V. S., Clegg, J. B., Weatherall, D. J., and Quattrin, N.,** Human globin gene expression: control of β, δ and γ chain production, *Cell,* 15, 437, 1978.

127. **Grosveld, G. C., Koster, A., and Flavell, R. A.,** A transcription map for the rabbit β-globin gene, *Cell,* 23, 573, 1981.

128. **Hofer, E. and Darnell, J. E.,** The primary transcription unit of the mouse β-major globin gene, *Cell,* 23, 585, 1981.

129. **Miyake, T., Kung, C. K. H., and Goldwasser, E.,** The purification of human erythropoietin, *J. Biol. Chem.,* 252, 5558, 1977.

130. **Sasaki, R., Yanagawa, S. I., and Chiba, H.,** Isolation of erythropoietin by monoclonal antibody, *Biomed. Biochim. Acta,* 42, S202, 1983.

131. **Jacobs, K., Shoemaker, Ch., Rudersdorf, R., Neill, S. D., Kaufman, R. J., Mufson, A., Seehra, J., Jones, S. S., Hewick, R., Fritsch, E. F., Kawakita, M., Shimizu, T., and Miyake, T.,** Isolation and characterization of genomic and cDNA clones of human erythropoietin, *Nature (London),* 313, 806, 1985.

132. **Lai, P. H., Everett, R., Wang, F. F., and Goldwasser, E.,** The primary structure of human erythropoietin, *Abstract TH-275, 13th Int. Congr. Biochem.,* Amsterdam 1985.

133. **Iscove, N. N.,** The role of erythropoietin in regulation of population size and cell cycling of early and late erythroid precursors in mouse bone marrow, *Cell Tissue Kinet.,* 10, 323, 1977.

134. **Krantz, S. B.,** Response of polycythemia vera marrow to erythropoietin in vitro, *J. Lab. Clin. Med.,* 71, 999, 1968.

135. **Necheles, T. F., Sheehan, R. G., and Meyer, H. J.,** Studies on the control of hemoglobin synthesis: nucleic acid synthesis and normoblast proliferation in the presence of erythropoietin, *Ann. N.Y. Acad. Sci.,* 149, 449, 1968.

136. **Paul, J. and Hunter, A.,** DNA-synthesis is essential for increased hemoglobin synthesis in response to erythropoietin, *Nature (London),* 219, 1362, 1968.

137. **Izak, G. and Karsai, A.,** The effect of erythropoietin on synchronized rabbit hemopoietic tissue in culture, *Blood,* 39, 814, 1972.

138. **Boll, J.,** Studien zur humoralen Regulation des erythropoietischen Proliferationsspeichers beim Menschen, *Klin. Wochenschr.,* 56, 187, 1978.

139. **Chang, S. C.-S., Sikkema, D., and Goldwasser, E.,** Evidence for an erythropoietin receptor protein on rat bone marrow cells, *Biochem. Biophys. Res. Commun.,* 57, 399, 1974.

140. **Lafferty, M. D., Ackerman, G. A., Dunn, C. D., and Lange, R. D.,** The ultrastructural immunocytochemical demonstration of erythropoietin receptors on developing erythrocytic cells of fetal mouse liver, *Exp. Hematol.,* 8, 1063, 1980.

141. **Graber, S. E., Bomboy, J. D., Jr., Salmon, W. D., Jr., and Krantz, S. B.,** Effect of erythropoietin preparations on cyclic AMP and cyclic GMP levels in rat fetal liver cell cultures, *J. Lab. Clin. Med.,* 90, 162, 1977.

142. **Chiuini, F., Della Torre, G., Fanio, G., and Viti, A.,** Early increase of cyclic adenosine monophosphate level induced by erythropoietin on rabbit bone marrow cell suspensions, *Acta Haematol.,* 61, 251, 1979.

143. **White, L., Fisher, J. W., and George, W. J.,** Rôle of erythropoietin and cyclic nucleotides in erythroid cell proliferation in fetal liver, *Exp. Hematol.,* 8, Suppl. 8, 168, 1980.

144. **Fanio, G., Della Torre, G., Menchetti, G., Secca, T., and Marsili, V.,** Cyclic nucleotides and hemoglobin concentration in rabbit bone marrow cell suspensions stimulated by purified erythropoietin, *Cell Biochem. Function,* 2, 119, 1984.

145. **Rodgers, G. M., Fisher, J. W., and George, W. J.,** Elevated cyclic GMP concentrations in rabbit bone marrow culture and mouse spleen following erythropoietic stimulation, *Biochem. Biophys. Res. Commun.,* 70, 287, 1976.

146. **Fedorov, N. A., Ermil'chenko, G. V., Malykhina, J. S., and Borisov, B. N.,** Cyclic guanosine-3',5'-monophosphate-mediator of the action of the hormone erythropoietin, *Dokl. Akad. Nauk SSSR,* 233, 985, 1977.

147. **White, L. E. and George, W. J.,** Increased concentrations of cyclic GMP in fetal liver cells stimulated by erythropoietin, *Proc. Soc. Exp. Biol. Med.,* 166, 186, 1981.

148. **Yigit, R. and Fisher, J. W.,** Activation of erythroid colony forming cells (CFU-E) by cyclic AMP in adult mouse bone marrow and fetal mouse liver cultures, *Haematologica,* 68, 301, 1983.

149. **Brown, J. E. and Adamson, J. W.,** Studies on the influence of cyclic nucleotide on in vitro hemoglobin synthesis, *Br. J. Haematol.,* 35, 193, 1977.

150. **Brown, J. E. and Adamson, J. W.,** Modulation of in vitro erythropoiesis. The influence of β-adrenergic agonists on erythroid colony formation, *J. Clin. Invest.,* 60, 70, 1977.

151. **Przala, F., Gross, D. M., Beckman, B., and Fisher, J. W.,** Influence of albuterol on erythropoietin production and erythroid progenitor cell activation, *Am. J. Physiol.,* 236, H422, 1979.

152. **Fedorov, N. A. and Ermil'chenko, G. V.,** Effect of cyclic nucleotides and hormones of homologous methylation in nuclear homogenates of rat bone marrow, *Biokhimiya,* 45, 1048, 1980.

153. **Gross, M. and Goldwasser, E.,** On the mechanism of erythropoietin-induced differentiation. V. Characterization of the RNA formed as a result of erythropoietin action, *Biochemistry,* 8, 1795, 1969.

154. **Gross, M. and Goldwasser, E.,** On the mechanism of erythropoietin-induced differentiation. IX. Induced synthesis of 9S RNA and of hemoglobin, *J. Biol. Chem.,* 246, 2480, 1971.

155. **Maniatis, G. M., Rifkind, R. A., Bank, A., and Marks, P. A.,** Early stimulation of RNA synthesis by erythropoietin in cultures of erythroid precursor cells, *Proc. Natl. Acad. Sci. U.S.A.,* 70, 3189, 1973.

156. **Ramirez, F., Gambino, R., Maniatis, G. M., Rifkind, R. A., Marks, P. A., and Bank, A.,** Changes in globin messenger RNA content during erythroid cell differentiation, *J. Biol. Chem.,* 250, 6054, 1975.

157. **Sahr, K. and Goldwasser, E.,** The effects of erythropoietin on the biosynthesis of translatable globin mRNA, in *Regulation of Hemoglobin Biosynthesis,* Goldwasser, E., Ed., Elsevier, New York, 1983, 153.

158. **Marks, P. A., Rifkind, R. A., Bank, A., Tevada, M., Maniatis, G. M., Reuben, R. C., and Fibach, E.,** Erythroid differentiation and the cell cycle, in *Growth Kinetics and Biochemical Regulation of Normal and Malignant Cells,* Drewinko, B. and Humphrey, R. M., Eds., Williams & Wilkins Co., Baltimore, Md., 1977, 329.

159. **Conkie, D., Harrison, P. R., and Paul, J.,** Cell-cycle dependence of induced hemoglobin synthesis in Friend erythroleukemia cells temperature-sensitive for growth, *Proc. Natl. Acad. Sci. U.S.A.,* 78, 3644, 1981.

160. **Leder, A., Orkin, S., and Leder, P.,** Differentiation of erythroleukemic cells in the presence of inhibitors of DNA synthesis, *Science,* 190, 893, 1975.

161. **Tabuse, Y., Kawamura, M., and Furusawa, M.,** Induction of hemoglobin synthesis in Friend leukemia cells without the necessity of mitosis, *Differentiation,* 7, 1, 1976.

162. **Koury, M. J., Bondurant, M. C., Duncan, D. T., Krantz, S. B., and Hankins, W. D.**, Specific differentiation events induced by erythropoietin in cells infected in vitro with the anemia strain of Friend virus, *Proc. Natl. Acad. Sci. U.S.A.*, 79, 635, 1982.

163. **Nigon, W. and Godet, J.**, Genetic and morphogenetic factors in hemoglobin synthesis during higher vertebrate development: an approach to cell differentiation mechanisms, *Int. Rev. Cytol.*, 46, 79, 1976.

164. **Boll, I., Mersch, G., Schoen, S., Göttke, U., Boxheimer, D., and Lucke, G.**, Hormoneinwirkung auf die Proliferationskinetik humaner Knochenmarkkulturen, *Klin. Wochenschr.*, 46, 608, 1968.

165. **Byron, J. W.**, Nature of the erythropoietin-independent response of CFU-S to steroids, *Exp. Hematol.*, 8(Suppl. 8), 160, 1980.

166. **Golde, D. W., Bersch, N., Lippman, M. E., and Friend, Ch.**, Detection of glucocorticoid receptors on Friend erythroleukemic cells, *Proc. Natl. Acad. Sci. U.S.A.*, 76, 3515, 1979.

167. **Kurtz, A., Jelkmann, W., and Bauer, C.**, Insulin stimulates erythroid colony formation independently of erythropoietin, *Br. J. Haematol.*, 53, 311, 1983.

168. **Izak, G.**, Erythroid cell differentiation and maturation, *Prog. Hematol.*, 10, 1, 1977.

169. **Adamson, J. W., Popovic, W. I., and Brown, J. E.**, Hormonal control of erythropoiesis, in *Hematopoietic Cell Differentiation*, Golde, D. W., Metcalf, D., Cline, M. J., and Fox, C. F., Eds., Academic Press, New York, 1978, 53.

170. **Lewis, J. P., Moores, R. R., Neal, W. A., Garver, F. A., Lutchov, C. L., Zucali, J. R., and Miraud, E. A.**, Inhibition of Friend virus (FVP)-induced murine erythropoiesis with prostaglandin (PGF2α): potentiation and inhibition of erythropoietin and prevention of both with PGD2, *Exp. Hematol.*, 9, 540, 1981.

171. **Sieff, C., Bicknell, D., Caine, G. L., Robinson, J., Lam, G., and Greaves, M.**, Changes in cell surface antigen expression during hemopoietic differentiation, *Blood*, 60, 703, 1982.

172. **Lesley, J., Hyman, R., Schulte, R., and Trotter, J.**, Expression of transferrin receptor on murine hematopoietic progenitors, *Cell. Immunol.*, 83, 14, 1984.

173. **Stohlman, F., Jr., Howard, D., and Beland, A.**, Humoral regulation of erythropoiesis. XII. Effect of erythropoietin and iron on cell size in iron deficiency anemia, *Proc. Soc. Exp. Biol.*, 113, 986, 1963.

174. **Rutherford, T. R. and Weatherall, D. J.**, Deficient heme synthesis as the cause of noninducibility of hemoglobin synthesis in a Friend erythroleukemia cell line, *Cell*, 16, 415, 1979.

175. **Sassa, S.**, Heme biosynthesis in erythroid cells. Distinctive aspects of the regulatory mechanism, in *Regulation of Hemoglobin Biosynthesis*, Goldwasser, E., Ed., Elsevier, New York, 1983, 359.

176. **Wilczynska, A., Ponka, P., and Schulmann, H. R.**, Transferrin receptors and iron utilization in DMSO-inducible and -uninducible Friend erythroleukemia cells, *Exp. Cell Res.*, 154, 561, 1984.

177. **Pelicci, P. G., Tabilio, A., Thomopoulos, P., Titeux, M., Vainchenker, W., Rochant, H., and Testa, U.**, Hemin regulates the expression of transferrin receptors in human hematopoietic cell lines, *FEBS Lett.*, 145, 350, 1982.

178. **Gusella, J. F., Weil, S. C., Tsiftsoglou, A. S., Volloch, V., Neumann, J. R., Keys, C., and Housman, D. E.**, Hemin does not cause commitment of murine erythroleukemia (MEL) cells to terminal differentiation, *Blood*, 56, 481, 1980.

179. **Dean, A., Erard, F., Schneider, A. B., and Schechter, A. N.**, Induction of hemoglobin accumulation in human K 562 cells by hemin is reversible, *Science*, 212, 459, 1981.

180. **Dabney, B. J. and Beaudet, A. L.**, Increase in globin chains and globin mRNA in erythroleukemia cells in response to hemin, *Arch. Biochem. Biophys.*, 179, 106, 1977.

181. **Ross, J. and Sautner, D.**, Induction of globin mRNA accumulation by hemin in cultured erythroleukemia cells, *Cell*, 8, 513, 1976.

182. **Granick, J. L. and Sassa, S.**, Hemin control of heme biosynthesis in mouse Friend-virus-transformed erythroleukemia cells in culture, *J. Biol. Chem.*, 253, 5402, 1978.

183. **Nudel, U., Saimon, J., Fibach, E., Terada, M., Rifkind, R., Marks, A. P., and Bank, A.**, Accumulation of α- and β-globin messenger RNAs in mouse erythroleukemia cells, *Cell*, 12, 463, 1977.

184. **Rovera, G., Vartikar, J., Connolly, G. R., Magarian, C., and Dolby, T. W.**, Hemin controls the expression of the β minor globin gene in Friend erythroleukemic cells at the pretranslational level, *J. Biol. Chem.*, 253, 7588, 1978.

185. **Bonanou-Tzedaki, S. A., Sohi, M. N., and Arnstein, H. R. V.**, The effect of hemin on RNA synthesis and stability in differentiating rabbit erythroblasts, *Eur. J. Biochem.*, 144, 589, 1984.

186. **Bonanou-Tzedaki, S. A., Sohi, M., and Arnstein, H. R. V.**, Regulation of erythroid cell differentiation by hemin, *Cell Differ.*, 10, 267, 1981.

187. **Glass, J., Lavidor, L. M., and Robinson, S. H.**, Studies of murine erythroid cell development. Synthesis of heme and hemoglobin, *J. Cell Biol.*, 65, 298, 1975.

188. **Tsiftsoglou, A. S., Nunez, M. T., Wong, W., and Robinson, S. H.**, Dissociation of iron transport and heme biosynthesis from commitment to terminal maturation of murine erythroleukemia cells, *Proc. Natl. Acad. Sci. U.S.A.*, 80, 7528, 1983.

189. **Beru, N., Sahr, K., and Goldwasser, E.**, Inhibition of heme synthesis in bone marrow cells by succinylacetone: effect on globin synthesis, *J. Cell. Biochem.*, 21, 93, 1983.

Chapter 2

FROM THE PROERYTHROBLAST TO THE RETICULOCYTE

I. INTRODUCTION

The processes of differentiation of proerythroblasts and maturation of erythroid cells have been the subject of much early work. Three types of methods have been used to characterize and study the various stages of erythroblasts arising from the proerythroblasts:

1. Methods based on the fractionation of bone marrow cells; such procedures yield comparatively large amounts of material and are suitable for biochemical studies. They have the disadvantage that the cell population are not separated sharply and that there is therefore a considerable overlap.
2. Methods based on cytophotometry of single cells. These procedures are as precise as the criteria used to characterize them, but are limited to a few components such as the content of water, nucleic acids, and total protein and hemoglobin.
3. In recent times automated cell sorting has been used. So far this method suffers from the defect that only a minority of the cells subjected to sorting are classified and the majority are rejected, so that the yield is small and may not be representative.[1]

A new, highly promising system permitting the study of erythroid cell differentiation and maturation during culture in vitro, beginning with the late (CFU-E) progenitors, has been worked out.[2] It allows both cytological and biochemical investigations.

As an introduction, a brief survey of the dimensional and morphological changes which occur during differentiation and maturation will be given.

II. DIMENSIONAL AND MORPHOLOGICAL CHANGES

In the progression from the proerythroblast to the late orthochromatic erythroblast there are morphological and dimensional changes which serve to classify the cell types (for excellent reviews see Bessis[3] and Wickramasinghe et al.[4]). Several criteria help to distinguish cell types, among which the most useful are cell size and nuclear diameter; both can be used for cell sorting. The proerythroblasts are large cells with a diameter >20 μm and have big round nuclei, large nucleoli, dispersed chromatin, and a thin rim of basophilic cytoplasm. The basophilic erythroblasts are somewhat smaller (10 to 16 μm); they contain little condensed chromatin. The early polychromatic erythroblasts range in diameter from 8 to 10 μm with a nuclear diameter of >6 μm and contain a moderate quantity of condensed chromatin; their cytoplasm stains grey-greenish. The later polychromatic erythroblasts differ only in degree from the early ones; the main distinction is a nuclear diameter of <6 μm, which is based on the observation that the majority of these polychromatic cells have diploid DNA and do not incorporate tritiated thymidine.[4] Orthochromatic erythroblasts have a diameter of <9 μm and are characterized by their pyknotic nucleus which is often eccentrically located.

The change in the staining of the cytoplasm reflects two types of processes: (1) decrease in activity of transcription and translation with a consequent decline in RNA content, which is the main basophilic material of the cell, and (2) increase in hemoglobin, which as a basic protein is acidophilic.

Cell organelles undergo dramatic changes during erythroid differentiation. The number of ribosomes declines continuously. As far as the mitochondria are concerned, their number as judged both by electronmicroscopy as well as by cytochrome oxidase activity per cell, decreases with each cell division so that the late stages contain only one third or less of the original CFU-E progenitor (Nijhof and Wierenga[2]). At the same time there is an increase in the total cytochrome oxidase activity of the cell culture, indicating continuing active production of mitochondria as long as the cells divide.

Ultrastructural studies indicate that after the last mitosis there is a decrease in the rough endoplasmatic reticulum and the Golgi apparatus,[5-7] as well as in lysosomes, so that the reticulocyte is practically devoid of these organelles. The number of mitochondria decreases somewhat during the transition from intermediate to orthochromatic erythroblasts, and greatly with nuclear expulsion. This change is accompanied by a fivefold increase in the volume density of autophagosomes, many of which contain mitochondria.[6] In other work, actual expulsion of mitochondria accompanying that of the nucleus was observed (Nijhof and Wierenga,[8,9] Simpson and Kling,[10] and Gasko and Danon[11]). Thus the conclusion is inescapable that there are several mechanisms of degradation of mitochondria: one by autophagy operating in late erythroblasts and bone marrow reticulocytes, another by expulsion, and finally, one by means of lipoxygenase and ATP-dependent proteolysis, effective during the maturation of the reticulocytes (see Chapter 12). The changes of the cell membrane including those of the various receptors and the cytoskeleton are discussed in the appropriate chapters (see Chapters 4 and 10).

III. BIOCHEMICAL CHANGES

A. Microspectrophotometric Studies

Microspectrophotometry was used in several studies.[12-17] A bimodal distribution of DNA was found, classified as 2-N and 4-N, corresponding to the process of reduplication in the proliferating cells (Figure 2), whereas in the orthochromatic erythroblast stage only 2-N cells were found as a reflection of the stopped DNA synthesis.[14,15] In cell-kinetic studies of synchronized populations of CFU-E precursor cells, performed by means of flow cytometry, these results were confirmed. The DNA content of reticulocytes and erythrocytes of chicks is equal, as is to be expected[18,19] for nondividing nucleated cells.

During the entire course of development there is a continuous decrease of total RNA, beginning with the basophilic erythroblast (Figure 3). The rate of synthesis in the nucleus decreases steeply at the stage of the proerythroblast, indicating the shutdown of transcription and export of both mRNA and rRNA. These changes correspond to the well-documented loss of basophilia and the reduction of the number of ribosomes. On the other hand, the proteins show a biphasic change; there is a decrease in cytoplasmic proteins up to the stage of the polychromatic erythroblast, probably owing to the reduction of cell size which is superseded by an increase in the last stages, which is due to the accumulation of hemoglobin. The protein content of the nucleus declines parallel to increasing pyknosis.

B. Studies on Fractionated Bone Marrow

Much more detailed information was obtained in studies of fractionated bone marrow.[20] By the use of a velocity sedimentation technique it was possible to obtain a satisfactory fractionation, especially of the later stages of differentiation, and to separate erythroblasts before and after final cell division.[20-22] In agreement with cytospectrophotometrical work, it was found that hemoglobin concentration, which accumulates slowly to about one third of its final complement at the final erythroblast stage, doubles during the last cell division and passage to the orthochromatic stage. The synthesis of globin paralleled closely the concentration of globin mRNA at all stages of differentiation.[23] A reciprocal change with a decline to one third or to one fifth was observed with glucose-6-phosphate dehydrogenase, 6-phosphogluconate dehydrogenase, purine nucleoside phosphorylase, and adenosine

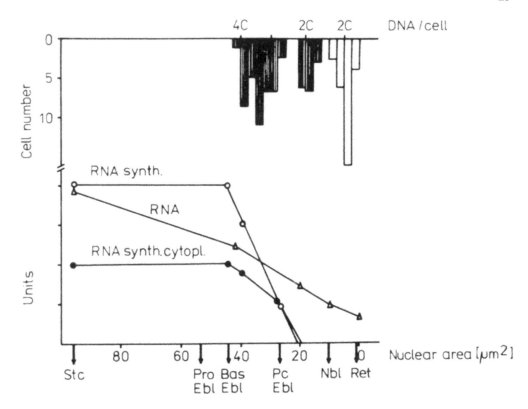

FIGURE 2. Changes in DNA per cell; content and synthesis of RNA during erythroid differentiation. Stc, stem cell; Pro, proerythroblast; Bas, basophilic; Pc, polychromatic, Ebl, erythroblast; Nbl, normoblast; Ret, reticulocyte. (Adapted from Grasso, J. A., et al., *Proc. Natl. Acad. Sci. U.S.A.*, 50, 134, 1963.)

deaminase[20] (Figure 4). Carbohydrate metabolism enzyme changes will be discussed in more detail further on.

The decline of the two purine metabolism enzymes is in contrast to changes that occur during early differentiation as determined in dimethyl sulfoxide-induced murine erythroleukemic cells.[24] In this study it was found that purine salvage pathway enzymes remained constant after induction, whereas two key enzymes of *de novo* synthesis, i.e., phosphoribosyldiphosphate aminotransferase (EC 2.42.14) and ribose-5-phosphate aminophospherase (EC 6.3.4.7.), declined. These changes occurred synchronously with the accumulation of hemoglobin and are probably connected with the restriction of proliferation occurring after induction.

Carboanhydrase was found to rise continuously with progressive differentiation, although the overall increase was not as great as that of hemoglobin.[20] In dimethyl sulfoxide-induced murine erythroleukemic cells, a clear-cut accumulation of carboanhydrase paralleling that of hemoglobin was observed.[25-27]

Enzymes involved in the defense against the toxicity of intermediate oxygen reduction products, in particular catalase, have also received attention. Catalase and superoxide dismutase increase during late differentiation and maturation, whereas glutathione peroxidase and reductase undergo a decline, which is particularly marked at the stage of orthochromatic erythroblasts.[28] With this in mind, the exceptional case of the maturational change of catalase in the red cells of ducks should be mentioned. In mature duck erythrocytes, the catalase activity is about three orders of magnitude lower than in human erythrocytes. It is about one order of magnitude higher in the bone marrow and reaches about the same value in the reticulocytes of peripheral blood in phenylhydrazine-induced anemia.[29] A biological half-life of 1.5 days for the maturational decline of catalase was calculated, which corresponds with estimates in other organs.

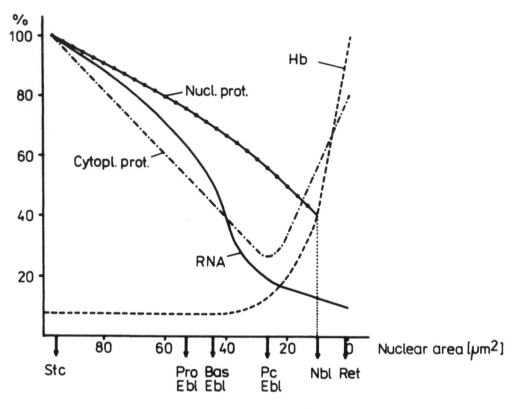

FIGURE 3. Changes in proteins, RNA, and hemoglobin during erythroid differentiation. Stc, stem cell; Pro, proerythroblast; Bas, basophilic erythroblast; Pc, polychromatic erythroblast; Ebl, erythroblast; Nbl, normoblast; Ret, reticulocyte. (Adapted from Grasso, J. A., et al., *Proc. Natl. Acad. Sci. U.S.A.*, 50, 134, 1963.)

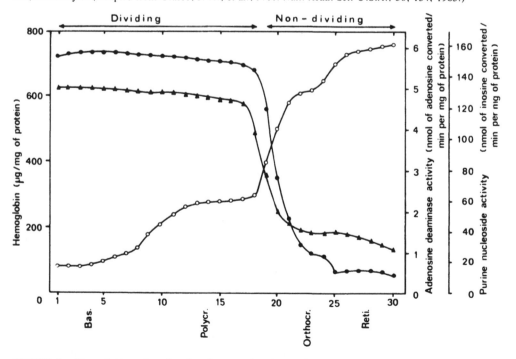

FIGURE 4. Hemoglobin, adenosine deaminase, and purine nucleosidase in fractionated rabbit bone marrow. Bas., basophilic erythroblast; Polycr., polychromatic erythroblast; Orthocr., orthochromatic erythroblast. (Adapted from Denton, M. J., et al., *Biochem. J.*, 146, 205, 1975.)

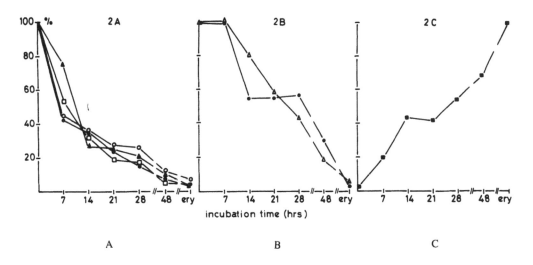

FIGURE 5. Activities of glycolytic enzymes at different stages of erythroid differentiation in vitro. (A) Hexokinase (●); phosphofructokinase (○); pyruvate kinase (□); and enolase (▲). (B) Glucose-6-P dehydrogenase (△) and aldolase (●). (C) 2,3-P₂G mutase (■). (From Nijhof, W., et al., *Blood*, 64, 607, 1984. With permission.)

During erythroid development in bone marrow, a decrease of overall ribonuclease activity was described commensurate with the changes of other enzymes, which decline steeply after the last division.[30-32] A further drop was noted after the extrusion of the nucleus. Evidence was provided that the ribonuclease activity is caused by several electrophoretically distinct enzymes, and that differential changes of the enzyme activities occur among various subcellular fractions. It is tempting to ascribe to the ribonucleases a role in this process, in view of the fact that total ribosomal population decreases during cell differentiation. However, one should keep in mind the occurrence of a ribonuclease inhibitor in red cells, the developmental changes of which have not yet been defined, and which may mask some activities.[33,34]

C. Changes in Activities and Isoenzyme Patterns of Glycolytic Enzymes

Although nucleated erythroid cells as well as reticulocytes derive most of their energy from respiration and have a highly active citrate cycle, they also exhibit much higher glycolytic capacity, and in keeping with this, high activities of glycolytic enzymes. In a study on hexokinase in density-fractionated bone marrow cells of anemic rabbits, the specific activity of the enzyme was found to be 150-fold higher in the basophilic erythroblast as compared with mature circulating erythrocytes.[8] Most of the decrease took place during the transition from the polychromatic to the orthochromatic stage. Concomitantly, the pattern of the isoenzymes changed. In another study[35] of in vitro maturation of late committed progenitor cells of erythropoiesis, CFU-E cells (isolated from mouse spleens) the time course of six glycolysis enzymes (hexokinase, phosphofructokinase, pyruvate kinase, enolase, aldolase, and 2,3 diphosphoglycerate mutase, as well as of glucose-6-phosphate dehydrogenase) during a 48-hr culture period. Within this time the cell morphology changed markedly. After 7 hr of incubation the pronormoblasts dominated with 70%; the distribution shifted after 48 hr to 84% basophilic erythroblasts. After 21 and 28 hr, polychromatic erythroblasts reached a relative peak of about 50% and orthochromatic erythroblasts made their appearance. After 48 hr, two thirds of the cells were reticulocytes; the remainder were mostly orthochromatic erythroblasts. During in vitro differentiation, all enzymes studied except diphosphoglycerate mutase exhibited a drastic decline by one order of magnitude or more (Figure 5).

Two patterns were observed for the enzyme decrease: hexokinase, phosphofructokinase, pyruvate kinase, and enolase declined to 25 to 35% of initial values after two cell divisions, i.e., after 14 hr, while aldolase and glucose-6-phosphate dehydrogenase remained constant during the early period. The authors assume that the first group of enzymes are no longer synthesized after the first cell division. There are comparatively minor changes in enzyme activities between the stages of basophilic and orthochromatic erythroblasts, which is in agreement with studies on fractionated rabbit bone marrow cells.[18,20] A similar course of events was observed for lactate dehydrogenase.[36] In contrast, it was found that phospho-fructokinase exhibited, in the late stages of erythroid differentiation of rabbits, an increase in activity after an early decline.[37,38]

2,3 Diphosphoglycerate mutase is nearly absent in the CFU-E cells but increases continuously during differentiation. There is a coincidence between the synthesis of hemoglobin, which starts after one cell division,[39] and the formation of mutase, in keeping with the close relationship between 2,3 bisphosphoglyceric acid (2,3-DPG) and hemoglobin in the bone marrow of rabbits.[37,38]

During differentiation there are also changes in the patterns of the isoenzymes. In the CFU-E cells the I-A subtype of hexokinase predominates with a shift during differentiation in favor of the I-B subtype. One should keep in mind that in the work referred to,[35] only cytosolic enzyme activities were determined, while the mitochondrial, nuclear, and membrane fractions were discarded. As will be discussed later, this circumstance renders statements, particularly concerning hexokinase, somewhat dubious, since this enzyme is partly bound to the mitochondria with a disproportional representation of its isoenzymes.

Pyruvate kinase exhibits in man at least three types of isoenzymes, the L-type (liver), the R-type (erythrocytes), and the M-type, which is present in most other tissues including leukocytes. The R-isoenzyme of pyruvate kinase exhibits immunological identity with the L-type; both isoenzymes are encoded by the same gene, from which two slightly differing mRNAs originate.[40] Both the M-type as well as the R-type (liver) and their hybrids were observed in CFU-E cells. The M-type disappeared after one cell division. All the later stages contained only the R-type, in accord with the observations of Kahn et al.[41] In another study the presence of M-type pyruvate kinase in human proerythroblasts was demonstrated by immunofluorescence.[42]

During differentiation of erythroblasts from human fetal liver, similar changes occur with the M-isoenzyme declining and the R-enzyme persisting.[43] It was also found that two human erythroleukemia cell lines, K 562 and HEL-60, contain high M-pyruvate kinase activity as well as small but definite amounts of the R-isoenzyme, in keeping with their erythroid character.

As differentiation progresses drastic changes in the pattern of lactate dehydrogenase isoenzymes occur according to studies on fractionated bone marrow cells of rabbits.[36] Basophilic erythroblasts contain considerable levels of M-subunits, which occur both as M_4-tetramer and in the form of various hybrids. When the formation of M-subunits stops, the percentage of H_4-isoenzyme increases to account for more than 90% of total lactate dehydrogenase activity. The occurrence of the M_4-isoenzyme in hyperactive erythroid tissue and in nucleated bird erythrocytes had been noted earlier.[44,45]

Malate dehydrogenase appears to be the only citrate cycle enzyme that has been investigated. Setchenska and Arnstein[46] found changes both in the activity and number of isoenzymes during differentiation of erythroid cells in rat bone marrow. The activity of the enzyme decreased to one tenth in the progression from basophilic erythroblast to the reticulocyte and changed little during its maturation. In the cytosol of erythroblasts, two isoenzymes were found but only one in reticulocytes.

The changes of NADH-cytochrome b_5 reductase show interesting features. This enzyme, found predominantly in the endoplasmic reticulum of Friend erythroleukemic cells, is mostly located in the cytosol of mature erythrocytes. Differentiation appears to go hand in hand with its release from the endoplasmic reticulum by proteolytic action, which removes the hydrophobic tail of the polypeptide chain of the enzyme.[47]

In view of the derivation of granulocytes, platelets, and erythrocytes from a common stem cell, it is of interest to compare isoenzyme pattern and enzyme activities between the three types of cells. Both platelets and granulocytes contain K-type pyruvate kinase. In the human erythroleukemic cell line[39] K 562, predominantly M-type pyruvate kinase was in accord with the classification of the cells as being in a developmental stage before CFU-E. As far as enzyme activities, granulocytes of man, dog, and various rodents exhibit 30- to 300-fold greater activity of glycolytic enzymes such as phosphohexose isomerase, aldolase, and lactate dehydrogenase[48,49] as compared to erythrocytes; furthermore, thrombocytes resemble red cells. It would appear that the activities in granulocytes are comparable to those of the erythroid series progenitors and that both red cells and thrombocytes undergo similar developmental changes.

It is apparent so far from this review that only a small number of enzymes or other cell components have been studied and that these were mostly chosen on the basis of methodical convenience. Until now, investigations are of a descriptive character and not mechanistically oriented.

IV. PYCNOSIS OF THE NUCLEUS AND ITS REACTIVATION

A. Pycnosis

Pycnosis of the nucleus is a general phenomenon of erythroid differentiation. In mammalian cells it precedes enucleation, whereas in mature erythrocytes of nonmammalian species the nucleus is retained in an inert, highly condensed state. For this reason the studies of the mechanism of pycnosis have been limited to nucleated red cells, mainly those of the avian species. It is generally assumed that the major erythrocyte-specific histone, H5, which is rich in lysine, serine, and alanine, may be responsible for the condensing and packaging of the DNA, similar to the role of sperm-specific basic proteins (for a review see Neelin[50]). Early studies had indicated that H5 is synthesized during the terminal stages of differentiation.[51] However, in later work it was found that this histone is already present, although in lower amounts, in the early dividing erythroblast.[52,53]

During reactivation of pycnotic chick erythrocyte nuclei in chick/HeLa heterokaryons, there was an exchange of the chick specific H5, for the H1 of the HeLa cell.[54] The histones including H5 undergo phosphorylation, but at the terminal stage of cell development — in the nondividing reticulocyte — H5, while being synthesized, is completely dephosphorylated.[55,56] It was assumed that after dephosphorylation H5 interacts strongly with DNA and thereby brings about the condensation of the chromatin of the mature nonmammalian erythrocyte. This simple scheme does not explain the absence of nucleoli, the loss of internal nuclear matrix,[57,58] or the loss of nonhistone proteins from the condensed nuclei.[59]

The pycnotic nuclei still retain some RNA synthesis[60,61] which could be specified in later work.[62] According to this work about 15% of RNA polymerase II, 11% of ribonucleotide transferase, and 5% of RNA-depending terminal adenyltransferase are still present.

B. Reactivation of the Pycnotic Nucleus

Deeper insight in the processes of maturational changes could be obtained by the study of the "reactivation reaction" which can be induced in heterokaryons of chick erythrocytes with proliferating mammalian cells, obtained by fusion (for reviews see Harris[63] and Appels and Ringertz[64]). In such a system the dormant chick nuclei undergo reactivation, the amount of highly condensed chromatin decreases, nucleoli make their appearance, specific proteins are taken up,[65] and the nuclei swell and resume synthesis of DNA and RNA. At a later stage of reactivation, new chick proteins are synthesized in the heterokaryons.

One of the key proteins taken up from the host cytoplasm is RNA polymerase I, which mediates the transcription of the genes coding for rRNA.[65] It is found in reforming nucleoli of the chick nuclei. At the same time the synthesis of chick rRNA can first be detected.[66]

A RNA polymerase II, which accounts for the synthesis of the mRNAs, is also selectively taken up and is responsible for the entire synthesis of the chick proteins, estimated at around 30 to 40 polypeptides.[67] Reactivation goes hand in hand with marked changes of the antigenicity of the nucleus by acquisition of nucleoplasmic and nuclear envelope antigens from their host.[68,69] The reactivation of the chicken erythrocyte nuclei results in the expression of adult chicken globin genes[70] to a variable extent, depending on the mammalian parental cell. The level of globin synthesis was high in heterokaryons with rat myoblasts but very low in other types of heterokaryons.[71]

The initial steps of reactivation, such as increase in mass, nuclear expansion, and chromatin decondensation, as well as initiation of synthesis of RNA and DNA, are independent of the presence of the host cell nucleus. They occur in the same manner if nuclei of chicken erythrocytes are fused with denucleated cytoplasts.[72,73] The importance of host cell cytoplasm was shown in experiments in which inactive nuclei of chicken erythrocytes failed to bind a monoclonal antibody against a DNA-"tight-binding" protein, or a human antiserum against small nuclear RNPs, but did so after fusion with cytoplasts. The development of the antibody-binding sites was not affected by α-amanitin or cycloheximide, inhibitors of RNA and protein synthesis, indicating that the reactivation was not dependent on new formation of RNA or protein.

In earlier work, irradiation with UV-light to produce lesions in the DNA, prior to the preparation of heterokaryons, failed to affect either characteristic morphological changes or protein accumulation.[64] From all of these data it may be concluded that the cytoplasm of the host contains proteins which enter the chick nucleus, and that neither synthesis of DNA nor synthesis of mRNA, be it in the nucleus of the chick erythrocyte or in the host, is needed for the reactivation reaction. On the basis of these results, one may surmise that the stores of nucleus-specific proteins, or of factors mediating the uptake by the nucleus, are gradually depleted during differentiation and are absent in the orthochromatic erythroblasts of mammals or the mature erythrocytes of submammalian species.

C. Enucleation

Enucleation, a process specific for mammalian eyrthroid cells, has been studied time and time again, mostly on murine cells.[74-77] A classic description with a convincing documentation is presented in the monograph by Bessis.[3] At 37°C within a period of 10 to 30 min, active movements of the erythroblast may be observed with the formation of numerous cytoplasmic projections.[75] The nucleus is located in one of them. It is expelled after a few convulsions. There is an obvious similarity to the events which may be observed during the postmitotic separation of two daughter cells. In other studies[10,78] it was found that the nuclei migrate to the plasmalemma with an assembly of mitochondria and vesicles. In later work, a decrease in the cell mitochondria — and their actual expulsion — was demonstrated by electron microscopy,[8,9] which went hand in hand with a steep decline in cytochrome oxidase activity. During enucleation, a constriction of the nucleus may be observed with a corresponding change in the plasma membrane.[79] Actin-like filaments appear to be instrumental in the movement of the nucleus.

The process of enucleation goes hand in hand with the remodeling of the cell surface.[80] A nonuniform distribution of concanavalin A receptors, thought to bind mostly glycophorin, is observed. The receptors accumulate in the region of the cell membrane surrounding the nucleus[81-82] and are lost with enucleation.[83] In an outstanding study by Geiduschek and Singer[84] on the molecular changes in the membrane of mouse erythroid cells, performed using double-fluorescence techniques, it was found that spectrin synthesis decreased drastically shortly before enucleation. Lectin receptors showed enrichment in the plasma membrane surrounding the extruding nucleus, which had assumed an eccentric position close to the cell membrane. During extrusion, the nucleus becomes enveloped by a domain of the membrane which progressively changed its composition and was closely juxtaposed with the nuclear membrane, so that finally, only a narrow cytoplasmic bridge remained. The authors identified three cell membrane domains: one contained the extruded portion of the nucleus which is devoid of spectrin; a second contained the remainder of the nucleus; and a third represented the structure of the reticulocyte-to-be in which the spectrin is segregated into a definite arrangement.

During enucleation, there are also changes in the order of membrane lipids. Using merocyanine 542, an impermeant fluorescent dye which binds preferentially to membranes with loosely packed lipid molecules, it was shown that mice erythroblasts in their late stages exhibit a patchy fluorescence. These domains aggregate to a single region of the cell surface which surrounds the nucleus as it is extruded.[85] There is a close topical correspondence between the domains stainable by the dye and the cell surface receptors which are eliminated with the nucleus.[83] The changes in the lipid order of the cell membrane could well be due to disruption of the spectrin network, which may cause loss of normal lipid compositional asymmetry. The extruded nucleus is found surrounded by a thin layer of cytoplasm, carrying with it both the dye-binding domains and the cell surface receptors, about to undergo rapid destruction.

All in all, there can be little doubt that enucleation represents an abortive cell division with active movements of both the nucleus and the cell membrane. The energy for this cell division is, in all likelihood, provided by the ATP-generating mitochondria.

REFERENCES

1. **Lesly, J., Hyman, R., Schulte, R., and Trotter, J.,** Expression of transferrin receptor on murine hematopoietic progenitors, *Cell. Immunol.,* 83, 14, 1984.
2. **Nijhof, W. and Wierenga, P. K.,** A new system for the study of erythroid cell differentiation, *Exp. Hematol.,* 12, 115, 1984.
3. **Bessis, M.,** *Living Blood Cells and Their Ultrastructure,* Springer-Verlag, Berlin, 1973.
4. **Wickramasinghe, S. N., Cooper, E. H., and Chalmers, D. G.,** A study of erythropoiesis by combined morphologic quantitative cytochemical and autoradiographic methods. Normal human bone marrow, vitamin B_{12} deficiency and iron deficiency anemia, *Blood,* 31, 304, 1968.
5. **Heynen, M. J. and Verwilghen, R. L.,** A quantitative ultrastructural study of normal rat erythroblasts and reticulocytes, *Cell Tissue Res.,* 224, 397, 1982.
6. **Heynen, M. J., Tricot, G., and Verwilghen, R. L.,** Autophagy of mitochondria in rat bone marrow erythroid cells. Relation to nuclear extrusion, *Cell Tissue Res.,* 239, 235, 1985.
7. **Hellenau, T., Eden, E., and North, R.,** The structure and composition of rat reticulocytes. I. The ultrastructure of reticulocytes, *Blood,* 20, 347, 1962.
8. **Magnani, M., Stocchi, V., Dacha, M., and Fornaini, G.,** Hexokinase in developing rabbit erythroid cells, *Biochim. Biophys. Acta,* 802, 346, 1984.
9. **Nijhof, W. and Wierenga, P. K.,** The isolation and characterization of the erythroid progenitor cell: CFU-E, *J. Cell Biol.,* 96, 386, 1983.
10. **Simpson, C. F. and Kling, J. M.,** The mechanism of denucleation in circulating erythroblasts, *J. Cell Biol.,* 35, 237, 1967.
11. **Gasko, O. and Danon, D.,** Endocytosis and exocytosis in membrane remodeling during reticulocyte maturation, *Br. J. Haematol.,* 28, 463, 1974.
12. **Thorell, B.,** Studies on the formation of cellular substances during blood cell production, *Acta Med. Scand.,* 120(Suppl. 200), 1, 1947.
13. **Thorell, B.,** Cytochemistry of red blood cell maturation, *Folia Haematol.,* Leipzig, 78, 7, 1961.
14. **Grasso, J. A. and Woodard, J. W.,** The relationship between RNA synthesis and hemoglobin synthesis in amphibian erythropoiesis: cytochemical evidence, *J. Cell Biol.,* 31, 279, 1966.
15. **Grasso, J. A., Woodard, J. W., and Swift, M.,** Cytochemical studies of nucleic acids and proteins in erythrocytic development, *Proc. Natl. Acad. Sci. U.S.A.,* 50, 134, 1963.

16. **Grasso, J. A. and Woodard, J. W.**, DNA synthesis and mitosis in erythropoietic cells, *J. Cell Biol.*, 33, 645, 1967.
17. **Yataganos, X., Gahrton, G., and Thorell, B.**, DNA, RNA and hemoglobin during erythroblast maturation. A cytophotometric study, *Exp. Cell Res.*, 62, 254, 1970.
18. **Augustin, H. W. and Rapoport, S.**, Über den Reifungsprozeß kernhaltiger Erythrozyten. Phosphatverteilung und Nukleinsäuregehalt, *Acta Biol. Med. Germ.*, 5, 600, 1960.
19. **Augustin, H. W. and Rapoport, S.**, Über den Reifungsprozeß kernhaltiger Erythrozyten. Morphologie, Dimensionen, Wasser und Hämoglobingehalt, *Acta Biol. Med. Ger.*, 6, 213, 1961.
20. **Denton, M. J., Spencer, N., and Arnstein, H. R. V.**, Biochemical and enzymic changes during erythrocyte differentiation, *Biochem. J.*, 146, 205, 1975.
21. **Denton, M. J. and Arnstein, H. R. V.**, Characterization of developing adult mammalian erythroid cells separated by velocity sedimentation, *Br. J. Haematol.*, 24, 1, 1973.
22. **Clissold, P. M., Arnstein, H. R. V., and Chesterton, C. J.**, Quantitation of globin mRNA levels during erythroid development in the rabbit and discovery of a new β-related species in immature erythroblasts, *Cell*, 11, 353, 1977.
23. **Stewart, A. G., Clissold, P. M., and Arnstein, H. R. V.**, The initiation of globin synthesis in differentiating rabbit-bone-marrow erythroid cells, *Eur. J. Biochem.*, 65, 349, 1976.
24. **Reem, G. H. and Friend, C.**, Purine metabolism in murine virus-induced erythroleukemic cells during differentiation in vitro, *Proc. Natl. Acad. Sci. U.S.A.*, 72, 1630, 1975.
25. **Kabat, D., Sherton, C. C., Evans, L. M., Bigley, R., and Koler, R. D.**, Synthesis of erythrocyte-specific proteins in cultured Friend leukemia cells, *Cell*, 5, 331, 1975.
26. **Scher, W., Parkes, J., and Friend, C.**, Increased carbonic anhydrase activity in Friend erythroleukemia cells during DMSO-stimulated erythroid differentiation and its inhibition by BrdU, *Cell Differ.*, 6, 285, 1977.
27. **Jeffery, S. and Spencer, N.**, Changes in carbonic anhydrase isoenzyme content accompanying differentiation in rabbit erythroid cells, *FEBS Lett.*, 95, 323, 1978.
28. **Russanov, E. M., Kirkova, M. D., Setchenska, M. S., and Arnstein, H. R. V.**, Enzymes of oxygen metabolism during erythrocyte differentiation, *BioSci. Rep.*, 1, 927, 1981.
29. **Rapoport, S., Hartwig, A., and Gross, J.**, Abhängigkeit der Katalaseaktivität von der Reifung roter Blutzellen der Ente, *Acta Biol. Med. Ger.*, 32, 601, 1974.
30. **Hulea, S. A., Denton, M. J., and Arnstein, H. R. V.**, Ribonuclease activity during erythroid cell maturation, *FEBS Lett.*, 51, 346, 1975.
31. **Hulea, S. A. and Arnstein, H. R. V.**, Interacellular distribution of ribonuclease activity during erythroid cell maturation, *Biochem. Soc. Trans.*, 3, 911, 1976.
32. **Hulea, S. A. and Arnstein, H. R. V.**, Intracellular distribution of ribonuclease activity during erythroid cell development, *Biochim. Biophys. Acta*, 476, 131, 1977.
33. **Rost. G.**, Eigenschaften und Vorkommen eines Ribonuklease Hemmstoffes im stromafreien Hämolysat roter Blutkörperchen, *Acta Biol. Med. Ger.*, 3, 276, 1959.
34. **Burka, E. R.**, Erythroid cell RNase: activation by urea and localization to the cell membrane, *J. Clin. Invest.*, 50, 60, 1971.
35. **Nijhof, W., Wierenga, P. K., Staal, G. E. J., and Jansen, G.**, Changes in activities and isoenzyme patterns of glycolytic enzymes during erythroid differentiation in vitro, *Blood*, 64, 607, 1984.
36. **Setchenska, M. S. and Arnstein, H. R. V.**, Changes in the lactate dehydrogenase isoenzyme pattern during differentiation of rabbit bone-marrow erythroid cells, *Biochem. J.*, 170, 193, 1978.
37. **Narita, M., Yanagawa, S. I., Sasaki, R., and Chiba, H.**, Synthesis of 2,3-bisphosphoglycerate synthase in erythroid cells, *J. Biol. Chem.*, 256, 7059, 1981.
38. **Narita, H., Ikura, K., Yanagawa, S., Sasaki, R., Chiba, H., Saimyoji, H., and Kumagai, N.**, 2,3-Bisphosphoglycerate in developing rabbit erythroid cells, *J. Biol. Chem.*, 255, 5230, 1980.
39. **Jansen, G., Koenderman, L., Hennekam, R. C. M., Beemer, F. A., Cats, B. P., and Staal, G. E. J.**, Red cell pyruvate kinase deficiency associated with hydrops fetalis and intrauterine disseminated intravascular coagulation, in Enzymological Aspects *of Erythroid Cell Differentiation*, Thesis, University of Utrecht, Utrecht, the Netherlands, 1984, 71.
40. **Marie, J., Simon, M. P., Dreyfus, J. C., and Kahn, A.**, One gene, but two messenger RNAs encode liver L and red cell L' pyruvate kinase subunits, *Nature*, 292, 70, 1981.
41. **Kahn, A., Marie, J., Garreau, H., and Sprengers, E. D.**, The genetic system of L-type pyruvate kinase forms in man, *Biochim. Biophys. Acta*, 523, 59, 1978.
42. **Takegawa, S., Fujii, H., and Miwa, S.**, Change of pyruvate kinase isozymes from M- to L-type during development of the red cell, *Br. J. Haematol.*, 54, 467, 1983.
43. **Max-Audit, T., Testa, U., Kechemir, D., Titeux, M., Vainchenker, W., and Rosa, R.**, Pattern of pyruvate kinase isozymes in erythroleukemic cell lines and in normal human erythroblasts, *Blood*, 64, 930, 1984.

44. **Vesell, E. S. and Bearn, J. C.,** Variations in the lactic dehydrogenase of vertebrate erythrocytes, *J. Gen. Physiol.,* 45, 553, 1962.

45. **Starkweather, W. H., Cousineau, L., Schoch, H. K., and Zarafonetis, C. J.,** Alterations of erythrocyte lactate dehydrogenase in man, *Blood,* 26, 63, 1965.

46. **Setchenska, M. S. and Arnstein, H. R. V.,** Changes in malate dehydrogenase isoenzymes during differentiation of rabbit bone marrow erythroid cells, *Int. J. Biochem.* 10, 817, 1979.

47. **Lostanlen, D. and Kaplan, J.-C.,** Expression of NADH-cytochrome b₅ reductase during dimethyl sulfoxide-induced differentiation of Friend erythroleukemia cells, *FEBS Lett.,* 143, 35, 1982.

48. **Lindena, J., Sommerfeld, U., Höpfel, C., and Wolkersdorfer, R.,** Enzyme activities in blood cells of man and dogs after separation on a discontinuous Percoll gradient, *Enzyme,* 29, 100, 1983.

49. **Lindena, J., Sommerfeld, U., Höpfel, C., and Wolkersdorfer, R.,** Enzyme activities in rabbit, guinea pig, rat and mouse blood cells after separation on a discontinuous Percoll gradient, *Enzyme,* 29, 229, 1983.

50. **Neelin, J. M., Callahan, P. X., Lamb, D. C., and Murray, K.,** The histones of chicken erythrocyte nuclei, *Can. J. Biochem.,* 42, 1743, 1964.

51. **Purkayastha, R. and Neelin, J. M.,** Comparison of histones from avian erythroid tissues by zone electrophoresis, *Biochim. Biophys. Acta,* 127, 468, 1966.

52. **Appels, R. and Wells, J. R. E.,** Synthesis and turnover of DNA-bound histone during maturation of avian red blood cells, *J. Mol. Biol.,* 70, 425, 1972.

53. **Moss, B. A., Joyce, W. G., and Ingram, V. M.,** Histones in chick embryonic erythropoiesis, *J. Biol. Chem.,* 248, 1025, 1973.

54. **Appels, R., Bolund, L., and Ringertz, N. R.,** Biochemical analysis of reactivated chick erythrocyte nuclei isolated from chick/HeLa heterokaryons, *J. Mol. Biol.,* 87, 339, 1974.

55. **Sung, M. T., Harford, J., Bundman, M., and Vidalakas, G.,** Metabolism of histones in avian erythroid cells, *Biochemistry,* 16, 279, 1977.

56. **Sung, M. T.,** Phosphorylation and dephosphorylation of histone V(H5): controlled condensation of avian erythrocyte chromatin, *Biochemistry,* 16, 286, 1977.

57. **LaFond, R. E. and Woodcock, C. L. F.,** Status of the nuclear matrix in mature and embryonic chick erythrocyte nuclei, *Exp. Cell Res.,* 147, 31, 1983.

58. **LaFond, R. E., Woodcock, H., Woodcock, C. L., Kundahl, E. R., and Lucas, J. J.,** Generation of an internal matrix in mature avian erythrocyte nuclei during reactivation in cytoplasts, *J. Cell Biol.,* 96, 1815, 1983.

59. **Jones, R., Okamura, C., and Martin, T. E.,** Loss of nonhistone proteins in pyknosis, *J. Cell Biol.,* 86, 235, 1980.

60. **Attardi, G., Parnas, H., and Attardi, B.,** Pattern of RNA synthesis in duck erythrocytes in relationship to the stage of cell differentiation, *Exp. Cell Res.,* 62, 11, 1970.

61. **Schechter, N. M.,** Studies of RNA polymerase in mature avian erythrocytes, *Biochim. Biophys. Acta,* 308, 129, 1973.

62. **Longacre, S. S. and Rutter, W. J.,** Nucleotide polymerases in the developing avian erythrocyte, *J. Biol. Chem.,* 252, 273, 1977.

63. **Harris, H.,** *Cell Fusion,* Harvard University Press, Cambridge, Mass., 1970.

64. **Appels, T. and Ringertz, N. R.,** Chemical and structural changes within chick erythrocyte nuclei introduced into mammalian cells by cell fusion, *Curr. Top. Dev. Biol.,* 9, 137, 1975.

65. **Scheer, U., Lanfranchi, G., Rose, K. M., Franke, W. W., and Ringertz, N. R.,** Migration of rat RNA polymerase I into chick erythrocyte nuclei undergoing reactivation in chick-rat heterokaryons, *J. Cell Biol.,* 97, 1641, 1983.

66. **Bramwell, M. E.,** Detection of chick rRNA in the cytoplasm of heterokaryons containing reactivated chick red cell nuclei, *Exp. Cell Res.,* 112, 63, 1978.

67. **Zuckerman, S. H., Linder, S., and Ringertz, N. R.,** Transcription of chick genes by mammalian RNA polymerase II in chick erythrocyte-mammalian cell heterokaryons, *J. Cell. Physiol.,* 113, 99, 1982.

68. **Jost, E., d'Arcy, A., and Ely, S.,** Transfer of mouse nuclear envelope specific proteins to nuclei of chick erythrocytes during reactivation in heterokaryons with mouse A9 cells, *J. Cell Sci.,* 37, 97, 1979.

69. **Nyman, U., Lanfranchi, G., Bergman, M., and Ringertz, N. R.,** Changes in nuclear antigens during reactivation of chick erythrocyte nuclei in heterokaryons, *J. Cell. Physiol.,* 120, 257, 1984.

70. **Linder, S., Zuckerman, S. H., and Ringertz, N. R.,** Reactivation of chicken erythrocyte nuclei in heterokaryons results in expression of adult chicken globin genes, *Proc. Natl. Acad. Sci. U.S.A.,* 78, 6286, 1981.

71. **Linder, S., Zuckerman, S. H., and Ringertz, N. R.,** Patterns of chick gene activation in chick erythrocyte heterokaryons, *J. Cell Biol.,* 95, 885, 1982.

72. **Dupuy-Coin, A. M., Ege, T., Bouteille, M., and Ringertz, N. R.,** Ultrastructure of chick erythrocyte nuclei undergoing reactivation in heterokaryons and enucleated cells, *Exp. Cell Res.,* 101, 355, 1976.

73. **Woodcock, C. L. F., LaFond, R. E., Woodcock, H., Baldwin, L. A., and Bhorjee, J. S.,** Reactivation of avian erythrocyte nuclei in mammalian cytoplasts. A dominant role for pre-existing cytoplasmic components, *Exp. Cell Res.,* 154, 155, 1984.
74. **Albrecht, M.,** Studien zur Frage der Erythroblastenentkernung an Kulturen von Meerschweinchenknochenmark, *Acta Haematol.,* 6, 83, 1951.
75. **Bessis, M. and Bricka, M.,** Aspect dynamique des cellules du sang. Son étude par la microcinématographie en contraste de phase, *Rev. Hematol.,* 7, 407, 1952.
76. **Rind, H.,** Kinetik der Erythroblastenentkernung mit Mikrofilmdemonstration (Phasenkontrast), *Folia Haematol. (Leipzig),* 74, 262, 1956.
77. **Awai, M., Okada, S., Takebayashi, J., Kubo, T., Inoué, M., and Seno, S.,** Studies on the mechanism of denucleation of the erythroblast, *Acta Haematol.,* 39, 193, 1968.
78. **Skutelsky, E. and Danon, D.,** An electron microscopic study of nuclear elimination from the late erythroblast, *J. Cell Biol.,* 33, 625, 1967.
79. **Repasky, E. A. and Eckert, B. S.,** A reevaluation of the process of enucleation in mammalian erythroid cells, *Prog. Clin. Biol. Res.,* 55, 679, 1981.
80. **Wraith, D. C. and Chesterton, C. J.,** Cell-surface remodeling during mammalian erythropoiesis, *Biochem. J.,* 208, 239, 1982.
81. **Skutelsky, E. and Farquhar, M. G.,** Variation of distribution of ConA receptor sites and anionic groups during red cell differentiation, *J. Cell Biol.,* 71, 218, 1976.
82. **Schlegel, R. A., Phelps, B. M., Waggoner, A., Terada, L., and Williamson, P.,** Binding of merocyanine 540 to normal and leukemic erythroid cells, *Cell,* 20, 321, 1980.
83. **Schlegel, R. A., Phelps, B. M., Cofer, G. P., and Williamson, P.,** Enucleation eliminates a differentiation-specific marker from normal and leukemic murine cells, *Exp. Cell Res.,* 139, 321, 1982.
84. **Geiduschek, J. B. and Singer, S. J.,** Molecular changes in the membranes of mouse erythroid cells accompanying differentiation, *Cell,* 16, 149, 1979.
85. **Cofer, G. P., Williamson, P., and Schlegel, R. A.,** Plasma membrane lipid order of leukemic and normal immature avian erythroid cells, *Exp. Cell Res.,* 153, 32, 1984.

Chapter 3

THE BIOLOGY OF THE RETICULOCYTE

I. INTRODUCTION

The reticulocyte is a stage of maturation of erythroid cells well-defined by morphological and biochemical criteria. It is characterized by elimination of the nucleus in mammalia or its inactivation in nonmammalian red cells on the one hand, and on the other, by the presence of functional mitochondria and ribosomes. The transition to the erythrocyte occurs by way of complete degradation of mitochondria and ribosomes, and partial breakdown of the cell membrane with partial or complete loss of various receptors and active transport systems. In addition there is a selective decay of organelle and cytosol enzymes.

The reticulocyte derives its name from the characteristic network which appears when it is stained with basic dyes.[1] This network is not preexisting, but rather represents a precipitate of the ribosomal ribonucleic acid with basic dyes, yet still contains some tightly bound ribosomal proteins. It took nearly 70 years to establish the causal connection between the characteristic staining material "substantia reticulofilamentosa" on the one hand, and RNA on the other. In 1936 the occurrence of RNA in the reticulocyte was demonstrated cytophotometrically by Caspersson.[2] Indirect evidence that the staining material consisted of ribonuclear proteins[3] was followed in 1952 by chemical proof of the presence of RNA in the reticulocyte.[4] The substantia reticulofilamentosa was demonstrated to be an artifact.[5,8] Under well-defined conditions, there is proportionality between the amount of brilliant cresyl blue taken up by the reticulocyte and its content of RNA.[5,6]

Various basic dyes may be used for staining, such as brilliant cresyl blue, methylene blue, neutral red, and perhaps with higher sensitivity, the fluorescent dye acridine orange.[9] As pointed out by Bessis and Breton-Gorius,[8] the staining is postvital. As simple as the procedure appears, it has its limitations and pitfalls. For one, the recognition of the reticulocyte has a threshold of visibility. The appearance of the precipitate is variable and does not always present the characteristic network. Very mature reticulocytes may contain only one or two specks and thus easily escape detection. The existence of a threshold for the perception of a precipitate has the consequence that the RNA in mature erythrocytes, the concentration of which ranges from 10 to 35 mg/100 mℓ cells, remains undetected.[10] Also, keep in mind that the apparent size of the precipitate is proportional to the two-thirds power of the concentration. In plasma-free solutions the staining is poor, probably because lowering of the pH retards the penetration of the dye. Therefore, a phosphate buffer solution of pH 8 is recommended.[11]

Under the phase contrast microscope the unstained living reticulocyte is polylobulated. These lobuli are caused by slow movements of the protoplasm, which contracts in some areas while throwing out projections in others.[12] These movements decrease as the cell matures (for other morphologic features see Bessis[12]). With the scanning electron microscope, the multishaped and variable three dimensional form of the reticulocyte is clearly demonstrated (Figure 6).

Under normal conditions the maturation time of the reticulocyte is estimated to be about 3 days, with 1 day spent in peripheral blood. Such reticulocytes are highly mature as judged by their low content of RNA and mitochondria; the degraded state of the mitochondria; and by functional criteria such as low respiration, low capacity of the respiratory chain enzymes, uncoupling of mitochondrial respiration, and a low or absent Pasteur effect.[13,14] No reliable data exist on the biochemical characteristics of bone marrow reticulocytes in the normal steady state of erythropoiesis.

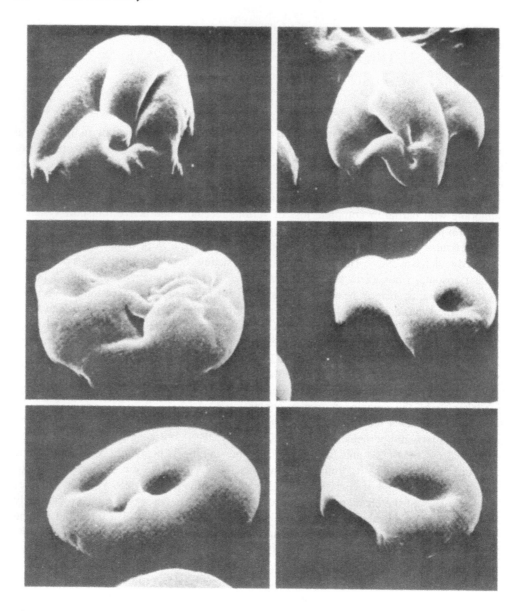

FIGURE 6.   Reticulocytes as seen with the scanning electron microscope. (From Bessis, M., *Living Blood Cells and Their Ultrastructure*, Springer-Verlag, Berlin, 1973. With permission.)

An increased number of reticulocytes in the peripheral blood is a sign of stimulated erythropoiesis. Usually, the reticulocyte count is expressed as a percentage of total red cells. This mode of expression may be misleading if the total number of red cells deviates greatly from the norm. Therefore it would be preferable to state the absolute number of reticulocytes, which reflects the activity of the bone marrow independent of alterations in the number of red cells per volume of blood. A reticulocyte index which provides an approximate correction based on the hematocrit has been proposed.[15]

II. NORMO- AND MACRORETICULOCYTES

The reticulocytes normally found in the peripheral blood are slightly larger than normal erythrocytes and contain about 90% of the hemoglobin concentration of erythrocytes.[16] In contrast, one finds under conditions of highly stimulated erythropoiesis reticulocytes which are double in size.[17] In experiments on rabbits with phenylhydrazine-induced anemia, cell size increased from a mean value of 67 μm^3 to peak values of 140 μm^3, whereas hemoglobin concentration declined from 20 mM to less than 13 mM Hb/$_4$ cells. Similar results were obtained in anemic rats and guinea pigs both after administration of phenylhydrazine and after bleeding.[18,19] Macroreticulocytes were also found after stimulation with erythropoie-tin.[19,20] All the authors of these and other publications[21-23] proposed that under the effect of erythropoietic stress, a generation in the differentiation sequence is skipped and cells which double their normal size are prematurely released from the bone marrow. An alternative hypothesis has been formulated by Ganzoni et al.,[24] who claimed that macroreticulocytes represent normal bone marrow reticulocytes, which once in the circulation, lose hemoglobin and water to become normocytes. The appearance of these cells in the circulation would therefore only be due to their shift from the bone marrow to the blood. This hypothesis is supported to some extent by other work[25-27] in which it was shown that a remodeling of the cell surface occurs with significant losses of membrane material. It is, however, hardly conceivable that shrinking of the macroreticulocyte to the extent necessary to convert them to normocytes could occur. Somewhat connected with this point is the question of the fate of macro-red cells. Many reports, both on experimental animals and on patients, indicate clearly that these are rapidly eliminated from the peripheral blood even before they can mature.[28-30] The most decisive type of experiment was performed in the following manner by Rapoport et al.[31] A normal rabbit received by exchange transfusion from anemic donor rabbits blood which was highly enriched in reticulocytes. A rapid loss of reticulocytes with a half-life of 18 hr and a corresponding shift of the Price-Jones curve was observed. These changes were considerably faster than expected from maturation time. In an independent study on the lifespan of rabbit phenylhydrazine-induced reticulocytes transfused into a normal host, it was found that half of the cell population disappeared within 2 days, with a lifespan one order of magnitude lower than the other half of the cells.[32] In rabbits, the macro-red cell differs qualitatively from the normocyte by its content of large amounts of lipoxygenase, and quantitatively by a larger concentration of RNA.[33] The macroreticulocyte has both increased rate of protein synthesis as well as ATP values which are nearly twice the normal concentration.[17,19,30,34,35] The differential count of bone marrow cells — with the great increase in the proportion of basophilic and polychromatic erythroblasts and the contrasting decrease in orthochromatic erythroblasts — would also suggest the skipping of the last division. All in all, the conclusion that macroreticulocytes and macrocytes resulting from macroreticulocytes are distinct in origin, maturation, and dwelling time in the peripheral blood seems justified.

What could be the mechanism for a skipped division? A reasonable hypothesis has been proposed by Stohlmann[36] and others.[37,38] It is assumed that hemoglobin, when present in a concentration of about 12.5 mM Hb/$_4$ cells, may block synthesis of nucleic acids and cell division.[36,39] In the normal steady state of erythropoiesis, the critical hemoglobin concentration is reached just before the emergence of the orthochromatic erythroblast. With high levels of erythropoietin, hemoglobin synthesis starts earlier and is accelerated. Thus, it is conceivable that the critical level of hemoglobin is reached earlier, i.e., at the polychromatic erythroblast stage, and results in an orthochromatic macroerythroblast. Support for this concept was provided by scanning microspectrophotometry.[38] This hypothesis would explain that administration of erythropoietin even in nonanemic animals causes the appearance of macroreticulocytes, and also, that lack of iron, necessary for the effectuation of the action of erythropoietin, suppresses macroreticulocytosis.

Another explanation for the appearance of macroerythroid cells may be proposed. It is conceivable that high erythropoietin concentrations may stimulate an excessive synthesis of membrane and cytoskeletal proteins with their complement of phospholipids out of proportion with the accumulation of globin. In this manner, large and relatively underhemoglobinized cells might arise. This hypothesis would circumvent the difficulty that a skipped division should lead to an underproduction of cells. This remains to be tested.

III. STAGES OF RETICULOCYTE MATURITY

The differences between normo- and macroreticulocytes should be distinguished from the maturational changes within either cell type. Unfortunately, little is known concerning their progression in the normal reticulocyte. In recent years, however, criteria have become known which permit classification of macroreticulocytes with regard to their maturity. The most immature reticulocytes have their mRNA for lipoxygenase completely masked.[40] They are found in the bone marrow of normal and anemic rabbits and in the topmost 1 to 5% fraction of lowest density of the peripheral blood at the height of anemia. The next stage of maturation is represented by reticulocytes with unmasked mRNA for lipoxygenase, the mitochondria of which are, however, not yet susceptible to the attack by lipoxygenase. These are found in the upper 15% of density-fractionated peripheral blood of anemic rabbits. A third stage of maturation is characterized by susceptibility of the mitochondria to lipoxygenase and subsequent ATP-dependent proteolysis. The fourth and most mature stage is represented by reticulocytes the mitochondria of which have already undergone the deleterious effects of lipoxygenase, with ongoing ATP-dependent proteolysis.[41,42]

These properties undergo qualitative changes and thus appear suitable for an objective, nonarbitrary classification. All other distinctions appear to present a continuous series of quantitative changes. Their use in defining maturational stages must therefore have an element of arbitrariness. This restriction applies to density per se, and correspondingly, to hemoglobin concentration, which increases gradually throughout the stages of reticulocyte maturation. It also applies to the continuous decrease of RNA.[43]

The number of ribosomes and RNA decreases during maturation, but there is evidence for a special type, "Line 2" reticulocytes, according to Borsook et al.[19] (see also References 44 to 51), that have an extremely high RNA concentration. The same pattern is observed with respect to respiration and cytochrome oxidase, and one would expect it to be reflected in the number of mitochondria.

The activity of a variety of enzymes apparently does not reflect early maturational changes but does undergo drastic diminution in the last stages of maturation. There are three possible exceptions: DNase and glucokinase are found only in the early stages of anemia in which one would expect to find the most immature reticulocytes. Inorganic pyrophosphatase exhibits a high peak in the presumed Line 2 cells.[49]

Morphologically, mitochondria represent a progression from a preponderance for orthodox and condensed forms to forms showing various degrees of deterioration.[14,52] Number of mitochondria and ribosomes and percentage of polysomes decrease (Figures 7 and 8).

Reticulocytes in the peripheral blood under normal steady-state conditions represent a highly advanced stage of maturation with low oxygen consumption and a high degree of uncoupling. In man, even during anemia with reticulocytosis, reticulocytes usually appear to be quite mature.[13] Additionally, there may be species differences between man and rabbit. As indicated in Table 1, the activities of cytochrome oxidase and of the succinate-cyt c system are in human reticulocytes about one order of magnitude lower than in stress reticulocytes of rabbits, whereas the activities of the $NADH_2$-cyt c reductase system are comparable.

IV. THE INFLUENCE OF LOCAL ENVIRONMENT

There are various indications that the local environment influences in subtle ways the

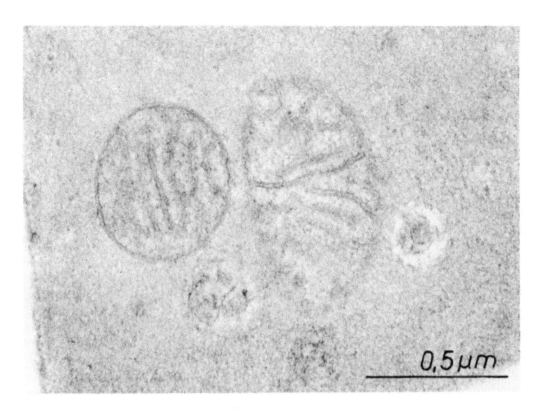

FIGURE 7. Mitochondria of human reticulocytes — orthodox forms. (Final magnification × 75,000.) (From Richter-Rapoport, S. K. N., et al., *Acta Biol. Med. Ger.*, 36, 53, 1977. With permission.)

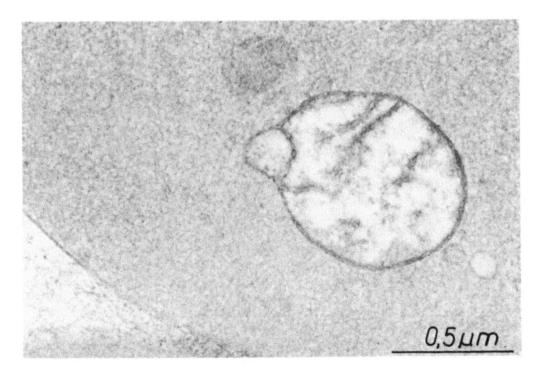

FIGURE 8. Mitochondrium of human reticulocyte — degraded form. (Final magnification × 65,000.) (From Richter-Rapoport, S. K. N., et al., *Acta Biol. Med. Ger.*, 36, 53, 1977. With permission.)

Table 1
COMPARISON OF RETICULOCYTES IN
MAN AND RABBIT

	Man	Rabbit
Cyt c oxydase	24 ± 5	270 ± 40
NADH$_2$-cyt c reductase	240 ± 30	330 ± 40
O$_2$ consumption	30 ± 4	110 ± 9
Reticulocyte (%)	28.5 ± 3.4	30.6 ± 2.4

Note: Values in 10^{-3} μM/min·mℓ cells.[13]

characteristics of erythroid cells including reticulocytes. With intensive loss or destruction of red cells, the capacity for erythropoiesis of red bone marrow increases about fivefold[53,54] by formation of new erythropoietic islands. Furthermore, previously dormant fatty bone marrow is activated. There is evidence that the new erythropoietic sites produce a relatively immature red cell population even during recovery from anemia. A sensitive indicator of activated sties in man appears to be the creatine concentration of red cells. The most striking example was observed in a case of osteopetrosis with practically complete obliteration of bone marrow. The erythropoietic sites were located in liver and spleen. The red cells produced were somewhat larger (102 μm^3) than normal, with a hemoglobin concentration lower by 10% and a reduced lifespan. The creatine concentration in all density-separated fractions were severalfold higher than normal, with a calculated peak value of 50 mM creatine per liter for the reticulocytes.[55,56] Chronic hemolytic anemias such as those caused by pyruvate kinase mutants, with extensive extramedullary erythropoiesis, exhibit a similarly high creatine concentration.[55] The red cells of infants born at term exhibit a higher creatine concentration than those of adults; this difference is even more pronounced in prematures.[57-60] It is reasonable to assume that the local environment of erythropoiesis is related to the higher creatine concentration. A single blood donation of volunteers was followed by a surprisingly large and long-lasting effect on the creatine concentration of red cells.[58] It was interpreted as a reflection of activation of new erythropoietic sites in the bone marrow.

V. THE CELL ORGANELLES

A. Mitochondria

Information on reticulocyte mitochondria is of general interest since it offers the opportunity to study the aging and degradation process of these organelles. Because of maturational changes, reticulocytes mitochondria are extremely heterogeneous. In density, i.e., age-fractionated reticulocytes, the most immature cells contain mostly orthodox and condensed forms of occasionally vacuolized mitochondria.[14,50,52,61-63] Many mitochondria exhibit a few cristae and a loose matrix structure. In more mature cells, degraded forms predominate, particularly involving the cristae which are often destroyed with the persistence of some remnants. Vacuoles are found in the matrix which join to form vesicles.[14,64] Some ferritin granules are found, possibly as an expression of inhibition of heme synthesis. The degradation of the mitochondria is associated with swelling. The average area in electron microscopy sections is 0.07 to 0.15 μm^2, similar to that of liver mitochondria; their number is strongly related to the maturity of the cell. Also, there is a strict correlation between the activity of cytochrome oxidase and the number of mitochondria (Figure 9). The amount of mitochondrial protein was calculated to vary between 18 to 30 mg/mℓ reticulocytes in anemic rabbits.[65]

Isolated mitochondria are likewise multiform. There appears to exist a correlation between the specific activity of succinate dehydrogenase and the form of the mitochondria. Those with the highest activity are predominantly the condensed type; those with intermediate

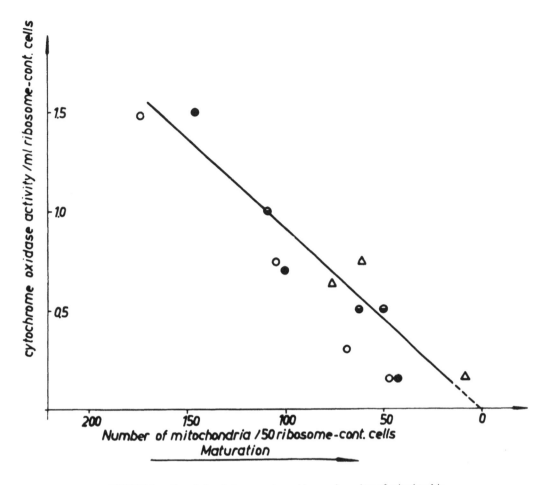

FIGURE 9.   Correlation between cyt c oxidase and number of mitochondria.

activity, the orthodox type; and those with the lowest activity exhibit mostly degraded forms.[66] These changes may be related to the progressive effects of lipoxygenase and ATP-dependent proteolysis (see Chapters 12 and 13).

Recently, attempts were undertaken to separate mitochondria according to maturity and based on density differences. It was found that the density of the mitochondria isolated from fractionated rabbit reticulocytes decreased with progressive maturity. In a complementary experiment, mitochondria were separated according to density from a mixed population of reticulocytes with corresponding results (Figure 10). The higher density of immature reticulocyte mitochondria is most plausibly explained by their condensed state.[67]

The chemical composition of reticulocyte mitochondria shows interesting differences compared to those of liver (Table 2),[68] with about double the content of cytochromes b, c, and c_1. The cytochrome a content per milligram mitochondrial protein is also the highest among mitochondria so far described.[65]

The large amount of phospholipids consist mainly of phosphatidyl choline, phosphatidyl-ethanolamine, sphingomyelin, cardiolipin, and phosphatidylinositol (Table 3).[69] Phosphatidylserine is practically absent. Lysophosphatides are found in minute amounts, indicating that phospholipase A_2 activity may be limiting, compared to phospholipase D-type enzymes. The high sphingomyelin content is unusual; it is nearly one order of magnitude larger than that of liver mitochondria. There is good evidence that the sphingomyelin content is not due to contamination with cell membranes. The distribution of fatty acids in phospholipids also

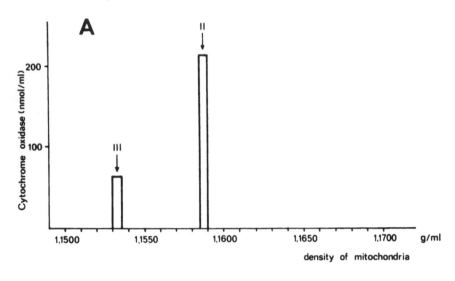

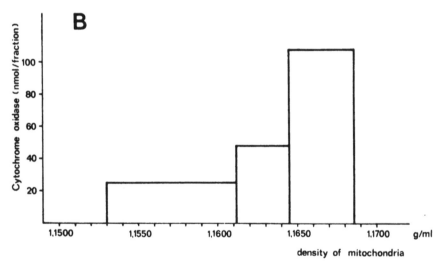

FIGURE 10.   Density differences between mature and immature rabbit reticulocyte mitochondria. (A) Density-separated mitochondria from a mixed population; (B) densities of mitochondria from immature and mature reticulocytes.

Table 2
COMPARISON OF RABBIT RETICULOCYTE
AND RAT LIVER MITOCHONDRIA[68]

Component	Protein (μmol/mg)	Reticulocyte/liver quotient
Cytochrome oxidase	0.3—0.6	2.4
Cytochrome b	0.58	1.8
Cytochrome c_1	0.42	2.2
Cytochrome c	0.65	1.8
CoQ	1.45	1.0
Phospholipid	0.37[a]	2.6
NAD$^+$	0.16	0.04
NADP	0.47	0.09

[a]   mg/mg Protein.

Table 3
DISTRIBUTION OF
PHOSPHOLIPIDS IN
MITOCHONDRIA OF RABBIT
RETICULOCYTES[69]

Phospholipid	Phospholipid-P (%)
Phosphatidylcholine	39.0 ± 4.5
Phosphatidylethanolamine	31.4 ± 3.6
Sphingomyelin	12.5 ± 5.3
Cardiolipin	8.9 ± 2.0
Phosphatidylinositol	7.6 ± 2.6
Lysophosphatidylcholine	0.6 ± 0.4

shows some characteristic features. Phosphatidylcholine, phosphatidylethanolamine, and phosphatidylinositol show great similarity in composition, with the distinction that the hexodecanoic acid content is highest in phosphatidyl choline and lowest in phosphatidylinositol, while the reverse is true for octadecanoic acid. The sum of the C_{18}-fatty acids amounts to about 50% or more of the total; unsaturated fatty acids predominate. Cardiolipin differs strongly from the other phospholipids, containing only unsaturated C_{18}-fatty acids, with the bulk (60%) consisting of linoleic acid. The fatty acid composition of phospholipids so far discussed does not differ greatly from that of other types of mitochondria. Sphingomyelins are characterized by high concentration of unsaturated fatty acids with 22- or 24-C atoms. A comparatively high percentage of free fatty acids is found; this appears to be correlated with the proportion of degraded forms.[68]

In studies on subcellular fractions of rabbit reticulocytes, mitochondria exhibited an active incorporation of fatty acids into phospholipids. The acylation processes characteristic for mitochondria from other tissues, namely acylation of sn-glycerol 3-phosphate and lysophospholipids, proceeds actively.[70] The authors drew the conclusion that the degradation of mitochondria during maturation apparently is not related to their inability either to form phosphatidic acid or to repair their phospholipids by reacylation.

Isolated mitochondria are extremely depleted of both NAD and NADP.[65,71] This loss is probably due to their permeability for pyridine nucleotides and to the high activity of NAD^+-glycohydrolase on the outer cell membrane.[72] At the physiological intracellular NAD^+ concentration of about 0.3 mM, there is an active uptake of NAD so that the mitochondria contain pyridine nucleotides within the same range as the mitochondria of other cell types. It may be concluded that in the intact reticulocyte, the mitochondria contain a similar complement of pyridine nucleotides.[71]

Respiratory characteristics of the reticulocyte mitochondria correspond on the whole to those of other animal cells. They contain a full complement of citric cycle enzymes, glutamate dehydrogenase, and aspartate aminotransferase (GOT). The enzyme pattern resembles that of other mitochondria, with malate dehydrogenase being the highest, followed by aspartate aminotransferase and succinate dehydrogenase. Noticeable is the low activity of succinate dehydrogenase, probably owing to inactivation by lipoxygenase, as well as the absence of NAD-dependent isocitrate dehydrogenase.[65] Isocitrate dehydrogenase was higher than in liver mitochondria but similar to heart mitochondria;[66] glutamate dehydrogenase, on the other hand, shows the opposite pattern: it is low in heart reticulocyte mitochondria compared to liver reticulocyte mitochondria.[65] Oxaloacetate is a particularly good substrate of both mitochondria and intact reticulocytes.[65,73,74] Mitochondria from rabbit reticulocytes contain considerable amounts of hexokinase, as will be discussed later in this chapter.

In rat reticulocyte, monoamine oxidase, a marker for the outer membrane of mitochondria, has been demonstrated, which is a clear indication for the functionality of this structure.[75,76]

B. Lysosomes and Other Organelles

Morphologic work indicates that there are few if any lysosomes in reticulocytes. They disappear after the last cell division. However, some autophagosomes are found.[14,52,62,63] The claim that functional lysosomes persist in reticulocytes[77] is contradicted by both morphological evidence and further biochemical studies.[78-80] In a study of normal individuals and patients with homozygous β-thalassemia, the activities of enzymes indicative of lysosomes, i.e., β-*N*-acetylglucosaminidase, β-galactosidase, and β-glycuronidase, were assessed.[78] Their activities, particularly of β-*N*-acetylglucosaminidase, were very low and entirely cell-membrane bound. In thalassemic cells, some activity was found in the cytosol which was linearly correlated with the number of normoblasts. This may have been due to lysosomes. The author suggests, in agreement with others,[81] that a fusion of the lysosomes with the inner side of the cell membrane may occur during maturation. This explains localization of enzyme activities usually found in lysosomes in the cell membrane of mature red cells. Fusion of the lysosomes to the cell membrane may also explain the masked conditions of most of the enzyme activities of the cell membrane, although other explanations are not excluded. One may conclude that lysosomes play an insignificant part in the maturational changes of reticulocytes. On the other hand, a specific role of endosomes within acid pH has been demonstrated with respect to the endocytotic cycle of transferrin (see Chapter 10).

After the last mitosis there is a drastic decrease in the amount of rough endoplasmic reticulum and of the Golgi apparatus.[62,82] There appear to be species differences, since in mouse reticulocytes small amounts of both endoplasmic reticulum and Golgi apparatus have been described.[8] If one considers the NADPH-cytochrome c reductase as a marker enzyme for the endoplasmic reticulum, this fraction amounts to about 1/100 of the value in liver cells.[70] A strict correlation between the activities of CDP-choline-1,2 diglyceride phosphatidyl choline transferase and NADPH-cytochrome c reductase was observed, indicating that both activities reflect the presence of traces of the endoplasmic reticulum.

VI. ENZYMES

There are many differences in activity and type of enzymes between the reticulocyte and mature erythrocyte. They include the disappearance or drastic reduction of those enzymes which are located in organelles that are degraded during maturation. Also in this category are enzyme activities and constituents of the respiratory chain which is situated at the inner membrane of the mitochondria, as well as enzyme activities of the outer membrane, and many of the mitochondrial matrix enzymes. The same is true for most of the enzymes of the endoplasmic reticulum, the Golgi apparatus, lysosomes, and the cell membrane. Some enzymes like NADH-cytochrome b_5 reductase become cytosolic by splitting off their hydrophobic anchor and acquire or retain a function, in this case, as a methemoglobin reductase.[83,84] Some decreases in enzyme activity are due to the lability of one or more isoenzymes and are thus connected with a simplification of the isoenzyme pattern. In addition to gene-determined changes in the profile of isoenzymes, there may be posttranslational modifications. A clear-cut example for such a mechanism has been provided by Johnson et al.[85] These workers found three isoenzymes of hypoxanthine-guanine phosphorbosyl transferase in human erythroid cells. One isoenzyme predominates in the reticulocyte, the IP of which is identical with the lymphoblast enzyme. During maturation of the reticulocyte, two more acidic isoenzymes appear that correspond to dimers containing one or two shortened peptide chains. It is not unlikely that similar effects are responsible for the microheterogeneity of hexokinase and galactose-1-phosphate uridyl transferase,[86] among others.

The following discussion is selective and is focused on (1) the enzymes related to glycolysis, and (2) the hydrolases. Other groups of enzymes are discussed in the context of the subject matter of other chapters, which include protein synthesis enzymes, the proteases, amino acid metabolism enzymes, heme synthesis, and the cAMP system.

A. Glycolysis Enzymes

The cytosol of the reticulocyte as compared to the erythrocyte is characterized by much greater activity of some glycolysis enzymes and of the oxidative pentose phosphate pathway.

Several studies have dealt with glycolytic enzyme changes during the transition from the reticulocyte to the erythrocyte stage, and have also considered the dynamics in the course of experimental anemia.

The greatest difference was found with respect to hexokinase, as to be expected from its important role in the control of glycolytic flux. This enzyme declines by more than one order of magnitude during maturation.[87-90] Changes in enzymatic activity are closely correlated with other indexes of maturity such as cytochrome oxidase, aspartate aminotransferase, inorganic pyrophosphatase, RNA, and reticulocyte count during the course of a bleeding anemia in rabbits.[45-51]

In mature erythrocytes hexokinase is found entirely in the cytosol.[91] Gellerich and Augustin[92-94] and Schlame et al.[95] obtained clear-cut evidence that about one half of the hexokinase in reticulocytes was bound to the outside of the external membrane of the mitochondria with an uneven distribution of the isoenzymes. They found a higher ATP affinity for the bound hexokinase and much lower inhibition by G-6-P. They suggested a regulatory role for the bound hexokinase on the basis of kinetic differences compared to cytosolic enzyme, and found some indications for the preferential use of mitochondrially generated ATP by bound hexokinase in a test system with isolated mitochondria. One part of the hexokinase appears to be localized more internally than the bulk in the mitochondria and tends to be more tightly bound. A functional role for this fraction could not be established. This work was confirmed on both human and rabbit reticulocytes.[96] This group of workers found 60% of total hexokinase to be located in the mitochondria with a preponderance of isoenzyme Ia, while in the cytosolic fraction the isoenzyme Ib dominated with 62%.[97-99] A similar compartmentation of hexokinase was also found in red cells in man.[97,100] There was a rapid decrease in enzyme activity in the declining phase of an experimental anemia of rabbits as well as during in vitro incubation of reticulocytes. The cytosolic isoenzyme Ib was preferentially and rapidly degraded by an ATP- and ubiquitin-dependent process, whereas bound hexokinase was protected. Nevertheless, in mature erythrocytes, which of course no longer contain mitochondria, there is also less of the Ia isoenzyme which is entirely cytosolic. The isoenzyme Ib has completely disappeared.

The liver type of glucokinase with low affinity for glucose was found in young reticulocytes during the early stages of anemia.[94,101] It was suggested that this enzyme represents a property of the Line 2 cells.[19] Glucokinase and hexokinase isoenzymes II and III disappear completely during the later stages of anemia and in maturation.

Phosphofructokinase, an important enzyme which shares control of the glycolytic flux with hexokinase, exhibited a lesser degree of change compared to hexokinase in the course of bleeding anemia of rabbits.[102] Its activity increased threefold with a maximal ratio of five between the density-separated youngest and oldest fractions. Although there was, as a whole, a good correlation with aspartate aminotransferase and RNA, enzyme activity declined more slowly during the recovery period from anemia than the other indexes. Some evidence was found for the occurrence of an additional enzyme in reticulocytes that is not found in erythrocytes.

Changes in pyruvate kinase activity, an enzyme of great importance for the regulation of the concentrations of glycolytic intermediates, resemble those of hexokinase during the course of an experimental anemia in rabbits.[103] A steep increase in enzyme activity during the development of the anemia was observed. At its peak, reticulocytes exhibited tenfold activity of pyruvate kinase, compared to mature erythrocytes exhibiting a rapid decline during the recovery period. The changes were well correlated with the other indexes of immaturity such as aspartate aminotransferase, cytochrome oxidase, and RNA. Similar trends in maturity-dependence of pyruvate kinase were observed by other workers.[104]

Among the enzymes that work near equilibrium during glycolysis, aldolase and glyceraldehyde phosphate dehydrogenase show a distinct three- to fivefold difference in the activity between reticulocyte and erythrocyte stages.[105,106] Lactate dehydrogenase activity, which declines by one order of magnitude at the polychromatic erythroblast stage (see Chapter 2), undergoes no further changes during the transition from reticulocyte to erythrocyte.

Of all the glycolytic enzymes, only hexokinase is bound to mitochondria to any significant extent, whereas the percentage of all others found in isolated mitochondria does not exceeed 2%.[107] Glucose-6-phosphate dehydrogenase is a well-known indicator of erythrocyte age. Increased activities have been reported for reticulocytes.[88,108,109]

Among the other enzymes connected with the metabolism of glucose, two are noteworthy: glycerol-3-phosphate dehydrogenase and phosphoglycolate phosphatase. The dehydrogenase is found only in reticulocytes, and with even greater activity in the red cells of cord blood in man.[110] The phosphoglycolate phosphatase was studied from the perspective of its possible influence on the level of 2,3-DPG.[111] Phosphoglycolate is a powerful activator of bisphosphoglycerate mutase/phosphatase and it was assumed that increased activity of the enzyme degrading it might be reflected in the levels of 2,3-DPG. Phosphoglycolate phosphatase increased in anemia but a definite relationship to 2,3-DPG could not be established.

B. Hydrolases

Enzymes degrading either RNA or DNA have been observed in various types of red cells. Studies relating enzyme activity to the maturation process were mostly carried out on the rabbit. The results are difficult to assess in a quantitative manner because of several complicating factors. The enzymes appear to be located in various cell fractions, with most of their activity masked. This is caused partly by a well-defined inhibitor of ribonuclease in the cytosol. Furthermore, the enzymes are highly sensitive to variations of ionic strength. In addition, there is the problem of the existence of an endogenous substrate for the enzymatic attack, the accessibility of which depends on experimental conditions. Lastly, there are difficulties in comparing results obtained by various methods for the determination of enzymatic activity. In this section the enzymes and their inhibitors are reviewed only to the extent that it is germane to the maturational process.

So far, two ribonucleases have been clearly defined in rabbit reticulocytes. A pyrimidine-specific endoribonuclease with a pH optimum of 7.5 corresponding in some properties to Type I pancreatic ribonuclease has been described.[112-115] The pyrimidine-specific RNases of rabbit reticulocytes and pancreas differ considerably from each other in molecular structure and in the mechanism of cleavage. The reticulocyte enzyme has a molecular mass of 25 kD but is not a dimer of pancreatic RNase. The reticulocyte RNase is also more sensitive to temperatures above 65°C and to reduction of its disulfide bridges. Clusters of pyrimidine bases are split by this enzyme preferentially. The primary cleavage product of uracil oligonucleotides is a trinucleotide. This enzyme is found in ribosomes and cannot be washed away by 0.5 M NH$_4$Cl. This enzyme may be responsible for the self-degradation of ribosomes which can be observed in vitro. It attacks only single-stranded sections of the RNA. It is practically inactive at low ionic strength; maximal activity is found at 0.15 M KCl. It is likely that enzyme activity with an optimum pH of 8.2 refers to the same protein.[116] This ribonuclease has a pronounced dependency on maturation, and only traces of any activity are left in erythrocytes.

The occurrence of a membrane-bound acid phosphodiesterase with an optimum pH of 6.5 was reported in rabbit reticulocytes.[117,118] It is a base-nonspecific diesterase which produces 3'-nucleotides. This enzyme also shows a pronounced decline at maturation to 1/30 its previous level. An enzyme from rat reticulocytes with an optimum pH of 6.5 to 7.5 may also belong to the same class.[119] Both the enzymes from rabbit and rat reticulocytes are largely masked by RNase inhibitors.

A further enzyme activity corresponding to a pyrimidine-specific endonuclease has been described in the cytosol which is similar in its properties to ribosomal RNase.[120] It is uncertain whether it represents a third type of enzyme or differs only in location from ribosomal ribonuclease. This enzyme activity also declines with maturation. Pyrimidine-5'-nucleotidase, which mediates one of the last steps in the breakdown of RNA, is greatly increased up to fivefold in reticulocyte-rich red cell populations in man.[121] An endoribonuclease, which is activated by $ppp(A2'p)_nA$, has been obtained from rabbit reticulocytes. It appears to be identical with a polynucleotide binding protein previously described. It cleaves predominantly the U–U and the U–A bonds of rRNA. Like other ribonucleases it requires Mg^{2+} and a high ionic strength, 50 to 150 mM KCl, for its activity. It is not inhibited by RNase inhibitor.[122]

The presence of inhibitory proteins for RNases in various tissues is well-known. An inhibitory activity in the cytosol was also described in rabbit reticulocytes and in small amounts in erythrocytes of various species.[123-125] The inhibitory protein was partially purified.[129] A similar inhibitory protein has been demonstrated in rat reticulocytes.[119] Inhibitory activity declines with maturation. The inhibitor appears to contain essential and accessible SH groups.

In bone marrow and in reticulocytes of anemic rabbits, stroma-bound DNase activity has been found.[117] The pH optimum is at 5.0, but there may be additional enzymes with higher pH optima. The enzyme is independent of Mg^{2+}. Enzyme activity is nearly completely masked in hemolysates because of an inhibitory effect of hemoglobin which apparently forms complexes with the DNA, which is thereby protected from enzymatic attack. Similar enzymes have been described in the red cells of amphibia and snakes where their activity is also completely masked.[126,127] The biological significance of the enzyme may be connected with the expulsion of the nucleus in mammals, which is disintegrated after its expulsion, or with the breakdown of mitochondria DNA.

There are a variety of protein phosphatases in reticulocyte cytosol.[128-133] At least five different phosphatases belonging to Types I and II (according to the Cohen nomenclature) were identified. One of their functions may be the dephosphorylation of the initiation factor eIF-2 (see Chapter 11).

Enzymes degrading phospholipids may be presumed to play an important role in both the degradation of mitochondria and cellular membrane. Unfortunately, however, there is scant information on their type and activity in the reticulocyte. Indirect evidence for their action is the occurrence of free fatty acids in both mitochondria and the cytosol, in the decline of the phospholipid content during maturation, and in the appearance of degraded forms of mitochondria. Phospholipase A_2 activity was detected in mitochondria and cytosol of rabbit reticulocytes.[134] In mature rat erythrocytes a phospholipase A_2 has been described[135] and a membrane-bound lysophospholipase was observed in human erythrocytes.[136]

High concentrations of inorganic pyrophosphatase have been found in various studies.[90,137,138] ITPase activity in rabbit reticulocyte cytosol is sixfold higher than in rabbit erythrocytes.[139]

VII. LOW MOLECULAR COMPOUNDS

The concentration profiles of low molecular compounds reflect both the greater metabolic activity and diversity of the reticulocyte as well as the species characteristics of mature erythrocytes (Table 4). Generally, the concentration of free nucleotides is considerably higher in the reticulocyte. Also, many additional nucleotides are found in the reticulocyte. In rabbits and rats the ATP concentration of about 3 mM is about 3 times higher in the reticulocyte than in the erythrocyte and accounts for more than 90% of total nucleotides. More than 90% of adenine nucleotides are found in the cytosol.[140]

GTP in the rat reticulocyte is one order of magnitude higher than in the erythrocyte. UTP concentration is 0.8 mM and CTP is 0.6 mM in the reticulocyte; nothing is found in the mature adult cell. The same holds for UDP acetylhexosamine. In cattle, mature erythrocytes contain less than 1/5 the ATP concentration. Compared to rabbit or rat erythrocytes, the relative differences of the nucleotides are similar. The reticulocytes contain three times as much ATP and small amounts of the nucleotide triphosphates that are absent in mature cells. An interesting feature is the behavior of 3-ribosyluric acid. It is the major nucleotide in the adult mature erythrocyte of cattle with a concentration of about 1 mM. It has a concentration in reticulocytes of less than 0.2 mM and may well be nil.

There are interesting patterns in glycolytic intermediates that are related to oxidative metabolism and to evolutionary adaptations in reticulocytes. This consideration refers particularly to 2,3-DPG. There is a definite tendency for higher levels of G-6-P which, as will be discussed later (see Chapter 7), may be connected with the Pasteur effect, i.e., the partial suppression of glucose consumption by respiration, which involves an inhibition of hexokinase by glucose-6-phosphate. 2,3-DPG does not appear to show consistent changes. It has been reported to be higher in reticulocyte-rich blood of man but lower in rabbits, rats, and pigs. Glycerol-3-phosphate was found in tenfold higher concentration in reticulocytes, indicating the activity of glycerine phosphate dehydrogenase, which is almost absent in mature erythrocytes.

The concentration of glutathione has been reported to be considerably higher in reticulocytes than in erythrocytes of several species. The high concentration of creatine in human reticulocytes — which is more than tenfold that of erythrocytes — is perhaps the best indicator of erythropoiesis activity. The reticulocytes of experimental animals such as rabbits and chickens also have elevated creatine concentrations, but to a lower extent than those of man, on account of their passive permeability for creatine.

In density-separated human red cells, high concentration gradients of the polyamines putrescine, spermine, and spermidine were observed. In rabbit reticulocytes a manyfold greater concentration of spermidine was found compared to human red cells. With this in mind, it is interesting that the polyamines activate the casein kinase II of red cells.[141] Even in species with a high concentration of K$^+$ in mature erythrocytes, the K$^+$ concentration of reticulocytes is distinctly higher than that of erythrocytes (see Chalfin[142]); however, if calculated on the basis of cell water, the difference is much smaller. These observations are explained by the lower hemoglobin concentration of reticulocytes and are in full agreement with a Donnan distribution. Reticulocytes of dogs, the mature erythrocytes of which have a low K$^+$ concentration of about 6 mM, contain 100 mM of K$^+$ — nearly identical with the concentration of K$^+$-rich body cells.[143]

Total magnesium in reticulocytes may exceed 14 mM but is only 2 to 3 mM in erythrocytes.[144] Based on the content of RNA-P in reticulocytes, about 9 mM magnesium is ribosome-bound with an Mg^{2+}/RNP ration of 0.3. If the more than 3 mM concentration of ATP in reticulocytes is also taken into account, it appears likely that the level of free Mg^{2+} does not differ greatly between reticulocytes and erythrocytes and is less than 1 mM.

VIII. TIME COURSE OF EXPERIMENTAL ANEMIA

The time course of experimental anemia reveals significant changes in the characteristics of reticulocytes and of the erythrocytes resulting from them. The most extensive complex study was carried out on rabbits.[45-51]

Table 4
SELECTIVE COMPARISON OF THE LOW MOLECULAR COMPOUNDS OF RETICULOCYTES, FETAL/NEONATAL, AND ADULT RED CELLS[a]

Compound	Species	Nucleotides (mM)			Ref.
		Reticulocyte	Fetal/neonatal	Adult	
ATP	Man	1.8—1.9	1—1.2	1.1—1.3	196,178
		(20%	1.6	1.3	168
		reticulocytes)	1.5		177
			(1.9 Premature)		177
			1.6	1.3	197
	Rabbit	3	2	1.2—1.5	17,198,199,200,201
	Rat	3—4	3.5	1.2	202,203
	Dog		2.0	1.2	204,205
	Pig		2.1	1.9	206
	Cattle	0.6	0—0.4	0.2	207
	Sheep		1.1	0.6	208
			0.6	0.2	209
	Goat		0.8	0.3	210
	Chicken		1.7	1.0	194
			4—5 (1—5 Days)		
	Turkey		2.8	1.0	194
	Duck		1.5—2	3.0	194
			4—5 (1—5 Days)		
GTP	Rat	1.2	0.5	0.1	202,203
	Cattle	0.08	0.09 (GTP + UTP)	0	207
UTP	Rat	0.8	0.8	0	202,203
	Cattle	0.06		0	207
CTP	Rat	0.6	0.4	0	202,203
	Cattle	0.04		0	207
UDP-Acetyl hexoseamine	Rat	0.15	0.3	0—0.05	203
	Cattle	0.2	0.18	0.01	207
3-Ribosyluric acid	Cattle	<0.2	0 (Fetal)	0.6—1.0	207,209
2,3-DPG	Man	5.7	3.1—3.6	4.5	178,179,211,212
		8.0	6.0		
	Rabbit	5.0—7.0	0.2	8.0—9.0	192,201,212,213
	Rat	3.5	0.1—0.4	5.6—6.0	202,214,215,216
	Guinea pig		0.7—1.1	6.0—8.0	212,214,216,217
	Hamster		1.4	5.7	214
	Mouse		0.8—1.0	6.1	214,218
	Horse		3.0—4.0	5.0—6.8	219,220
	Dog		0.8	4.4	221
			1.8 (Newborn)	6.5	204,205
					204,205
	Pig	7.0	3.0	7.0—10.0	222,219,206
			<2.0		
			<1.0		
	Cattle		0.6	0.05	219
	Sheep	0.4	8.0 (3 Days)	0.2	223,219,212
			(1.3 Fetal)		208
				0.01	223,193
				0.03	193
	Goat		1.5	0.3	224,219
			8.0 (3 Days)		
			9.0		210
	Chicken		4.0—5.0	0	194

Table 4 (continued)
SELECTIVE COMPARISON OF THE LOW MOLECULAR COMPOUNDS OF RETICULOCYTES, FETAL/NEONATAL, AND ADULT RED CELLS[a]

Compound	Species	Nucleotides (mM)			Ref.
		Reticulocyte	Fetal/neonatal	Adult	
	Turkey		IP5[b]: 0.5—0.8	4.5	
			4.0—5.0	0	194
	Duck		IP5[b]: 0.6	4.0	
			4.0—5.0	0	194
Gl-6-P	Man		IP5[b]: 0.3—0.4	4.5	
			45—70	25	179,178
	Rabbit	300		50	201
	Rat	450	800—1000	50	202,203
Glycerol-3-phosphate					
	Rabbit	200		20	201
P-rib-PP	Man		13	6	225
Glutathione	Man		2.4	1.7	166
			3.6	2.2	165
			3.9	2.4	167
	Goat		1.8	2.4	210
	Sheep	GSH +[c]>3		2.2	209
		GSH −[c]1.5		0.5	
Creatine	Man	3—10	0.6—0.75	0.4—0.5	56,57,59
		(→ 50)	(11.0)		211,226
					55,227
	Rabbit	0.5—1.9		0.13	228
	Rat	0.5		0.2	229
	Chicken	0.2		0.08	230
Putrescine (μM)	Man		0.7	0.5	231
				0.2	232
Spermidine (μM)	Man		36	20	231
	Rabbit	1000			233,141
Spermine (μM)	Man		44	31	231
				6	232

[a]   Only constituents that show drastic and characteristic differences from the adult type of cell ($+50\%$; -30%) are listed.

[b]   Inositol pentaphosphate.

[c]   GSH +, genetically GSH-rich; GSH −, GSH-poor red cells.

The red cells were fractionated on various days during the development of anemia which was produced by 7 days of bleeding. Their dimensional and biochemical characteristics were determined (Figure 11). The biochemical criteria were chosen so as to represent the ribosomes (RNA), mitochondria (cytochrome oxidase, aspartate aminotransferase, and 5-ALA synthase), and the cytosol (pyrophosphatase). Even one day after the beginning of bleeding there was an outpouring of reticulocytes, which were normal-sized, contained about 4 mg RNA per milliliter cells, and exhibited low respiration and low activities of all enzymes studied. On the 4th day, or even on the 3rd, a nearly pure population of cells was found, which are characterized by somewhat larger volume and a high hemoglobin content. They contained extremely high amounts of RNA, as much as 25 mg RNA per milliliter cells, and exhibited very high respiration and activity of mitochondrial enzymes as well as of the pyrophosphatase of the cytosol. They lacked lipoxygenase activity. They correspond to the Borsook[19,44] "Line 2" cells. Thereafter, from the 5th to 6th day on, macroreticulocytes appeared with volumes about two times larger than the normocytes. These cells are char-

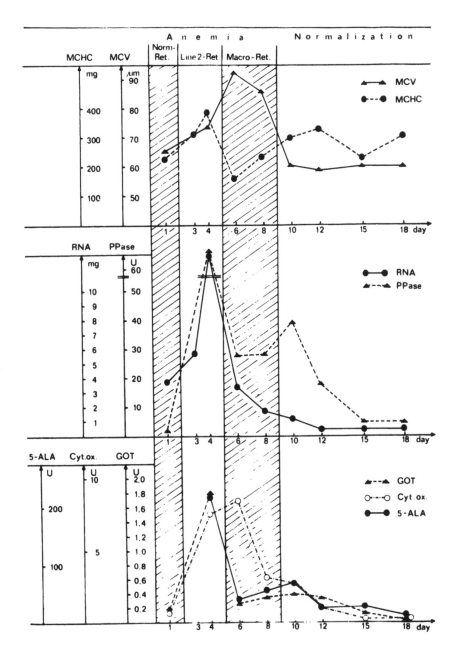

FIGURE 11.   Changes in cellular parameters and enzymes during the course of bleeding anemia of rabbits. PPase = inorganic pyrophosphate; 5-ALA = 5-ALA synthase; GOT = aspartate aminotransferase.

acterized by a low cellular hemoglobin concentration of about 12 mM Hb/$_4$, an RNA content of about 5 mg/mℓ cells, and a respiration about half as large as that of the "Line 2" cells. The macroreticulocytes exhibit high concentrations of lipoxygenase. During recovery a population of cells with a nearly normal hemoglobin concentration of 18.5 mM Hb/$_4$ and fairly high cytochrome oxidase activity was produced. They may originate from newly activated erythropoietic sites. Thus it would appear that at least three, and possibly four, distinct populations of red cells may be differentiated. This circumstance and the differences in the time courses of maturation of the various constituents explain discrepancies between the criteria used for characterization of reticulocytes or their maturation. In particular there is no direct correlation between the number of ribosomes and the number of mitochondria in the various types of reticulocytes. Most mitochondrial constituents such as cytochrome oxidase, succinate dehydrogenase, isocitrate dehydrogenase,[73,74] 5-ALA synthase, aspartate transaminase, malate dehydrogenase, and glutaminase[145] exhibit a plateau of high activity at all RNA concentrations in excess of 2 mg/mℓ cells.[146] Within a narrow range of RNA concentrations, below a critical level of 2 mg/mℓ cells, there is a steep decline of all these enzymes. The dissociation between the fate of mitochondria and ribosomes is also clearly demonstrated in experiments on in vitro maturation (see Chapter 12). A selective degradation of immature mitochondria without change in the reticulocyte count could be achieved. Similarly, differences between the behavior of cytosolic enzymes and of membrane constituents compared to ribosomes or mitochondria can be observed. Also, it would appear that the reticulocytes appearing in the peripheral blood during the course of bleeding anemia of rabbits differ in their creatine concentration.[147]

IX. THE EGRESS OF RETICULOCYTES FROM THE BONE MARROW

The egress of reticulocytes from the bone marrow tissue into the blood capillaries poses a number of unresolved questions. Under normal circumstances, why do only the most mature reticulocytes enter the circulation? Under conditions of stress, what causes a shift so that immature reticulocytes may pass through the capillary wall? Are there any special properties of blood capillaries in the bone marrow compared to those elsewhere?

The signal for reticulocyte passage into the bone marrow capillaries has not been defined so far. The movement appears to occur via diapedesis with pseudopods of the reticulocytes passing through small openings in the capillary wall. The number of the dimensions of the openings may vary, resulting in a dynamic state. There is some influence of bone marrow pressure or cell packing in bone marrow sinuses on the rate of cell egress from the bone marrow.[150,151] It is possible that the effects of adrenalin and other catecholamines on releasing reticulocytes into the circulation in a short time may be based upon the same phenomenon. Whether increased blood pressure or blood flow changes the dynamics of the capillary openings or exercises a direct effect on the reticulocytes is uncertain. The failure of immature reticulocytes to pass the capillary wall may be related to the stickiness of their surface and/ or to their lesser deformability.[152] It is well documented that deformability of red cells progresses with maturation. On the other hand, the ability to form pseudopods is greater in younger reticulocyte, which may be connected to both their larger energy supply and to the incompleteness of their membrane skeleton.[153]

A possible factor may be the adhesion of immature red cells to fibronectin, a glycoprotein associated with the extracellular matrix of many tissues including the bone marrow. It is tightly bound to the surface of various types of cells. The ability to adhere to fibronectin was demonstrated for reticulocytes in man, rabbit, and mouse. During maturation in vivo, this property disappears, presumably on account of the loss of a fibronectin-adhesive component of the reticulocyte membrane. The speculation was advanced that the loss of fibronectin adhesion is associated with the release of red cells into the circulation. On the other hand, two considerations militate against this hypothesis: one is the fact that during anemia, reticulocytes still exhibiting fibronectin-binding ability leave the bone marrow; the second is the demonstration that splenectomy inhibits the loss of this property so that even mature erythrocytes in the circulation can still attach to fibronectin. Apparently the spleen is one of the sites in which the ability to adhere to fibronectin is lost.

Which of the various factors involved underlies the facilitated egress of immature reticulocytes is at present a matter of speculation. It is certain that there is increased blood flow through the anemic bone marrow which serves to supply iron and nutrients to the stimulated bone marrow.[154]

Overall, the regulation of red cell entry into the circulation appears to be both finely regulated and easily disturbed, possibly as a result of the subtle balance of weak forces.

X. THE FETAL AND NEONATAL RED CELL

Erythropoiesis and the red cells of the human fetus and the newborn have been the object of extensive studies. The cells of this period of life differ in several respects from those produced postnatally and in adult life and resemble reticulocytes in some properties. For this reason a short survey is given, mainly from a comparative point of view. During the fetal period the production of red cells occurs mainly in the liver and to a smaller extent in the spleen and thymus. Beginning with about the 20th week, the bone marrow becomes increasingly more important as the site of erythropoiesis and becomes the exclusive site shortly after birth. During the last 2 months of pregnancy erythropoiesis is highly active, with a daily red cell production rate of about 4%,[155] i.e., about four times greater than in adult life, partly related to the rapid increase of red cell mass commensurate with growth of the fetus, and partly due to the reduced lifespan of the fetal red cell which is only about 60 to 80 days,[156] the reason for which is not known.

The red cells produced in man after the 24th week of gestation are predominantly macrocytic with cell volumes of about 125 μm^3, with an increasing proportion of normocytes. The average volume at term birth is about 108 μm. The red cell population during fetal life is very heterogeneous for several reasons: (1) the coexistence of macro- and normocytic cells, (2) the diversity of erythropoietic sites, and (3) the stimulatory effect of erythropoietin. For this reason the percentage of reticulocytes is high before and at birth and drops to almost zero after the 4th day of life. The level of erythropoietin — which is high before birth — falls to an undetectable level after 3 days.[157,158] Newborn erythrocytes are less deformable than those of adults.[159] The amount of phospholipid and cholesterol per cell is higher with a different pattern of phospholipids and distribution of fatty acids.[160,161]

A. Enzymes

Among the characteristic features of newborn red cells is higher sensitivity to oxygen injury and tendency for the formation of Heinz bodies.[162-164] For this reason the enzymes that defend against toxicity of oxygen reduction products have been extensively studied. The activity of glutathione reductase, if measured without addition of FAD, is 50 or even 100% higher in fetal red cells but less so in its presence.[165-168] The glutathione peroxidase activity in neonatal and fetal erythrocytes is definitely lower, amounting to only about one half of the value in red cells of adults.[165,167,169-172] With a lower level of catalase[171,173] and of NADH-dependent methemoglobin reductase,[174] the greater sensitivity of the fetal and newborn red cells to oxygen injury seems explicable. On the other hand the activity of the superoxide dismutase does not differ in the cells of newborns and adults.[171,175]

Glutathione synthesis enzymes show differences in fetal and newborn erythrocytes compared to adults; this, however, is not reflected in the level of glutathione. The first enzyme, 5-glutamylcystein synthase, is about 30% higher in activity, whereas the second one, glutathione synthase, is one third lower.[165,170]

Glycosis of newborn red cells tends to be lower than in adults.[176,177] This circumstance may well be connected with the lower activity of phosphofructokinase, which has been found consistently,[168,177-182] and seems to be particularly pronounced in the older cells of the erythrocyte population at birth.[183] On the other hand, elevated levels of several other glycolysis enzymes have been found which were particularly pronounced for hexokinase, aldolase, phosphoglycerate kinase, phosphoglycerate mutase, and enolase[168,177-179,183,184] (Table 5). These data make evident the important role of phosphofructokinase for the glycolytic flux rate.

Table 5
ENZYMES RELATED TO GLYCOLYSIS IN FETAL/NEONATAL
AND ADULT RED CELLS IN MAN
μM/min·mℓ cells

Enzyme	Fetal/neonatal	Adult	Ref.
Hexokinase	0.24—0.34	0.13	168, 178, 179
PFK	0.85	1.50	179
	1.40	2.00	178
	1.9	2.3	168
	1.7	2.2	177
Aldolase	0.40	0.25	179
	0.60—0.70	0.40	178
Phosphoglycerate kinase	40—45	30	178, 179
	42	24	168
	50	35	177
Phosphoglycerate mutase	10.5	7.5	179
	21	14	168
	8.5	6	184
Enolase	8		168
	5.0	2.5	179
	3.5	1.60	178
	(5.50 Premature)		178
Glucose-6-phosphate	3.3	2.0	178, 179
dehydrogenase	(4.3 Premature)	2.1	178
	5.3 (Fetal)		184

Several fetal and neonatal cell enzymes show differences in isoenzyme pattern compared to adults.[185-188] The studies were complicated particularly with phosphofructokinase, because it is controlled by a three-genic system, which would lead one to expect as many as 15 isoenzyme species.[189] Earlier evidence would indicate that the phosphofructokinase of cord blood differs in its isoenzyme pattern from that of adult blood and that one or more of the fetal isoenzymes are more labile than the others (for a review see Travis[178]).

The activity of galactokinase has been reported to be three times higher in cord blood red cells as compared to adults, and to exhibit significant differences in its Michaelis constant for galactose and in its thermostability.[190] These results would also indicate the existence of a specific fetal isoenzyme.

Glucose-6-phosphate dehydrogenase consistently has been found to be elevated in fetal cells to an extent that is disproportionate to the percentage of reticulocytes.[178,179] Higher activities have also been described for several other enzymes. The activities of pyrimidine synthesis enzymes are much greater in newborn erythrocytes than in adult erythrocytes with both orotidine 5-phosphate decarboxylase and orotate phosphoribosyl transferase being more than threefold higher. Reticulocytosis may be a contributing factor, since in a patient with 12% reticulocytosis, the enzymes were 10- and 20-fold higher than in mature erythrocytes.[191] On the other hand, low activities have been reported for carbonic anhydrase and acetylcholinesterase.

B. Low Molecular Compounds

The profile of the low molecular compounds (see Table 4) of fetal and neonatal red cells reflect on the one hand their youthfulness, and on the other characteristic features of their ontogenis and evolutionary adaptation. There is a definite tendency towards higher ATP levels in fetal or neonatal cells in almost all the species so far examined. The presence of other nucleotides, either absent or in low concentration in mature erythrocytes, also may be taken as an indication of the youthfulness of neonatal cell populations. The elevated levels

Table 6
PYRUVATE KINASE AND GLYCERATE-2,3-P$_2$ SYNTHASE IN RED CELLS OF SPECIES WITH DRASTIC PERINATAL CHANGES IN 2,3 BISPHOSPHOGLYCERATE

Species	Pyruvate kinase (IU) Fetal/neonatal	Adult	Glycerate-2,3-P$_2$ synthase (IU) Fetal/neonatal	Adult	Ref.
	2.6	1.8			179
Man	3.8 (5.0 fetal)	2.4		2—3	178
Rabbit	60—70	5	20	20	199
Rat	57	4	0	7—8	215
	50—60	5	0	7—8	234
	50	5	0	7	214
Guinea pig	5.5	2	4	4	214
Mouse	50—60	10	3	7—8	214, 218
Hamster	25	8	2,5	5	214
Dog	17	2			205

Note: The data for man are representative of species with switching of fetal to adult hemoglobin without major changes in 2,3 bisphosphoglycerate.

of glucose 6-phosphate may be related to lower activities of phosphofructokinase, but possibly also to the presence of respiring cells, i.e., reticulocytes.

The changes of 2,3-DPG during fetal and postnatal life have been the subject of numerous studies. Diverse patterns have been found, which can be interpreted only if considered in the framework of the change of the oxygen function of the blood, particularly the properties of hemoglobins, in the transition from fetal to extrauterine existence. In man, a switch from fetal hemoglobin (with a high affinity for oxygen and low binding to 2,3-DPG) to the adult type of hemoglobin occurs, the oxygen affinity of which is modulated by 2,3-DPG. On the other hand, in many mammals there is a direct switch from synthesis of embryonic hemoglobin to that of adult hemoglobin during fetal life. In these species the high oxygen affinity needed for intrauterine life is achieved by a low concentration of 2,3-DPG (Table 6). Examples are the rabbit, rat, guinea pig, hamster, mouse, dog, and pig. Low 2,3-DPG may result from the high activity of pyruvate kinase, particularly in proportion to phosphofructokinase, this being the case in the rabbit, rat, mouse, hamster, dog, and to a lesser degree in the guinea pig. Another mechanism contributing to low 2,3-DPG may be the small activity of glycerate-2,3-P$_2$ synthase such as found in the rat. Reticulocytes of adult rabbits and red cells of neonatal rabbits therefore show a contrasting pattern with respect to 2,3-DPG content, owing to different constellations of pyruvate kinase and glycerate-2,3-P$_2$ synthase activity.

Both fetal and adult hemoglobin in ruminants is unresponsive to 2,3-DPG.[192] Fetal hemoglobin in ruminants has a high oxygen affinity. The adaptive lowering of oxygen affinity at birth is achieved in an indirect manner by a transient increase of 2,3-DPG, the concentration of which may rise as high as 8 mM in sheep and goats. DPG decreases the oxygen affinity of hemoglobin via the Bohr effect by a more acid intracellular pH, which is caused by Donnan equilibrium. Glycerate-2,3-P$_2$ synthase in red cells of newborn lambs was found to be highly active but nearly absent in adult sheep. The transient rise of 2,3-DPG has been demonstrated to be primarily due to an increase of pH, and to some extent, an increase in glucose concentration — factors which lead to a 50-fold concentration of fructose-1,6-bisphosphate.[193]

The phosphate compounds in submammalian species exhibit remarkable changes during development (for reviews see Issacks and Harkness[194] and Bartlett[195]). ATP, GTP, 2,3-DPG, and inositol pentophosphate may play a role in regulating oxygen transport and hemoglobin function in various species with dramatic ontogenetic changes.

Glutathione level is about 50% higher in fetal erythrocytes compared to adults, but perhaps less so in neonatal cells in man. The reverse has been found in goats.

Creatine concentration shows a negative correlation with gestational age but even at term is on the average about 50% higher than in adult red cells. The polyamines, putrescine, spermidine, and spermine, which may well be good markers for red cell age, are distinctly higher in fetal and neonatal red blood cells. However, it should be kept in mind that the presence of any type of nucleated cell, be it leukocytes and lymphocytes, or possibly erythroblasts, would vitiate the results because of their high content of polyamines.

REFERENCES

1. **Ehrlich, P.**, Bericht über einige Beobachtungen am anämischen Blut, *Berl. Klin. Wochenschr.*, 18, 43, 1881.
2. **Caspersson, T.**, Über den chemischen Aufbau der Strukturen des Zellkerns, *Scand. Arch. Physiol.*, 73(Suppl. 8), 98, 1936.
3. **Dustin, P., Jr.**, Contribution à l'étude histophysiologique et histochimique des globules rouges des vertébrés, *Arch. Biol.*, 55, 285, 1943.
4. **Holloway, B. W. and Ripley, S. H.**, Nucleic acid content of reticulocytes and its relation to uptake of radioactive leucine in vitro, *J. Biol. Chem.*, 196, 695, 1952.
5. **Coutelle, R., Rapoport, S., and Lindigkeit, R.**, Die Substantia reticulo-filamentosa, *Naturwissenschaften*, 42, 127, 1955.
6. **Coutelle, R. and Rapoport, S.**, Über eine kolorimetrisch-quantitative Bestimmungsmethode der Substantia reticulo-filamentosa bei experimentell erzeugten Anämien, *Folia Haematol. (Leipzig)*, 73, 326, 1955.
7. **Rosenthal, S., Künzel, W., and Wagenknecht, C.**, Charakterisierung der Mikrosomenäquivalente in roten Blutkörperchen von Kaninchen und die Änderung ihrer Zusammensetzung während der Reifung, *Folia Haematol. (Leipzig)*, 78, 135, 1962.
8. **Bessis, M. and Breton-Gorius, J.**, Le réticulocyte. Colorations vitales et microscopie électronique. *Nouv. Rev. Fr. Hematol.*, 4, 77, 1964.
9. **Kosenow, W.**, Über den Strukturwandel der basophilen Substanz junger Erythrocyten im Fluoreszenzmikroskop, *Acta Haematol.*, 7, 360, 1952.
10. **Coutelle, Ch., Rosenthal, S., Gross, J., Davis, H., and Uerlings, I.**, Leitkriterien der Retikulozytenreifung. VI. Verhalten der RNS- und Retikulozytenwerte sowie des Ribosomengehaltes in peripheren erythroiden Zellpopulationen verschiedener Dichte im Verlaufe einer Entblutungsanämie beim Kaninchen, *Acta Biol. Med. Ger.*, 31, 781, 1973.
11. **Coutelle, R., and Rapoport, S.**, Der Einfluß von pH und Ionenmilieu auf die Aufnahme von Brilliantkresylblau durch Kaninchenretikulocyten, *Folia Haematol. (Leipzig)*, 74, 191, 1956.
12. **Bessis, M.**, *Living Blood Cells and Their Ultrastructure*, Springer-Verlag, Berlin, 1973, 122 and 137.
13. **Richter-Rapoport, S. K. N., Dumdey, R., Hiebsch, Ch., Thamm, R., Uerlings, I., and Rapoport, S.**, Charakterisierung von Retikulozyten des Menschen: Atmung, Pasteur-Effekt und elektronenmikroskopische Befunde an Mitochondrien, *Acta Biol. Med. Ger.*, 36, 53, 1977.
14. **Krause, W., David, H., Uerlings, I., and Rosenthal, S.**, Veränderungen der Mitochondrien-Ultrastruktur von Kaninchenretikulozyten im Reifungsprozeß, *Acta Biol. Med. Ger.*, 28, 779, 1972.
15. **Ganzoni, A. M.**, Die Bedeutung der Retikulozytenzahl für die Beurteilung einer Anämie, *Dtsch. Med. Wochenschr.*, 95, 2291, 1970.
16. **Killmann, S. A.**, On the size of normal human reticulocytes, *Acta Med. Scand.*, 176, 529, 1964.
17. **Rapoport, S., Guest, G. M., and Wing, M.**, Size, hemoglobin content, and acid-soluble phosphorus of erythrocytes of rabbits with phenylhydrazine-induced reticulocytosis, *Proc. Soc. Exp. Biol. Med.*, 57, 334, 1944.
18. **Brecher, G. and Stohlman, F., Jr.**, Reticulocyte size and erythropoietic stimulation, *Proc. Soc. Exp. Biol. Med.*, 107, 887, 1961.
19. **Borsook, H., Lingrel, J. R., Scaro, J. L., and Milette, R. L.**, Synthesis of haemoglobin in relation to the maturation of erythroid cells, *Nature*, 196, 347, 1962.
20. **Stohlman, F., Jr., Howard, D., and Beland, A.**, Humoral regulation of erythropoiesis. XII. Effect of erythropietin and iron on cell size in iron deficiency anemia, *Proc. Soc. Exp. Biol. Med.*, 113, 986, 1963.

21. **Erslev, A. J.,** Effect of erythropoietin on the uptake and utilization of iron by bone marrow cells in vitro, *Proc. Soc. Exp. Biol. Med.,* 110, 615, 1962.
22. **Krantz, S. B., Gallien-Lartique, O., and Goldwasser, E.,** The effect of erythropoietin upon heme synthesis by marrow cells in vitro, *J. Biol. Chem.,* 238, 4085, 1963.
23. **Borsook, H., Ratner, K., Tattrie, B., Teigler, D., and Lajtha, L. G.,** Erythropoietin and the development of erythrocytes, *Nature,* 217, 1024, 1968.
24. **Ganzoni, A. S., Hillman, R. S., and Finck, C. A.,** Maturation of the macroreticulocyte, *Br. J. Haematol.,* 16, 119, 1969.
25. **Come, S. E., Shohet, S. B., and Robinson, S. H.,** Surface remodelling of reticulocytes produced in response to erythroid stress, *Nature,* 236, 157, 1972.
26. **Come, S. E., Shohet, S. B., and Robinson, S. H.,** Surface remodelling vs. whole-cell hemolysis of reticulocyte produced with erythroid stimulation of iron deficiency anemia, *Blood,* 44, 817, 1974.
27. **Shattil, S. J. and Cooper, R. A.,** Maturation of macroreticulocyte membranes in vitro, *J. Lab. Clin. Med.,* 79, 215, 1972.
28. **Stohlmann, F.,** Shortened survival of erythrocytes produced by erythropoietin or severe anemia in rats, *Proc. Soc. Exp. Med.,* 107, 884, 1961.
29. **Neuberger, A. and Niven, J. F.,** Haemoglobin formation in rabbit — shortened lifetime after erythropoietin stimulation, *J. Physiol.,* 112, 292, 1961.
30. **Wilner-Zehavi, T., Isak, G., and Mager, J.,** Studies on the biochemical alterations of the protein synthesizing apparatus in maturing rabbit reticulocytes, *Blood,* 27, 319, 1966.
31. **Rapoport, S., Ababei, L., Hinterberger, U., Kahrig, C., Künzel, W., Gerischer-Mothes, W., Raderecht, H. J., Scheuch, D., and Viereck, G.,** Die Reifung von Kaninchenretikulozyten im nichtanämischen Empfangertier nach Austauschtransfusion, *Acta Biol. Med. Ger.,* 7, 589, 1961.
32. **Millette, R. L. and Glowacki, E. R.,** In vivo maturation of immature reticulocytes transfused in a normal rabit, *Nature,* 204, 1207, 1964.
33. **Rosenthal, S., Künzel, W., and Wagenknecht, C.,** Biochemische Charakterisierung der Ribosomen von Kaninchenretikulozyten und ihre Reifungsänderung, *Acta Biol. Med. Ger.,* 13, 281, 1964.
34. **Glowacki, E. R. and Millette, R. L.,** Polyribosomes and the loss of hemoglobin synthesis in the maturing reticulocytes, *J. Mol. Biol.,* 11, 116, 1965.
35. **Marks, P. A., Rifkind, R. A., and Danon, D.,** Polyribosomes and protein synthesis during reticulocyte maturation in vitro, *Proc. Natl. Acad. Sci. U.S.A.,* 50, 336, 1963.
36. **Stohlman, F., Jr.,** Some aspects of erythrokinetics, *Semin. Hematol.,* 4, 304, 1967.
37. **Tooze, J. and Davies, H. G.,** The occurrence and possible significance of haemoglobin in the chromosomal regions of mature erythrocyte nuclei of the newt. Triturus crislatus crislatus, *J. Cell Biol.,* 16, 501, 1963.
38. **Brecher, G., Prenant, M., Haley, J., and Bessis, M.,** Origin of stress macroreticulocytes from macronormoblasts, *Nouv. Rev. Fr. Hematol.,* 15, 13, 1975.
39. **Yataganas, X., Gahrton, G., and Thorell, B.,** DNA, RNA and hemoglobin during erythroblast maturation, *Exp. Cell Res.,* 62, 254, 1970.
40. **Thiele, B. J., Andree, H., Höhne, M., and Rapoport, S. M.,** Translational repression of lipoxygenase mRNA in reticulocytes, *Eur. J. Biochem.,* 129, 133, 1982.
41. **Rapoport, S., Schmidt, J., and Prehn, S.,** Maturation of rabbit reticulocytes: Susceptibility of mitochondria to ATP-dependent proteolysis is determined by the maturational state of reticulocytes, *FEBS Lett.,* 183, 370, 1985.
42. **Schmidt, J., Prehn, S., and Rapoport, S. M.,** Proteolysis during in vitro-maturation of rabbit reticulocyte, *Biomed. Biochim. Acta,* 44, 1429, 1985.
43. **Kim, H. D., Luthra, M. G., Hildenbrandt, G. R., and Zeidler, R. B.,** Pig reticulocytes. II. Characterization of density-fractionated maturing reticulocytes, *Am. J. Physiol.,* 230, 1676, 1976.
44. **Lord, B. I.,** Emergence of two erythroblast lines in erythropoiesis during growth and haemorrhage in the rat, *Nature,* 217, 1026, 1968.
45. **Rosenthal, S., Gross, J., Grauel, E. L., Papies, B., Schulz, W., Belkner, J., Botscharowa, L., Coutelle, C., Hawemann, M., Nieradt-Hiebsch, C., Müller, M., Opitz, M., Prehn, S., Schultze, M., Staak, K., and Wiesner, R.,** Leitkriterien der Retikulocytenreifung, in *6th Internationales Symposium über Struktur und Funktion der Erythrozyten,* Rapoport, S. and Jung, F., Eds., Abhandlungen der Deutschen Akademie der Wissenschaften zu Berlin, Akademie-Verlag, Berlin, 1972, 513.
46. **Gross, J., Papies, B., Grauel, E. L., Hartwig, A., and Rosenthal, S.,** Leitkriterien der Retikulozytenreifung. I. Dichte von roten Blutzellen während einer Entblutungsanämie des Kaninchens im Dextrangradienten, *Acta Biol. Med. Ger.,* 30, 617, 1973.
47. **Wiesner, R., Rosenthal, S., and Hiebsch, C.,** Leitkriterien der Retikulozytenreifung. II. Das Verhalten von Zytochromoxydase und Hemmstoff F der Atmungskette bei der Retikulozytenreifung, *Acta Biol. Med. Ger.,* 30, 631, 1973.

48. **Gerber, G., Schröder, K., and Rosenthal, S.,** Leitkriterien der Retikulozytenreifung. III. Verhalten von Hexokinase and Glukokinase roter Blutzellen während einer Entblutungsnämie des Kaninchens, *Acta Biol. Med. Ger.,* 30, 773, 1973.

49. **Gross, J., Staak, R., and Rosenthal, S.,** Leitkriterien der Retikulozytenreifung. IV. Verhalten der anorganischen Pyrophosphatase, *Acta Biol. Med. Ger.,* 30, 781, 1973.

50. **Papies, B., Gross, J., Coutelle, C., and Rosenthal, S.,** Leitkriterien der Retikulozytenreifung. V. Mittleres Volumen, Volumenverteilungskurven, mittlere Hämoglobinkonzentration und mittlerer Hämoglobingehalt der Zellen im Verlauf einer Entblutungsanämie beim Kaninchen, *Acta Biol. Med. Ger.,* 31, 543, 1973.

51. **Gross, J., Pietsch, L., and Rosenthal, S.,** Leitkriterien der Retikulozytenreifung: osmotische Fragilität von im Dextrandichtegradienten getrennten roten Blutzellen im Verlaufe einer Entblutungsanämie, *Folia Haematol.,* 100, 81, 1973.

52. **Gasko, O. and Danon, D.,** Deterioration and disappearance of mitochondria during reticulocyte maturation, *Exp. Cell Res.,* 75, 159, 1972.

53. **Hamstra, R. D. and Block, M. H.,** Erythropoiesis in response to blood loss in man, *J. Appl. Physiol.,* 27, 503, 1969.

54. **Bessis, M. and Brecher, G.,** A second look at stress erythropoiesis — unanswered questions, *Blood Cells,* 1, 409, 1975.

55. **Syllm-Rapoport, I. and Hilgenfeld, E.,** in preparation.

56. **Fehr, J. and Knob, M.,** Comparison of red cell creatine level and reticulocyte count in appraising the severity of hemolytic processes, *Blood,* 53, 966, 1979.

57. **Schmidt, G., Gross, J., Grauel, E. L., Ihle, W., and Syllm-Rapoport, I.,** Separation and characterization of red blood cells from newborns and infants during the first trimenon of life by means of a dextran density gradient. Density distribution curves of erythrocytes and reticulocytes, *Biol. Neonate,* 31, 42, 1977.

58. **Syllm-Rapoport, I., Daniel, A., Götze, W., Holzhütter, H.-G., Liebe, A., Liebe, K., and Hilgenfeldt, E.,** Red cell creatine in steady state and stimulated erythropoiesis, *Biomed. Biochim. Acta,* 42, 229, 1983.

59. **Syllm-Rapoport, I., Vogtmann, Ch., Daniel, A., and Holzhütter, H.,** Red cell creatine in term, and preterm, adequate and small for gestational age newborns. I. Statistical analysis of creatine in various groups of newborns, *Biomed. Biochim. Acta,* 42, 359, 1983.

60. **Holzhütter, H.-G., Syllm-Rapoport, I., and Daniel, A.,** A mathematical model for the cell age-dependent decline of creatine in human red blood cells, *Biomed. Biochim. Acta,* 43, 153, 1984.

61. **Guggenheim, S. J., Bonkowsky, M. L., Harris, J., and Webster, L. T.,** The preparation and characterization of mitochondria from developing red blood cells, *J. Lab. Clin. Med.,* 69, 357, 1967.

62. **Heynen, M. J. and Verwilghen, R. L.,** A quantitative ultrastructural study of normal rat erythroblasts and reticulocytes, *Cell Tissue Res.,* 224, 397, 1982.

63. **Heynen, M., Tricot, G., and Verwilghen, R. L.,** Autophagy of mitochondria in rat bone marrow erythroid cells. Relation to nuclear extrusion, *Cell Tissue Res.,* 239, 235, 1985.

64. **Simpson, C. F. and Kling, J. M.,** The mechanism of denucleation in circulating erythroblasts, *J. Cell. Biol.,* 35, 237, 1967.

65. **Grecksch, G., Wiswedel, I., and Augustin, W.,** Enzymatic characterization of rabbit reticulocyte mitochondria, in *Abhandlungen der Akademie der Wissenschaften der DDR,* Akademie-Verlag, Berlin, 1973, 587.

66. **Schulz, W., Neymeyer, H. G., and Rosenthal, S.,** Mitochondrien aus Kaninchenretikulozyten. I. Präparation und zellphysiologische Reinheitskriterien, *Acta Biol. Med. Ger.,* 26, 439, 1971.

67. **Belkner, J., Schewe, T., and Rapoport, S.,** unpublished.

68. **Schewe, T., Wiesner, R., and Schulz, W.,** Mitochondrien aus Kaninchen-retikulozyten. III. Zytochrom- und Phospholipid-Gehalt, *Acta Biol. Med. Ger.,* 28, 1, 1972.

69. **Lutze, G., Kunze, D., Reichmann, G., Wiswedel, I., and Augustin, H. W.,** Phospholipidzusammensetzung und Fettsäuremuster der isolierten Phospholipide von Mitochondrien aus Kaninchenretikulozyten, *Acta Biol. Med. Ger.,* 36, 1403, 1977.

70. **Augustin, W., Zborowski, J., Baranska, J., Wiswedel, I., and Wojtczak, L.,** Synthesis of phospholipids in mitochondria and other membrane fractions of rabbit reticulocytes, *Biochim. Biophys. Acta,* 489, 298, 1977.

71. **Steinbrecht, I. and Augustin, W.,** Studies on NAD$^+$ permeability in intact mitochondria from rabbit reticulocytes, *Acta Biol. Med. Ger.,* 36, 555, 1977.

72. **Hofmann, E. C. G.,** Abbau und Synthese des DPN in den roten Blutkörperchen des Kaninchens, *Biochem. Z.,* 327, 273, 1955.

73. **Sarkar, S. R. and Rapoport, S.,** A study of isocitric-dehydrogenase of rabbit reticulocytes and erythrocytes. I. Properties, the effects of various ions and SH-reagents, *Acta Biol. Med. Ger.,* 11, 323, 1963.

74. **Rapoport, S. and Sarkar, S. R.,** Isocitric-dehydrogenase of rabbit reticulocytes and erythrocytes. II. The influence of physical and chemical agents on the liberation of the enzyme; the maximal activity of the enzyme in the cell, *Acta Biol. Med. Ger.,* 11, 335, 1963.

75. **Quiring, K., Kaiser, G., and Gauger, D.**, Monoamine oxidase activity in rat erythrocytes: evidence for its localization in reticulocyte mitochondria, *Experientia*, 32, 1132, 1976.

76. **Quiring, K. and Hubertus, S.**, Monoamine oxidase in rat reticulocytes: subcellular localization and identification of isoenzymes, *Naunyn Schmiedebergs Arch. Pharmacol.*, 300, 273, 1977.

77. **Kornfeld, S. and Gregory, W.**, The identification and partial characterization of lysosomes in human reticulocytes, *Biochim. Biophys. Acta*, 177, 615, 1969.

78. **Yatziv, S., Kahane, I., Abeliuk, P., Cividalli, G., and Rachmilewitz, E. A.**, "Lysosomal" enzyme activities in red blood cells of normal individuals and patients with homozygous β-thalassaemia, *Clin. Chim. Acta*, 96, 67, 1979.

79. **Van Renswoude, J., Bridges, K. R., Harford, J. B., and Klausner, R. D.**, Receptor-mediated endocytosis of transferrin and the uptake of Fe in K 562 cells: identification of a nonlysosomal acidic compartment, *Proc. Natl. Acad. Sci. U.S.A.*, 79, 6186, 1982.

80. **Veldman, A., Kroos, M. J., van der Heul, C., and van Eijk, H. G.**, Are lysosomes directly involved in the iron uptake by reticulocytes?, *Int. J. Biochem.*, 16, 39, 1984.

81. **Rapoport, S. M., Rosenthal, S., Schewe, T., Schultze, M., and Müller, M.**, The metabolism of the reticulocyte, in *Cellular and Molecular Biology of Erythrocytes*, Yoshikawa, H. and Rapoport, S., Eds., University of Tokyo Press, Tokyo, 1974, 93.

82. **Hellenau, T. and North, R.**, The structure and composition of rat reticulocytes. I. The ultrastructure of reticulocytes, *Blood*, 20, 347, 1962.

83. **Lostanlen, D. and Kaplan, J.-C.**, Expression of NADH-cytochrome-b_5-reductase during dimethyl sulfoxide-induced differentiation of Friend erythroleukemia cells, *FEBS Lett.*, 143, 35, 1984.

84. **Chaury, D., Reghis, A., Pichard, A. L., and Kaplan, S. C.**, Endogenous proteolysis of membrane-bound red cell cytochrome-b_5-reductase in adults and newborns. Its possible relevance to the generation of the soluble methemoglobin reductase, *Blood*, 62, 894, 1983.

85. **Johnson, G. G., Ramage, A. L., Littlefield, J. W., and Kazazian, H. H., Jr.**, Hypoxanthine-guanine phosphoribosyltransferase in human erythroid cells: posttranslational modification, *Biochemistry*, 21, 960, 1982.

86. **Schapira, F., Gregori, C., and Banroques, J.**, Microheterogeneity of human galactose-1-phosphate uridyl transferase. Isoelectrofocusing results, *Biochem. Biophys. Res. Commun.*, 80, 291, 1978.

87. **Rapoport, S.**, The regulation of glycolysis in mammalian erythrocytes, *Essays Biochem.*, 4, 69, 1968.

88. **Sabine, J. C. and Kwok, L. W.**, Observation of four age-related parameters in the red cells of various disease states, *Br. J. Haematol.*, 15, 507, 1968.

89. **Hinterberger, U., Ockel, E., Gerischer-Mothes, W., and Rapoport, S.**, Größe und pH-Abhängigkeit der anaeroben Glykolyse und der Hexokinase-aktivität von Erythrozyten und Retikulozyten des Kaninchens, *Acta Biol. Med. Ger.*, 7, 50, 1961.

90. **Rubinstein, D., Ottolenghi, P., and Denstedt, O. F.**, The metabolism of the erythrocyte. XIII. Enzyme activity in the reticulocyte, *Can. J. Biochem.*, 34, 222, 1956.

91. **Magnani, M., Dachà, M., Stocchi, V., Ninfali, P., and Fornaini, G.**, Rabbit red blood cell hexokinase, *J. Biol. Chem.*, 255, 1752, 1980.

92. **Gellerich, F. N. and Augustin, H. W.**, Electrophoretic characterization and subcellular distribution of hexokinase isoenzymes in red blood cells of rabbits, *Acta Biol. Med. Ger.*, 38, 1091, 1979.

93. **Gellerich, F. N. and Augustin, H. W.**, Studies on the functional significance of mitochondrial bound hexokinase in rabbit reticulocytes, *Acta Biol. Med. Ger.*, 36, 571, 1977.

94. **Gellerich, F. N. and Augustin, W.**, Electrophoretic characterization and subcellular distribution of hexokinase isoenzymes in red blood cells of rabbits, *Acta Biol. Med. Ger.*, 38, 1091, 1979.

95. **Schlame, M., Gellerich, F. N., and Augustin, W.**, Localization of hexokinase in mitochondria from rabbit reticulocytes and its relation to mitochondrial ATP-formation studied by measurement of ^{32}P-fluxes, *Acta Biol. Med. Ger.*, 40, 617, 1981.

96. **Stocchi, V., Magnani, M., Canestari, F., Dachà, M., and Fornaini, C.**, Rabbit red cell hexokinase: evidence for two distinct forms and their characterization from reticulocytes, *J. Biol. Chem.*, 256, 7856, 1981.

97. **Fornaini, C., Dachà, M., Magnani, M., and Stocchi, V.**, Molecular forms of red blood cell hexokinase, *Mol. Cell. Biochem.*, 49, 129, 1982.

98. **Magnani, M., Stocchi, V., Dachà, M., and Fornaini, C.**, Rabbit red blood cell hexokinase: intracellular distribution during reticulocytes maturation, *Mol. Cell. Biochem.*, 63, 59, 1984.

99. **Magnani, M., Stocchi, V., Dachà, M., and Fornaini, G.**, Rabbit red blood cell hexokinase, *Mol. Cell. Biochem.*, 61, 83, 1984.

100. **Rijksen, G., Staal, G. E. J., Beks, P. J., Streefkerk, M., and Akkerman, J. W. N.**, Compartmentation of hexokinase in human blood cells. Characterization of soluble and particulate enzymes, *Biophys. Acta*, 719, 431, 1982.

101. **Gerber, G., Schröder, K., and Rosenthal, S.,** Leitkriterien der Retikulozytenreifung. III. Verhalten von Hexokinase und Glukokinase roter Blutzellen während einer Entblutungsanämie des Kaninchens, *Acta Biol. Med. Ger.,* 30, 773, 1973.

102. **Jacobasch, G., Kühn, B., and Grieger, M.,** Maturation dependence of PFK, in *7th Internationales Symposium über Struktur und Funktion der Erythrozyten,* Rapoport, S. and Jung, F., Eds., Abhandlungen der Akademie der Wissenschaften der DDR, Akademie-Verlag, Berlin, 1975, 555.

103. **Grieger, M., Jacobasch, G., and Gerth, C.,** Die Reifungsabhängigkeit der Pyruvatkinase roter Blutzellen, in *7th Internationales Symposium über Struktur und Funktion der Erythrozyten,* Rapoport, S. and Jung, F., Eds., Abhandlungen der Akademie der Wissenschaften der DDR, Akademie-Verlag, Berlin, 1975, 561.

104. **Sass, M. D., Vorsanger, E., and Spear, P. W.,** Enzyme activity as an indicator of red cell age, *Clin. Chim. Acta,* 10, 21, 1964.

105. **Maretski, D. and Rapoport, S.,** Glyzerinaldehyd-3-phosphat-Dehydrogenase aus Erythrozyten des Menschen. I. Isolierung und einige Eigenschaften, *Acta Biol. Med. Ger.,* 29, 207, 1972.

106. **Bernstein, R. E.,** Alterations in metabolic energetics and cation transport during aging of red cells, *J. Clin. Invest.,* 38, 1572, 1959.

107. **Letko, G., Bohnensack, R., and Augustin, W.,** Studies on binding of GAPDH in particulate fractions of rabbit reticulocytes, in *7th Internationales Symposium über Struktur und Funktion der Erythrozyten,* Rapoport, S. and Jung, F., Eds., Abhandlungen der Akademie der Wissenschaften der DDR, Akademie-Verlag, Berlin, 1973, 631.

108. **Kahrig, C. and Rapoport, S.,** Das Verhalten der Atmung und der Glukose-6-phosphat-Dehydrogenase roter Blutkörperchen verschiedenen Alters während einer Entblutungsanämie bei Kaninchen, *Acta Biol. Med. Ger.,* 6, 238, 1961.

109. **Brewer, G. and Powell, R. D.,** Hexokinase activity as a function of age of the human erythrocytes, *Nature,* 199, 704, 1963.

110. **Fessas, P., Anagnou, N. P., and Loukopoulos, D.,** Glycerol-3-phosphate dehydrogenase activity in the red cells of patients with thalassemia, *Blood,* 55, 564, 1980.

111. **Somoza, R. and Beutler, E.,** Phosphoglycolate phosphatase and 2,3-diphosphoglycerate in red cells of normal and anemic subjects, *Blood,* 62, 750, 1983.

112. **Prehn, S., Rosenthal, S., and Rapoport, S. M.,** The temperature-dependent enzymatic breakdown of rRNA of reticulocytes, *Eur. J. Biochem.,* 24, 456, 1972.

113. **Rosenthal, S., Jagemann, K., Prehn, S., and Heinemann, G.,** Über eine alkalische RNase aus Ribosomen von Kaninchenretikulozyten: Eigenschaften des gebundenen und abgelösten Enzyms; Einfluß ein- und zweiwertiger Kationen, *Acta Biol. Med. Ger.,* 18, 329, 1967.

114. **Rosenthal, S., Rapoport, S. M., and Heinemann, G.,** Über eine ribosomale RNase aus Kaninchenretikulozyten, *Acta Biol. Med. Ger.,* 13, 946, 1964.

115. **Rosenthal, S., Prehn, S., and Rapoport, S.,** Uridinfreisetzung aus ribosomaler RNS von Kaninchenretikulozyten durch alkalische RNase aus Retikulozytenribosomen und Pankreas, *Acta Biol. Med. Ger.,* 17, 667, 1966.

116. **Rowley, P. T. and Morris, J. A.,** Protein synthesis in the maturing reticulocyte, *J. Biol. Chem.,* 242, 1533, 1967.

117. **Lindigkeit, R. and Rapoport, S.,** Die Nukleasewirksamkeit in den Retikulozyten von Kaninchen, *Folia Haematol. (Leipzig),* 74, 251, 1956.

118. **Adachi, K., Nagano, K., Nakao, T., and Nakao, M.,** Purification and characterization of ribonuclease from rabbit reticulocytes, *Biochim. Biophys. Acta,* 92, 59, 1964.

119. **Goto, S. and Mizuno, D.,** Degradation of RNA in rat reticulocytes. I. Purification and properties of rat reticulocyte RNase, *Arch. Biochem. Biophys.,* 145, 64, 1971.

120. **Farkas, W. and Marks, P. A.,** Partial purification and properties of a ribonuclease from rabbit reticulocytes, *J. Biol. Chem.,* 243, 6464, 1968.

121. **Beutler, E. and Hartman, G.,** Age-related red cell enzymes in children with transient erythroblastopenia of childhood and with hemolytic anemia, *Pediatr. Res.,* 19, 44, 1985.

122. **Wreschner, D. H., Silverman, R. H., James, T. C., Gilbert, C. S., and Kerr., I. M.,** Affinity labelling and characterization of the ppp (A 2 p)$_n$ A-dependent endoribonuclease from different mammalian sources, *Eur. J. Biochem.,* 124, 261, 1982.

123. **Rost, G.,** Eigenschaften und Vorkommen eines Ribonuklease-Hemmstoffes im stromafreien Hämolysat roter Blutkörperchen, *Acta Biol. Med. Ger.,* 3, 276, 1959.

124. **Traub, P., Zillig, W., Milette, R. L., and Schweiger, M.,** Untersuchungen zur Biosynthese der Proteine. VII. Aktivität verschiedener Desoxyribonucleinsäuren und eines Ribonuclease-inhibitors aus Kaninchen-reticulocyten in einem zellfreien Proteinsynthese-System aus Escherichia coli. DNA-abhängige in vitro-Synthese ''früher Proteine'' des E.-coli-Phagen T 4, *Hoppe Seyler's Z. Physiol. Chem.,* 343, 261, 1966.

125. **Priess, H. and Zillig, W.,** Inhibitor für pakreatische Ribonuclease aus roten Blutzellen, *Hoppe Seyler's Z. Physiol. Chem.,* 348, 817, 1967.

126. **Rapoport, S., Leva, E., and Guest, G. M.,** Acid and alkaline phosphatase and nucleophosphatase in the erythrocytes of some lower vertebrates, *J. Cell. Comp. Physiol.,* 19, 103, 1942.

127. **Rosenthal, S., Ruprecht, A., and Zank, W.,** Demaskierung der DNasen von Froscherythrozyten, *Folia Haematol. (Leipzig),* 90, 142, 1968.

128. **Foulkes, J. G., Ernst, V., and Levin, D. H.,** Separation and identification of type 1 and type 2 protein phosphatase from rabbit reticulocyte lysates, *J. Biol. Chem.,* 258, 1439, 1983.

129. **Wollny, E., Watkins, K., Kramer, G., and Hardesty, B.,** Purification to homogeneity and partial characterization of a 56,000-dalton protein phosphatase from rabbit reticulocytes, *J. Biol. Chem.,* 259, 2484, 1984.

130. **Mumby, M. and Traugh, J. A.,** Dephosphorylation of translational components by phosphoprotein phosphatasen from reticulocytes, *Methods Enzymol.,* 60, 522, 1979.

131. **Crouch, D. and Safer, B.,** Purification and properties of eIF-2 phosphatase, *J. Biol. Chem.,* 255, 7918, 1980.

132. **Stewart, A. A., Crouch, D., Cohen, P., and Safer, B.,** Classification of an eIF-2 phosphatase as a type-2 protein phosphatase, *FEBS Lett.,* 119, 16, 1980.

133. **Grankowski, N., Lehmusvirta, D., Kramer, G., and Hardesty, B.,** Partial purification and characterization of reticulocyte phosphatase with activity for phosphorylated peptide initiation factor 2, *J. Biol. Chem.,* 255, 310, 1980.

134. **Parra, M. G., Schewe, T., and Rapoport, S. M.,** On the presence of a calcium-stimulated phospholipase A in the stromafree supernatant fluid of rabbit reticulocytes, *Acta Biol. Med. Ger.,* 34, 1075, 1975.

135. **Delbauffe, D., Paysant, M., and Polonovski, J.,** Phosphatidyl-glycerolphospholipase A des globules rouges de rat. I. Influence des effecteurs, *Bull. Soc. Chim. Biol.,* 50, 1431, 1968.

136. **Ferber, E., Munder, P. G., Kohlschütter, A., and Fischer, H.,** Lysolecithin-Stoffwechsel in Erythrocytenmembranen. Lysolecithin-Acylierung und Lysophospholipase in alternden Erythrocyten, *Eur. J. Biochem.,* 5, 395, 1968.

137. **Scheuch, D. and Rapoport, S.,** Biologische Dynamik der anorganischen PP-ase im Verlauf einer Entblutungsanämie, *Acta Biol. Med. Ger.,* 6, 23, 1961.

138. **Mai, A., Sandring, D., Belkner, J., Prehn, S., and Rapoport, S. M.,** In vitro-Reifung von Retikulozyten. Verhalten von RNS und anorganischer Pyrophosphatase, *Acta Biol. Med. Ger.,* 39, 217, 1980.

139. **Chern, C. J., MacDonald, A. B., and Morris, A. J.,** Purification and properties of a nucleoside triphosphate pyrophosphohydrolase from red cells of the rabbit, *J. Biol. Chem.,* 244, 5489, 1969.

140. **Augustin, W. and Gellerich, F. N.,** Studies on the regulation of mitochondrial ATP formation in rabbit reticulocytes, *Acta Biol. Med. Ger.,* 40, 603, 1981.

141. **Hathaway, G. M. and Traugh, J. A.,** Kinetics of activation of casein kinase II by polyamines and reversal of 2,3 bisphosphoglycerate inhibition, *J. Biol. Chem.,* 259, 7011, 1984.

142. **Chalfin, D.,** Differences between young and mature rabbit erythrocytes, *J. Cell. Comp. Physiol.,* 47, 215, 1956.

143. **Henriques, V. and Ørskov, S. L.,** Untersuchunger über die Schwankungen des Kationgehaltes der roten Blutkörperchen. I. Die Änderungen der Kaliumkonzentration in den Blutkörperchen nach einem Aderlass, nach Vergiftung mit Phenylhydrazin, und nach Einführung von destilliertem Wasser in die Blutbahn, *Skand. Arch. Physiol.,* 74, 63, 1936.

144. **Ginzburg, S., Smith J. G., Ginzburg, G. M., Readon, J. Z., and Aikawa, J. K.,** Magnesium metabolism of human and rabbit erythrocytes, *Blood,* 20, 722, 1962.

145. **Ababei, L.,** Über eine Erythrozyten-Glutaminase; ihr Verhalten bei der Reifung der roten Blutzellen, *Acta Biol. Med. Ger.,* 5, 630, 1960.

146. **Wiesner, R., Rosenthal, S., and Hiebsch, C.,** Leitkriterien der Retikulozytenreifung. II. Das Verhalten von Zytochromoxydase und Hemmstoff F der Atmungskette bei der Retikulozytenreifung, *Acta Biol. Med. Ger.,* 30, 631, 1973.

147. **Syllm-Rapoport, I., Dumdey, E., and Rapoport, S.,** Creatine during bleeding anemia of rabbits, *Acta Biol. Med. Ger.,* 36, 411, 1977.

148. **Bessis, M. and Breton-Gorius, J.,** Diapédèse des réticulocytes et des érythroblastes, *C. R. Acad. Sci.,* 251, 465, 1960.

149. **Tavassoli, M.,** Red cell delivery and the function of the marrow-blood barrier: a review, *Exp. Hematol.,* 6, 257, 1978.

150. **Chamberlain, J. K., Weiss, L., and Weed, R. I.,** Bone marrow sinus cell packing: a determinant of cell release, *Blood,* 46, 91, 1975.

151. **Dabrowski, Z., Zygula, Z., and Miszta, H.,** Do changes in bone marrow pressure contribute to the egress of cells from bone marrow?, *Acta Physiol. Pol.,* 32, 729, 1981.

152. **Celada, A., Stroy, S., Sivarajan, M., and Finch, C.,** Iron supply for erythropoiesis in the rabbit, *J. Clin. Invest.,* 74, 161, 1984.

153. **Leblond, P. F., LaCelle, P. L., and Weed, R. I.,** Cellular deformability: a possible determinant of the normal release of maturing erythrocytes from the bone marrow, *Blood,* 37, 40, 1971.

154. **Patel, V. P., Ciechanover, A., Platt, O., and Lodish, H. F.,** Mammalian reticulocytes lose adhesion to fibronectin during maturation to erythrocytes, *Proc. Natl. Acad. Sci. U.S.A.,* 82, 440, 1985.

155. **Oski, F. A. and Komazawa, M.,** Metabolis of the erythrocytes of the newborn infant, in *Current Problems in Pediatric Hematology,* Oski, F. A., Jaffé, E. P., and Miescher, P. A., Eds., Grune & Stratton, New York, 1975, 49.

156. **Pearson, H. A.,** Life-span of the fetal red blood cell, *J. Pediatr.,* 70, 166, 1967.

157. **Meberg, A.,** Haemoglin concentrations and erythropoietien levels in appropriate and small for gestational age infants, *Scand. J. Haematol.,* 24, 162, 1980.

158. **Stockman, J. A., III., Garcia, J. F., and Oski, F. A.,** The anemia of prematurity: factors governing the erythropoietin response, *New Engl. J. Med.,* 296, 647, 1977.

159. **Gross, G. P. and Hathaway, W. E.,** Fetal erythrocyte deformability, *Pediatr. Res.,* 6, 593, 1972.

160. **Crowley, J., Ways, P., and Jones, J. W.,** Human fetal erythrocyte and plasma lipids, *J. Clin. Invest.,* 44, 989, 1965.

161. **Neerhout, R. C.,** Erythrocyte lipids in the neonate, *Pediatr. Res.,* 2, 172, 1968.

162. **Gasser, C.,** Die hämolytische Frühgeborenenanämie mit spontaner Innenkörperbildung; ein neues Syndrom, beobachtet an 14 Fällen, *Helv. Paediatr. Acta,* 8, 491, 1953.

163. **Gasser, C.,** Heinz body anemia and related phenomena, *J. Pediatr.,* 54, 673, 1959.

164. **Borges, A. and Desforges, J. F.,** Studies of Heinz body formation, *Acta Haematol.,* 37, 1, 1967.

165. **Lestas, A. N. and Rodeck, C. H.,** Normal glutathione content and some related enzyme activities in the fetal erythrocytes, *Br. J. Haematol.,* 57, 695, 1984.

166. **Berry, D. H. and Brewster, M. A.,** Riboflavin, FAD, glutathione and active glutathione reductase in neonatal erythrocytes, *Biol. Neonate,* 30, 245, 1976.

167. **Konrad, P. N., Valentine, W. N., and Paglia, D. E.,** Enzymatic activities and glutathione content of erythrocytes in the newborn: comparison with red cells of older normal subjects and those with comparable reticulocytosis, *Acta Haematol.,* 48, 193, 1972.

168. **Witt, I., Müller, H., and Künzer, W.,** Vergleichende biochemische Untersuchungen an Erythrozyten aus Neugeborenen- und Erwachsenen-Blut, *Klin. Wochenschr.,* 45, 262, 1967.

169. **Emerson, P. M., Mason, D. Y., and Cuthbert, J. E.,** Erythrocyte glutathione peroxidase content and serum tocopherol levels in newborn infants, *Br. J. Haematol.,* 22, 667, 1972.

170. **Minnich, V., Smith, M. E., and Rajanasathit, C.,** Erythrocyte glutathione synthesis in the neonate, *Biol. Neonate,* 24, 128, 1974.

171. **Agostoni, A., Gerli, G. C., Beretta, L., and Bianchi, M.,** Superoxide dismutase, catalase and glutathione peroxidase activities in maternal and cord blood erythrocytes, *J. Clin. Chem. Clin. Biochem.,* 18, 771, 1980.

172. **Gross, R. I., Bracci, R., Rudolph, N., Schroeder, E., and Kochen, I. A.,** Hydrogen peroxide toxicity and detoxification in erythrocytes of newborn infants, *Blood,* 29, 481, 1967.

173. **Jones, P. E. H. and McCance, R. A.,** Enzyme activities in the blood of infants and adults, *Biochem. J.,* 45, 464, 1949.

174. **Ross, J. D.,** Deficient activity of DPNH-dependent methemoglobin diaphorase in cord blood erythrocytes, *Blood,* 21, 51, 1963.

175. **Yoshioka, T., Sugine, A., Shimada, T., and Utsumi, K.,** Superoxide dismutase activity in the maternal and cord blood, *Biol. Neonate,* 36, 173, 1979.

176. **Oski, F. A. and Smith, C. A.,** Red cell metabolism in the premature infant. III. Apparent inappropriate glucose consumption for cell age, *Pediatrics,* 41, 473, 1968.

177. **Gross, R. T., Schroeder, E. A. R., and Brounstein, S. A.,** Energy metabolism in the erythrocytes of premature infants compared to full term newborn infants and adults, *Blood,* 21, 755, 1963.

178. **Travis, S. F. and Delivoria-Papadopoulos, M.,** Red cell enzymopathies in the newborn. I. Evaluation of red cell metabolism, *Ann. Clin. Lab. Sci.,* 12, 89, 1982.

179. **Oski, F. A.,** Red cell metabolism in the newborn infant. V. Glycolytic intermediates and glycolytic enzymes, *Pediatrics,* 44, 84, 1969.

180. **Wang, W. and Mentzer, W. C.,** Differentiation of transient erythroblastopenia of childhood from congenital hypoplastic anemia, *J. Pediatr.,* 88, 784, 1976.

181. **Kahn, A., Boyer, C., Cottreau, D., Marie, J., and Boivin, P.,** Immunologic study of the age-related loss of activity of six enzymes in the red cells from newborn infants and adults — evidence for a fetal type of erythrocyte phosphofructokinase, *Pediatr. Res.,* 11, 271, 1977.

182. **Gahr, M., Meves, H., and Schröter, W.,** Fetal properties in red blood cells of newborn infants, *Pediatr. Res.* 13, 1231, 1979.

183. **Komazawa, M. and Oski, F. A.,** Biochemical characteristics of "young" and "old" erythrocytes of the newborn infant, *J. Pediatr.,* 87, 102, 1975.

184. **Lestas, A. N., Rodeck, C. H., and White, J. M.,** Normal activities of glycolytic enzymes in the fetal erythrocytes, *Br. J. Haematol.,* 50, 439, 1982.

185. **Chen, S. H., Anderson, J. E., Giblett, E. R., and Stamatoyannopoulos, G.**, Isozyme patterns in erythrocytes from human fetuses, *Am. J. Hematol.*, 2, 23, 1977.

186. **Bartels, H. and Vogel, J.**, Isoenzyme der Enolase in Erythrozyten Neugeborener und Erwachsener, *Z. Kinderheilkd.*, 111, 247, 1971.

187. **Gahr, M.**, Different biochemical properties of fetal and adult red cell hexokinase isoenzymes, *Hoppe Seyler's Z. Physiol. Chem.*, 361, 829, 1980.

188. **Franzke, R. and Jelkmann, W.**, Characterization of the pyruvate kinase which induces the low 2,3 DPG level of fetal rabbit red cells, *Pfluegers Arch.*, 394, 21, 1982.

189. **Vora, S.**, Isozymes of human phosphofructokinase in blood cells and cultured cell lines: molecular and genetic evidence for a trigenic system, *Blood*, 57, 724, 1981.

190. **Magnani, M., Cucchiarini, L., Stocchi, V., Dacha, M., and Fornaini, G.**, Adult and fetal galactokinases in human red blood cells, *Mech. Ageing Dev.*, 18, 215, 1982.

191. **Tax, W. J. M., Veerkamp, J. H., and Schretlen, E. D. A. M.**, Pyrimidine metabolism in erythrocytes of the newborn, *Biol. Neonate*, 35, 121, 1979.

192. **Bauer, Ch. and Jelkmann, W.**, Prenatal ontogeny of blood oxygen affinity and its determinants, *Acta Biol. Med. Ger.*, 40, 639, 1981.

193. **Noble, N. A., Jensen, C., Nathanielsz, A. M., and Tanaka, K. R.**, Mechanism of red cell 2,3-diphosphoglycerate increase in neonatal lambs, *Blood*, 62, 920, 1983.

194. **Isaacks, R. E. and Harkness, D. R.**, Erythrocyte organic phosphate and hemoglobin function in birds, reptiles and fishes, *Am. Zool.*, 20, 115, 1980.

195. **Bartlett, G. R.**, Phosphate compounds in vertebrate red blood cells, *Am. Zool.*, 20, 103, 1980.

196. **Jacobasch, G.**, unpublished data.

197. **Stare, C. and Cara, J.**, Adenosinphosphate im Blut Frühgeborener, *Biol. Neonate*, 3, 160, 1966.

198. **Jelkmann, W. and Bauer, C.**, Oxygen affinity and phosphate compounds of red blood cells during intrauterine development of rabbits, *Pfluegers Arch.*, 372, 149, 1977.

199. **Jelkmann, W. and Bauer, C.**, High pyruvate kinase activity causes low concentration of 2,3 diphosphoglycerate in fetal rabbit red cells, *Pfluegers Arch.*, 375, 189, 1978.

200. **Schweiger, H. G. and Rapoport, S.**, Der N-Stoffwechsel bei der Erythrozytenreifung: die N-Bilanz unter endogenen Bedingungen, *Hoppe Seyler's Z. Physiol. Chem.*, 313, 97, 1958.

201. **Okyama, H. and Minakami, S.**, Studies in erythrocyte glycolysis. V. Change of the glycolytic intermediate pattern of reticulocytes during maturation, *J. Biochem.*, 61, 103, 1967.

202. **Oswald, B. and Dassler, G.**, Die Konzentration von Nukleotiden und anderen säurelöslichen P-Verbindungen im Blut der Ratte während der Erythrozytenreifung und -alterung, *Acta Biol. Med. Ger.*, 18, 163, 1967.

203. **Bartlett, C. R.**, Phosphate compounds in rat erythrocytes and reticulocytes, *Biochem. Biophys. Res. Commun.*, 70, 1055, 1976.

204. **Mueggler, P. A., Jones, G., Peterson, J. S., Bissonnette, J. M., Koler, R. D., Metcalfe, J., Jones, R. T., and Black, J. A.**, Postnatal regulation of canine oxygen delivery: erythrocyte components affecting Hb function, *Am. J. Physiol.*, 238, H73, 1980.

205. **Black, J. A. and Mueggler, P. A.**, Molecular changes which contribute to canine postnatal anemia, *Acta Biol. Med. Ger.*, 40, 645, 1981.

206. **Kim. H. D. and Duhm, J.**, Postnatal decrease in the oxygen affinity of pig blood induced by red cell 2,3-DPG, *Am. J. Physiol.*, 226, 1001, 1974.

207. **Smith, R. C. and Teer, P. A.**, 3-Ribosyluric acid and nucleotide content of erythroid cells in phenylhydrazine-induced anemia in cattle, *Int. J. Biochem.*, 13, 509, 1981.

208. **Battaglia, F. C., McGaughey, H., Makowski, E. L., and Meschia, C.**, Postnatal changes in oxygen affinity of sheep red cells: a dual role of diphosphoglyceric acid, *Am. J. Physiol.*, 219, 217, 1970.

209. **Smith, J. E.**, Elevated erythrocyte glutathione associated with elevated substrate in high- and low-glutathione sheep, *Biochim. Biophys. Acta*, 496, 516, 1977.

210. **Agar, N. S. and Harley, J. D.**, In vitro metabolism of red blood cells from newborn and adult goats, *Biol. Neonate*, 28, 113, 1976.

211. **Opalinski, A. and Beutler, E.**, Creatine, 2,3 Diphosphoglycerate and anemia, *New Engl. J. Med.*, 285, 483, 1971.

212. **Chemtob, S., Gibb, W., and Bard, H.**, Relationship of 2,3 diphosphoglycerate and 2,3-diphosphoglycerate mutase in various mammals, *Biol. Neonate*, 38, 36, 1980.

213. **Narita, H., Yanagawa, S., Sasaki, R., and Chiba, H.**, Synthesis of 2,3-bisphosphoglycerate synthase in erythroid cells, *J. Biol. Chem.*, 256, 7059, 1981.

214. **Jelkmann, W. and Bauer, C.**, Regulation of red cell DPG metabolism in fetuses and adults, *Acta Biol. Med. Ger.*, 40, 661, 1981.

215. **Jelkmann, W. and Bauer, C.**, 2,3 DPG levels in relation to red cell enzyme activities in rat fetuses and hypoxic newborns, *Pfluegers Arch.*, 389, 61, 1980.

216. **Bartels, H., Bartels, R., Rathschlag-Schaefer, A. M., Röbbel, H., and Iüdders, S.,** Acclimatization of newborn rats and guinea pig to 3000 to 5000 m simulated altitudes, *Respir. Physiol.,* 36, 375, 1979.
217. **Bard, H. and Shapiro, M.,** Perinatal changes of 2,3 diphosphoglycerate and oxygen affinity in mammals not having fetal type hemoglobins, *Pediatr. Res.,* 13, 167, 1979.
218. **Gilman, J. G. and Jenkins, G. M.,** Developmental changes of mouse red cell pyruvate kinase, *Biomed. Biochim. Acta,* 42, 273, 1983.
219. **Comline, R. S. and Silver, M.,** A comparative study of blood gas tensions, oxygen affinity and red cell 2,3-DPG concentrations in fetal and maternal blood in the mare, cow and sow, *J. Physiol.,* 242, 805, 1974.
220. **Bunn, H. F. and Kitchen, H.,** Hemoglobin function in the horse: the role of 2,3 diphosphoglycerate in modifying the oxygen affinity of maternal and fetal blood, *Blood,* 42, 471, 1973.
221. **Dhinds, D. S., Hoversland, A. S., and Templeton, J. W.,** Postnatal changes in oxygen affinity and concentrations of 2,3 DPG in dog blood, *Biol. Neonate,* 20, 226, 1972.
222. **Watts, R. P. and Kim, H. D.,** Comparison of 2,3-diphosphoglycerate metabolism between fetal and postnatal pig red cells, *Biol. Neonate,* 45, 280, 1984.
223. **Jacobasch, G. and Rapoport, S.,** Phosphoglyzerinsäureveränderungen in Retikulozyten und Erythrozyten von Schafen, *Folia Haematol.,* 83, 283, 1965.
224. **Blunt, M. H., Kitchens, J. L., Mayson, S. M., and Huisman, T. H. J.,** Red cell diphosphoglycerate and oxygen affinity in newborn goats and sheep, *Proc. Soc. Exp. Biol. Med.,* 138, 800, 1971.
225. **Borden, M., Nyhan, W. L., and Bakay, B.,** Increased activity of adenine phosphoribosyltransferase in erythrocytes of normal newborn infants, *Pediatr. Res.,* 8, 31, 1974.
226. **Griffiths, W. J. and Fitzpatrick, M.,** The effect of age on the creatine in red cells, *Br. J. Haematol.,* 13, 175, 1967.
227. **Syllm-Rapoport, I., Daniel, A., Starck, H., Götze, W., Hartwig, A., Gross, J., and Rapoport, S.,** Creatine in red cells: transport and erythropoietic dynamics, *Acta Biol. Med. Ger.,* 40, 653, 1981.
228. **Syllm-Rapoport, I., Dumdey, E., and Rapoport, S.,** Creatine during bleeding anemia of rabbits, *Acta Biol. Med. Ger.,* 36, 411, 1977.
229. **Syllm-Rapoport, I., Daniel, A., Lun, A., Pohle, R., Möller, R., and Gross, J.,** Das Verhalten von Kreatin in roten Blutzellen und im Blutplasma bei erwachsenen Ratten nach Hypoxie-Exposition, *Acta Biol. Med. Ger.,* 39, 1021, 1980.
230. **Syllm-Rapoport, I., Daniel, A., and Dumdey, E.,** Kreatingehalt in Erythrozyten und Blutplasma des Huhns vor und nach Erythropoese-Stimulation durch Anämie, *Acta Biol. Med. Ger.,* 39, 1015, 1980.
231. **Hiramatsu, Y., Eguchi, K., Yonezawa, M., Hayase, R., and Sekiba, K.,** Alterations of red blood cells polyamines during pregnancy and neonatal period, *Biol. Neonate,* 40, 136, 1981.
232. **Cooper, K. D., Shukla, J. B., and Rennert, O. M.,** Polyamine distribution in cellular compartments of blood and in aging erythrocytes, *Clin. Chim. Acta,* 73, 71, 1976.
233. **Jackson, R. J., Campbell, E. A., Herbert, P., and Hunt, T.,** The preparation and properties of gel-filtered rabbit-reticulocyte lysate protein-synthesis systems, *Eur. J. Biochem.,* 131, 289, 1983.
234. **Gilman, J. G.,** Red cells of newborn rats have low bisphosphoglyceromutase and high pyruvate kinase activities in association with low 2,3 bisphosphoglycerate, *Biochem. Biophys. Res. Commun.,* 98, 1057, 1981.

Chapter 4

CELL MEMBRANE AND CYTOSKELETON

I. INTRODUCTION

In recent years cell membranes have been the object of intensive studies with remarkable success. The best characterized cell membrane is that of the erythrocyte. Where the cytoskeleton is concerned, the structure of the erythrocyte only permits the study of the membrane skeleton since it lacks intracellular fibers. But, again, our knowledge of the membrane skeleton of erythrocytes is advanced farthest among the cells studied so far. Membrane and membrane skeleton acquire their final structure during maturation of the erythroid cell.

In order to understand developmental changes, an overview of the composition and structure of the mature erythrocyte membrane is appropriate.

II. CELL MEMBRANE AND CYTOSKELETON OF THE ERYTHROCYTE

The cell membrane of erythrocytes contains 60 to 80% protein, 20 to 40% lipids, and 5 to 10% protein-bound carbohydrates. The proteins are highly heterogeneous; although more than 200 different species can be demonstrated, a few types predominate. Most of the membrane proteins share common features, such as a high proportion of non-polar amino acids which cluster in certain sections of the polypeptide chains and are responsible for the interaction with the hydrophobic milieu of the lipid bilayer (for reviews see Steck,[1] Marchesi,[2,3] Goodman and Shiffer,[4] and Cohen[5]). Erythrocyte membrane lipids are made up of up to 50 to 70% phospholipids, 20 to 30% cholesterol, and about 5 to 10% glycolipids. On account of the variety of fatty acid chains, it is estimated that there are 150 to 200 different lipids in the membrane. Lipid composition is variable and subject to external influences like nutrition. Quantitatively, the most important phospholipids are phosphatidylcholine, sphingomyelin, phosphatidylethanolamine, and phosphatidylserine, which account for about 95% of the total phospholipids. Phosphatidylinositides represent the remainder. The choline-containing phospholipids amount to about one half of the total in erythrocytes of all mammalian species.[6] The two layers of the bilayer differ distinctly with respect to their phospholipids. The choline-containing phospholipids, phosphatidylcholine and sphingomyelin, are predominant in the outer layer, whereas the inner layer consists mainly of phosphatidylethanolamine and phosphatidylserine. The carbohydrates of the cell membrane are present either in glycoprotein or glycolipid form. They are almost exclusively found on the outer surface of the cell and form a coating, the glycocalix. The glycolipids resemble sphingomyelin, being derivatives of ceramide; their polar group consists of a great variety of complex polysaccharides, which may contain more than 20 sugar residues.

Operationally, the distinction has been made between (1) peripheral proteins, which are located on either surface of the lipid bilayer and which can be detached from the membrane under mild conditions — such as increase in ionic strength — and (2) integral proteins, which partially or completely penetrate the lipid bilayer and require drastic means — such as detergents — for their detachment. A more functional classification would distinguish membrane proteins in the strict sense and membrane skeleton proteins, the latter corresponding to the inner peripheral proteins. The mature erythrocyte is devoid of any component of the cytoskeleton.

The lipid bilayer itself lacks mechanic stability and cannot maintain the asymmetry of the phospholipid distribution or the biconcave shape of the erythrocyte. These properties are dependent on the membrane skeleton proteins, which form a highly interconnected network that is anchored to the cell membrane by specific interactions with its integral membrane proteins.

A. The Membrane Proteins

There are three major membrane proteins: the glycophorins; the anion-transporter, band 3; and the monosaccharide transporter.

The glycophorins (A, B, and C) make up the bulk of the sialo-glycoproteins with glycophorin A accounting for about 3/4 of the total (for reviews see Marchesi[2,3] and Furthmair[7,8]). Glycophorin A is a dimer with each single peptide chain consisting of 131 amino acids to which 16 oligosaccharides are attached, making up 60% of the molecular mass. Many of the carbohydrate side chains are bound to threonine and serine residues. They are concentrated at the amino terminal end of the molecule, which faces the cell exterior.[9] This part of the glycophorin molecule is responsible for the MN blood group determinants.[7,10] The sugar moieties bind specific lectins. It is likely that other blood group determinants and receptors are found on the glycosylated end of the glycophorin molecule. The non-polar amino acids are concentrated in a region roughly midway between the amino-terminal and the carboxy-terminal part of the polypeptide chain. There is good evidence that the hydrophobic part of glycophorin A is buried within the lipid bilayer of the cell membrane. The C-terminal segment of glycophorin A is characterized by a large number of charged amino acids with a sequence of acid residues at the very end of the chain. It is estimated that there are about 250,000 copies of glycophorin A dimer per erythrocyte. In addition to glycophorin A, glycophorin B and C, which so far have not been well characterized, are present in smaller amounts.[7,8]

The anion transporter (band 3) is also a glycoprotein occurring as a tetramer in the membrane and is present in about one million copies per erythrocyte, the highest representation (for reviews see Steck[11] and Jennings[12]). Its amino acid sequence has been established,[13] consisting of two domains with the amino-terminal region of 43 kD facing the cytoplasm, an orientation which is opposite to that of glycophorin A.[14] The N-terminal end projecting into the cytoplasm is extraordinarily acidic; among the first 32 amino acids there are 6 aspartate and 12 glutamate residues. The terminal amino group is acetylated. It appears to be adapted to the strong electrostatic binding of various proteins such as hemoglobin and glyceraldehyde phosphate dehydrogenase at low ionic strength. The cytoplasmic extension has a most important function in connecting the membrane skeleton to the membrane via ankyrin. There is some evidence that band 3 and glycophorin A may be associated within the membrane. The C-terminal region, which carries the property of the inorganic anion transporter also contains one carbohydrate chain which amounts to 5 to 8% on a weight basis and is composed of mannose, galactose, and N-acetylglucosamine in the approximate ratios of 1:2:2.[15] This composition suggests that the oligosaccharides attached to the protein are the complex type, which are linked to an asparagine residue by an N–glycosidic bond.[16] The C-terminal region has a most complicated topology; it crosses the membrane at least five times, with the C-terminal end buried in the lipid bilayer. This conclusion is based on evidence from sequence studies of chymotryptic, peptic, and papain-split products.[17-20]

The monosaccharide transport protein, which physiologically transports D-glucose, is the main component of band 4.5. It has been purified in recent years and found to be a membrane-spanning glycoprotein. The molecular mass of the monomer is 55 kD.[21,22] It can be characterized by its property of binding cyto-chalasin. Its terminal end groups have been determined.[23,24] Recently, significant information on its orientation in the human erythrocyte membrane has been obtained.[25] It was found that it resembles the anion transporter to a large extent in its arrangement with the N terminus in the cytosol, whereas the C-terminal domain is on the outside of the cell membrane. It carries the carbohydrate groups. From the number and type of proteolytic fragments obtained with various proteases, it was concluded that the protein spans the membrane at least three times (see Figure 12). There is good evidence that the nucleoside transporter is a component of the proteins of band 4.5.[26,27]

B. The Proteins of the Membrane Skeleton

There are five proteins that make up the membrane skeleton; they are spectrin, actin, ankyrin, band 4.1, and band 4.9.

Spectrin is a water-soluble protein composed of two subunits of 240 kD and 220 kD designated as α-chain (band 1) and β-chain (band 2), respectively.[28-32] They occur in 200,000 copies each per erythrocyte, accounting for more than one half of the membrane skeleton on a weight basis. The two chains differ not only in molecular mass, but also in constitution. Only the β-subunit is phosphorylated and contains three phosphoserines and one phosphothreonine clustered at its C-terminal end.[29] Both subunits contain many acidic amino acids and appear to be comprised of many homologous helical segments.[31,32]

The N-terminal region contains binding sites for band 4.1 and actin.[33-35] There is good evidence that spectrin occurs as a tetramer in the cell, located at the cytoplasmic surface of the membrane.[36]

Actin is ubiquitous in eukaryotic cells. It has a monomer molecular mass of 42 kD (actin G), occurs in about 400,000 copies per erythrocyte, and is capable of forming double-helical filaments (actin F) with a diameter of 5 nm. The actin in the membrane skeleton of red cells is primarily a β-actin isotype similar to that of striated muscle, yet different in its properties.[37-39] In contrast to other cells, the actin filaments are short and are comprised of 10 to 20 monomers.[40,41]

At first ankyrin was not considered a constituent of the membrane skeleton because of its close attachment to band 3 protein. It represents a polypeptide with a 210 kD molecular mass[42-44] and occurs in 100,000 copies per erythrocyte. It has the important function of linking the membrane skeleton to the bilayer by connecting the spectrin to the amino-terminal cytosolic domain of band 3. Ankyrin binds to the spectrin β-chain and in turn is bound to the membrane by attachment to 10 to 20% of band 3 molecules. A provisional, structural model was established and two functional domains could be identified: one is responsible for the binding of band 3 and the other is responsible for the binding of spectrin.

Band 4.1 represents two globular proteins (4.1a and 4.1b) with molecular mass of about 80 kD, which comprise about 5% of the mass of the membrane skeleton (for review see Goodman et al.[45]). These two proteins are phosphorylated and appear to be sequence-related.[45-47] Each occurs in about 100,000 copies per red cell. The function of protein 4.1 appears to be the strengthening of the weak spectrin-actin interaction by formation of a ternary complex.[48] Two domains can be recognized at each end, one basic and one acidic. The basic domain contains the spectrin binding site, while the phosphates are found in the acidic domain. Band 4.1 proteins have their own affinity attachment site to the cell membrane, thus serving as an additional link between the membrane skeleton and the cytoplasmic membrane surface. Recent evidence makes it very likely that the 4.1 protein binds to the cytosolic domain of glycophorin.[49,50]

The presence of 4.1 protein has recently been demonstrated in avian erythrocytes.[51]

Band 4.9 protein is undoubtedly part of the membrane skeleton; however, its function as yet is not clearly defined. It has a molecular mass of about 50 kD and is a phosphoprotein. It is represented in about 100,000 copies in the erythrocyte. It has been suggested that it serves to stabilize short actin oligomers, since it binds to F-actin.[52,53]

Recently, tropomyosin has been purified and characterized from human red cells. It is a dimer with a molecular mass of about 60 kD and represents 1% of the membrane proteins.[54] In its properties, it resembles muscle tropomyosin more closely than other non-muscle species. It exhibits highly cooperative binding to actin, depending on the Mg^{2+} concentration within a physiological concentration range. The amount of tropomyosin may be just sufficient to cover the two grooves of the actin filaments and thus serve to stabilize them. A conceptual difficulty arises from the fact that a competition was observed between tropomyosin and spectrin for actin binding. On the other hand, this circumstance might provide the possibility for regulating the properties of the membrane skeleton.

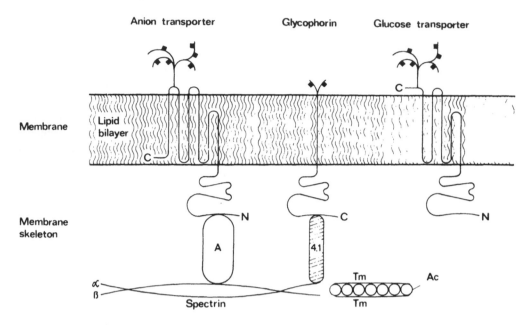

FIGURE 12.   Scheme of interconnections between the membrane and its skeleton. Depicted is half of a "unit",
i.e., one half of the spectrin tetramer and its connections. A, ankyrin; Ac, actin; Tm, tropomyosin.

Most of the work discussed has been performed on human erythrocytes. Comparative
studies have shown that some elements of the membrane skeleton of erythrocytes are similar
among various species, as judged by their molecular mass, whereas bands 3, 4.1a, and 4.1b
show substantial variations.[55]

Finally, the presence of an oligomeric protein, distinguished by its ability to form hollow
cylinders, should be noted.[56-60] It has received the name "cylindrin". This protein occurs
in the cytosol of erythrocytes in amounts exceeding 15 μg/mℓ cells, but is partly membrane
attached. It consists of at least five different polypeptides. The function of cylindrin is
unknown.

A schematic diagram (Figure 12) depicts the molecular organization of the cell membrane
and its skeleton (for a review see Branton et al.[61]). Figure 12 illustrates the actin-spectrin
network which provides the basic strong, flexible, elastic skeletal structure covering the
cytoplasmic surface of the erythrocyte membrane. Furthermore, it can be seen that the
spectrin molecule is multiply tethered to the membrane via ankyrin toward the middle of
the tetramers and via complexes of band 4.1 with actin at their ends. One can imagine a
unit cell of the membrane skeleton with a core of actin molecules forming a short filament
which is stabilized by its associations with spectrin, band 4.1, and possibly band 4.9. The
main anchor in the membrane is provided by band 3 and glycophorin.

It would be a simplification to assume that the membrane skeleton alone is responsible
for the characteristic shape of the red cell. It appears that it is the joint properties of membrane
skeleton and the lipid bilayer, to which it is attached, that account for it. The membrane
skeleton exerts a stabilizing and restrictive influence on both the mobility of the proteins as
well as on membrane phospholipids.[62-66] On the other hand, there is some evidence that the
bilayer influences the shape and stability of the whole system.[67,68] A recent model, the
protein-gel-lipid bilayer model, which takes into consideration physicochemical properties
such as the rubber elasticity of peptide chains, polymer-polymer affinity, and hydrogen ion
tension, describes in a quantitative fashion the shape of the red cell under various conditions;
these include biconcave and cup shape and crenation at low pH and invagination at high
pH values.[281] However, keep in mind that the system consisting of membrane and membrane

69

skeleton is not static but is subject to the influence of cell metabolism with regard to its composition and structure. It must be considered as in a steady state rather than at equilibrium.

III. CHANGES IN MEMBRANE AND CYTOSKELETON DURING DIFFERENTIATION AND MATURATION

There is ample evidence that the cell membrane, its skeleton, and the internal cytoskeleton undergo various changes during the course of the differentiation program. At least two processes are involved: (1) the appearance and assembly of red cell-specific components, and (2) the loss of constituents which are specific for the nondifferentiated precursors. It is as yet an insurmountable conceptual difficulty as to how these changes are coordinated without loss of the characteristic properties of a cell membrane. Only recently has it been possible to gain some comprehension of the sequence of events leading to the formation of the membrane skeleton, which by nature requires definite proportions among its constituents, as well as specific partners among the membrane proteins. The key to this understanding has been the realization that synthesis of the cytoskeletal proteins and their assembly is not coupled. Spectrin is synthesized very early in erythroid development, even before hemoglobin is made.[69-77] The accumulation of spectrin is stimulated by erythropoietin.[78] Another protein to be synthesized early is the anion transporter band 3.[79] On the other hand, glycophorin is synthesized fairly late, reaching a maximum at the intermediate erythroblast stage of differentiation.[80] Band 4.1 is probably the last component of the membrane skeleton to terminate synthesis, since it is still made by reticulocytes.[69]

The Lazarides and Moon[81] studies have provided a fairly clear picture of the assembly and topogenesis mechanism of the membrane skeleton during erythroid development. According to their work, assembly proceeds sequentially with each preceding component limiting the succeeding step.[82] The initial limitation is the development of cell membrane anchoring sites provided by the newly formed anion transporter. This step is followed by the association of ankyrin with these sites. Subsequently, the β-spectrin chains and then the α-spectrin chains are attached. The spectrins are synthesized in considerable excess of the amount required to form the membrane skeleton.

Individual nonassembled polypeptides undergo rapid proteolysis with a half-life of 45 min, whereas in an assembled state they are highly stable. Thus, the excessively produced building blocks are eliminated. The degradation of the unassembled α- and β- spectrin occurs by distinct intracellular pathways.[85] This scheme is supported by studies on mice with genetic defects causing hemolytic anemia.[84] They exhibit defects in the membrane skeleton of the red cells which are caused by deficient synthesis of ankyrin or β-spectrin.

In addition to the processes leading to assembly, other subtle changes in membrane organization take place. There is increasing restriction on the movement of membrane proteins.[85] The reorganization of the cell membrane preceding the expulsion of the nucleus has been described in Chapter 2. Loss of the nucleus and the ensuing reduction in cell size, with little loss of spectrin, results in a drastic increase in spectrin packing which is further increased by loss of membrane parts free of spectrin during the maturation of the reticulocyte.[86] Among others the transferrin receptors are lost by vesiculation.[87]

Hand in hand with this process of membrane skeleton solidification is the loss of lectin receptor mobility as well as ability for invagination and endocytosis.[88] It is interesting that the neonatal human erythrocyte exhibits domains of mobility and endocytosis of membrane receptors similar to the reticulocyte, properties which are absent in the mature erythrocyte of adults.[89,90]

As to be expected from our understanding of synthesis and cellular traffic of proteins, the components of the membrane skeleton, which are assembled on the cytosolic surface of the cell membrane, do not undergo processing after translation. Membrane proteins, on the other hand, are subject to modification prior to their insertion into the plasma membrane.

The biosynthesis of glycophorin corresponds to the well-studied model of virus-coded proteins. The polypeptide is synthesized carrying a signal sequence on membrane-bound ribosomes transferred as a nascent chain through the membrane of the endoplasmic reticulum, with the C-terminus remaining in the cytosol after which it is glycosylated.[91-93] During its passage, the carbohydrates are further modified — primarily in the Golgi apparatus — until the protein reaches the cell surface after 30 min, making its first appearance at the basophilic erythroblast stage.[94,95] Recently, it was shown that human erythroid cell differentiation is associated with increasing O–glycosylation of glycophorin A.[96]

A particularly interesting case is the anion transport protein. As mentioned before, a large N-terminal domain of this protein is cytosolic, in contrast to glycophorin. It was demonstrated in studies on mouse erythroblasts that the nascent anion transport protein is cotranslationally inserted into added microsomes in such a manner that the N-terminal domain remains in the cytosol, whereas the C-terminal region passes their membrane. No signal sequence could be detected; possibly one exists in the middle of the polypeptide chain. The glycosylation with a high-mannose side chain occurs, presumably by way of the dolichol-transfer system, during translation. Further modifications — such as removal of some mannoses and additional glycosylations by glycosyl transferases — occur in the Golgi apparatus until the protein appears on the cell surface after 30 to 45 min.[97,98]

The cytoplasm of eukaryotic cells contains a variety of filamentous proteins which are organized into three different three-dimensional networks: microtubules consisting of tubulins; microfilaments, mainly composed of actin; and intermediate-sized filaments. Five classes of polypeptides have been distinguished among the building blocks of intermediate filaments, which differ chemically, immunologically, and in their distribution among different cells. Usually, but not always, a single class of polypeptides is expressed in a given cell type. During differentiation, a switch from one type of polypeptide to another has been described for several nonerythroid cell types. In studies of human red cells and their progenitors, it was found that the early precursors expressed vimentin in a heterogeneous fashion, i.e., that only about one half of the immature erythroblasts exhibited it. During differentiation of mammalian cells vimentin disappears completely, apparently unrelated to the expulsion of the nucleus.[99] On the other hand, avian red cells retain their vimentin intermediate filaments.[100]

They also contain a marginal band of microtubules as a prominent cytoskeletal component even though the number of the microtubules decrease with maturation.[101,102] There are some indications that the microtubules serve even in the mature cells as a support for the elliptic cell shape.[103] This suggestion is corroborated by a report on the occurrence of microtubules in llama, which like all camelidae, have oval-shaped red cells.[104]

The active movements exhibited by reticulocytes lead one to expect the occurrence of myosin, which indeed has recently been found.[105,106] The mechanisms by which the early progenitors degrade or extrude the components of their cytoskeleton and associated proteins, like actin-binding proteins other than spectrin, remain to be explored.

IV. MEMBRANE ENZYMES

A multitude of enzymes are bound to the cell membrane of the mature erythrocyte and exhibit drastic changes during maturation of the reticulocytes, and as far as is known, during the preceding states of differentiation. Enzymes differ in their localization and arrangement in the membrane (Table 7). Among those that span the membrane are Na^+K^+ ATPase and $Ca^{2+}Mg^{2+}$ ATPase.

Imbedded in the external part of the bilayer are a variety of hydrolases such as acetylcholine esterase, nucleotidases, and glycosidases. They act on extracellular substrates and therefore may be called "ectoenzymes". Their function is uncertain. The cytosolic side of the mem-

Table 7
MEMBRANE-RELATED ENZYMES OF RED CELLS (SELECTION)

Location in cell	Enzyme	Ref.
Plasma-membrane-bound		
Transmembrane	Na^+K^+-ATPase	See text
	$Ca^{2+}Mg^{2+}$-ATPase	See text
Ectoenzymes	Acetylcholine esterase	244, 245
	Ecto ATP pyrophosphohydrolase[a]	246
	UTPase	247
	NAD, NADP glycohydrolase	248, 249, 250
	Acid phosphatase	251
	Glycosidases	252, 253
Cytosolic side of membrane	Adenylate cyclase	See text
	Protein kinases	See Chapter 11
	Phosphatidylinositol kinase	See Chapter 7
	Phosphatidylinositol-4-P kinase	See Chapter 7
	Diacylglycerol kinase	See Chapter 7
	Acyltransferases	254
	Methyltransferases	255
	Phosphatidic acid phosphatase	256
	Phosphatidylinositol-4-P phosphatase	257
	Phosphatidylinositol-4,5-Pl_2 phosphatase	258
	Polyphosphoinositide phosphodiesterase	259, 260
	Inositol-P_3 phosphatase	261
	Phospholipase A_2	262
	Proteases	See Chapter 12
	NADH-cyt c reductase	263
Unspecified	GTPase	264
	Lysophospholipase	265
	β-3-D-Galactosyl transferase	266
	N-Acetyl-D-galactosaminyl	267
	Transferase	268
	Neuraminidase	269
	Enzymes synthesizing dolichol derivatives	270
	Dolichol-PP-oligosaccharide transferase[b]	271
Cytosolic enzymes modifying	Protein kinases	See Chapter 10
membrane constituents	Protein phosphatases	See Chapters 4 and 11
	Phospholipase A_2	272, 273
	Phosphatidylinositol-4-P-phosphatase	274
	Transglutaminase	265
Artificial binding at low	Glyceraldehyde-3-phosphate dehydrogenase	276, 277, 282
ionic strength	Aldolase	278
	Phosphofructokinase	278
	Glutathionreductase	279
	Catalase	280

[a] Nucleated red cells only.
[b] Only in reticulocytes.

brane has an even greater variety of enzyme activities including several types of proteases, protein kinases, and enzymes instrumental in the synthesis and breakdown of phosphatidyl inositides. There are also several enzymes with, so far, unspecified membrane location. Table 7 also contains a selective list of cytosolic enzymes, the substrates of which are predominantly membrane constituents in the mature erythrocyte.In the reticulocyte, however, other substrates are also available. This is particularly true for protein kinases and -phosphatases as well as phospholipase A_2, which may have various locations in the cell, in the membrane, in the cytosol, and in the outer membrane of mitochondria. There are many reports on the binding of enzymes — in the first place, of carbohydrate metabolism — to the cytosolic side of cell membranes prepared by hypotonic hemolysis. This binding occurs by electrostatic interaction, primarily of the highly negatively charged N-terminal domain of the anion transporter, with cationic groups of cytosolic proteins including hemoglobin. Under isotonic conditions the binding is practically abolished. Nevertheless, it is possible that the still existing weak electrostatic attraction results in an inhomogeneous distribution of cytosolic proteins with a gradient from the membrane to the interior of the cell. A functional significance of such a distribution is most unlikely.

Some enzymes are universal in their occurrence among various species of erythrocytes, e.g., $Ca^{2+}Mg^{2+}$-ATPase; others are missing or very low in a restricted group of species, e.g., Na^+K^+-ATPase in the mature red cells of ruminants, while a third group seems to be confined to a few species, like UTPase. All in all it is difficult to envisage a function for the mature erythrocyte.

Among the enzymes showing a distinct maturation dependence are ATPases, proteases, protein kinases, the adenylate kinase system, and dolichol-PP-oligosaccharide transferase. They show drastic decreases in activity with maturation of the reticulocyte.

Several of the enzyme activities usually found in lysosomes appear to be membrane-associated in red cells. Their localization may arise by fusion of lysosomes with the membrane skeleton. One characteristic of many of the membrane enzymes is their masked state. They can be unmasked by a variety of manipulations, e.g., detergents, change of tonicity, or incubation at 37°C. It is unclear how their physiologic activation may occur.

V. CHANGES IN TRANSPORT

Mature erythrocytes differ greatly among various species in their ability to transport low molecular compounds. An extreme case is the red cells of pigs, which are virtually impermeable to glucose; very low permeabilities to glucose are also found in erythrocytes of sheep, rabbits, and guinea pigs.[107] Therefore, the question arises as to whether such differences are innate, genetically determined, or represent results of differences in maturation processes. Of course a combination of factors should also be considered.

Studies on rabbits and pigs clearly indicate that their reticulocytes exhibit high glucose permeabilities which decrease with maturation.[108-112] There is a striking parallelism between fetus reticulocytes and red cells and newborn animal reticulocytes and red cells, with respect to glucose permeability.[110,111,113]

Similar differences between neonatal and adult red cells of dogs and sheep were reported.[114] From cell kinetic studies it appears that the loss of transport activity after birth results from the replacement of glucose-permeable cells by glucose-impermeable erythrocytes. Of course, this explanation would not apply to glucose permeability loss observed during in vitro maturation of reticulocytes.[112] It is conceivable that the maturational changes are caused by a specific decay of the glucose transporter in the red cell membrane, possibly a structural, i.e., genetically determined, feature of the protein.

Nucleoside transport offers an interesting example of the interplay of maturational and genetic effects. Erythrocytes of adult sheep show genetic polymorphism in the transport of nucleosides.[115] The nucleoside-permeable cells of one phenotype possess a high affinity transport system for both purine and pyrimidine nucleosides, whereas the other phenotype lacks this system. This difference is controlled by two allelomorphic genes. Newborn lamb erythrocytes and sheep reticulocytes show manyfold higher inosine transport than mature red cells of adult sheep and also show no phenotype differences.[114,116] Inosine transport system properties of neonatal cells were found to be similar to those of the adult. Obviously, the genetic difference is only expressed at the end of the differentiation process, possibly due to a subtle difference, perhaps a point mutation, in the gene for the inosine transport protein. During maturation in vitro, both rate of transport and number of specific binding sites decline.

Creatine occurs in high concentrations in reticulocytes of man and shows strong maturational dependence of creatine transport.[117-119] Apparently there are two types of transport processes: one with high affinity, which represents an active transport, and a second with low-affinity transport by exchange diffusion. The high-affinity component is highly dependent on cell age and decreases with maturation.

Several functionally distinct systems have been described in mammalian erythrocytes for amino acid transport (for reviews see Christensen,[120] Young and Ellory,[121] and Guidotti et al.[122]). Three of these systems, denominated *ASC, Glu,* and *Gly,* are Na^+-dependent, the latter also requiring Cl^- for its activity. The distribution of amino acid carriers varies greatly among red cells of different species. For instance, the *Gly* system, which selectively transports glycine and sarcosine, is found in human and avian erythrocytes, but is absent in mature erythrocytes of rabbits, guinea pigs, and sheep.[123-129] On the other hand, the reticulocytes of these species possess a highly active *Gly* system with properties similar to those described in mature human and avian erythrocytes.[124,129] A strong maturation dependence has also been described for the transport of alanine mediated by the *ASC* system and for the transport of cationic amino acids of rabbit and man.[130-133] Similarly, it was found that only reticulocytes of guinea pigs — but not mature erythrocytes — exhibit the Na^+-dependent transport of L-alanine.[129]

Pigeon erythrocytes show many features which are comparable to mammalian reticulocytes.[130,134] Evidently, the fate of various amino acid transport systems differs both within one type of red cell and among different species. During maturation of rabbit reticulocytes, Na^+-dependent alanine transport decreases rapidly while transport for glycine and lysine decline much more slowly.[131,135] Some transport mechanisms are retained, e.g., the *Gly*-transporter in human erythrocytes and cationic amino acid transport in rabbits.

There have been several reports that membrane transport of various substrates such as glucose, nucleosides, and amino acids or substances related to these substrates decreases as a function of differentiation of erythroid cells.[136-139] In a careful study in which the unidirectional flux, independent of utilization, was tested, rather different conclusions were drawn. The depression in uptake previously found after dimethyl sulfoxide treatment of murine erythroleukemic cells was found to be spurious, if the decrease in surface area during differentiation was taken into account. Instead, increases in flux of up to 10- to 20-fold for a glucose-like substrate and 2- to 4-fold for amino acids were found.[140]

The transport of Na^+ and K^+ as well as Na^+K^+-ATPase exhibit great differences between reticulocytes and neonatal cells on the one hand and adult erythrocytes on the other in species with low K^+ concentration in mature red cells.[141-144] Sheep are a particularly interesting object of study since there is a dimorphism with respect to the concentration of monovalent cations in their mature erythrocytes. "High-K^+ sheep" have typical high K^+ and low Na^+ concentrations, whereas "low-K^+ sheep" show the reverse cation composition. This dimorphism is genetically determined. Also, thiol- and Cl-dependent passive K^+ transport, which is highly active in sheep reticulocytes, disappears during maturation. Concurrently Na^+K^+-ATPase and Na^+K^+-pump activity decrease.[145,146] The actual changes in Na^+K^+-ATPase during maturation were determined in rat reticulocytes by means of ouabain binding, with the result that reticulocytes exhibit about three times as many binding sites as mature erythrocytes with a correspondingly threefold influx and efflux of K^+. From these results one would surmise that passive permeability of the reticulocyte membrane is higher than that of the erythrocyte.[147] Similar observations were made on the red cells of dogs. The reticulocytes had several times more ouabain-binding sites than mature erythrocytes.[148] A specific carrier characterized by the inhibitory effects of furosamide and ethacrynic acid also declined, to one tenth, during maturation of rabbit reticulocytes.[149]

In line with the preceding conclusion, the reticulocyte exhibits a much higher — up to 40-fold — Ca^{2+} permeability than the erythrocyte.[150] Correspondingly there is also a greater turnover of calcium.[151] In keeping with these conditions, the share of calcium transport among ATP-consuming processes in reticulocytes is significant.[152] Increases in Ca^{2+}-ATPase activity, if tested in the presence of calmodulin, have been found in the young population of human red cells in contrast to earlier results obtained in the absence of calmodulin.[153] It is therefore significant that ten-times higher amounts of calmodulin were found in human reticulocytes as compared with erythrocytes.[154]

Little is known about Mg^{2+} transport of red cells. Its high concentration in reticulocytes has been referred to (see Chapter 3). An intriguing possibility connecting Mg^{2+} transport with the adrenergic receptor system should be mentioned and will be discussed in the next section of this chapter. Changes in iron transport are discussed in Chapter 10.

The diversity of transport properties in different species of mature erythrocytes leads one to doubt their importance for the metabolism of these cells. The situation is certainly different with respect to reticulocytes and earlier erythroid cells. In keeping with their intensive metabolism and high rate of protein synthesis, they require a large uptake of glucose and amino acids; accordingly, one finds high concentrations of several amino acids accumulated by active uphill transport in reticulocytes (see Chapter 8). The significance of the high creatine concentration is unclear as yet.

VI. RECEPTORS AND ANTIGENS

The membranes of all cells are provided with a variety of receptors having diverse functions: they serve cell-cell recognition and they are the targets of hormones and are instrumental in receptor-mediated transport. Three receptor systems will be discussed in this section: (1) the insulin receptor; (2) the adrenergic receptor; and (3) the blood group antigens. The iron transferrin receptor system is discussed in Chapter 10. Receptors for other hormones and substances are poorly defined as yet.

A. The Insulin Receptor

Insulin receptors have been demonstrated in the red cells of a variety of species including man, rabbit, rat, pig, and in avian species.[155-158] Their properties, as far as specificity and binding constants are concerned, do not differ from those of other cells. The insulin receptors are integral components of the membrane.[156] Recently great advances have been made in the characterization of insulin receptors including those of red cells. The insulin receptor is a glycoprotein consisting of two subunits: α of 135 kD and β of 95 kD. They form a heterotetramer, $\alpha_2\beta_2$. The α-subunit faces the outside of the cell and probably carries the binding site, whereas the β-subunit is a transmembranous polypeptide chain[159,160] which appears to be a protein kinase specifically phosphorylating tyrosyl residues.[161-163] With binding of insulin, the β-subunit undergoes autophosphorylation. Some authors maintain, however, that the subunit structure of the insulin receptors of red cells may differ from that of classic target cells.[164,165] The absence of covalent binding by insulin to insulin receptors of erythrocytes and reticulocytes has been demonstrated in contrast to adipocytes.[166]

The number of insulin receptors decreases drastically during differentiation and maturation. In a study on erythroleukemic cells, the receptor number per cell decreased from 30,000 in the early erythroblast to 10,000 in the normoblast and 200 in the reticulocyte to 45 in the mature erythrocyte.[167] Similar results were reported on normal bone marrow of rabbits and man.[168-170]

The well-known reversible decrease of functional receptors under the influence of high levels of insulin, so-called "down-regulation", is progressively diminished during differentiation. The receptor loss determined after 18 hr of incubation amounted to 60% in the most immature cells but declined to 30% in reticulocytes. It was completely absent in erythrocytes.[167] These results are in keeping with other studies which suggested that down-regulation requires protein synthesis, since inhibitors of protein synthesis suppressed the response. In another study, however, down-regulation was observed in human erythrocytes, which was reversed after 16 hr of incubation.[171] The same workers also found that down-regulation led to an increased affinity of the glucose transport system of erythrocytes.[172] The reason for the discrepancies is not obvious. Rabbit reticulocytes contain about one order of magnitude more insulin binding sites than erythrocytes,[168,173] with some indication that young red cells have a higher number of receptors than old ones. This is also true for human erythrocytes.[156,168,169,175] The difference in receptor number between reticulocytes and erythrocytes in rats is similar to that in rabbits.[174]

A comparison between reticulocytes, fetal red cells, and adult erythrocytes of pigs also demonstrated a tenfold number of high-affinity binding sites for insulin per cell between reticulocytes and adult erythrocytes and also a fivefold number in fetal red cells.[165] During prolonged incubation, up to 140 hr, of reticulocytes and fetal cells, a significant loss in insulin-binding capacity was observed, largely occurring within the first 2 days. It was also found that insulin receptors with predominantly high molecular weight were lost.

In a study of preterm and term newborn infants, the insulin-binding capacity of red cells of prematures was found to be considerably greater than term newborns, which in turn exceeded that of red cells of adults.[176]

One may ask what function the insulin receptors in erythroid cells might serve. It is not likely that the effects of insulin are exerted on glucose transport, as it is on muscle cells, since the permeability for glucose of all erythroid precursors is high. Accordingly, an effect of insulin glucose transport was not found in either reticulocytes or erythrocytes of rabbits.[173] Increased protein synthesis, measured as incorporation of leucine, was reported on bone marrow cells and reticulocytes of man, rabbit, and rats.[169] It is therefore conceivable that the growth-stimulating property of insulin, which it shares with growth factors, is recognized by erythroid cells and may serve a biological function. In keeping with this assumption it was found that rat reticulocytes possess specific receptors for insulin-like somatomedin peptides, which are less in number by about one order of magnitude in erythrocytes.[174]

B. The β-Adrenergic Receptor and the Adenylate Cyclase System

In recent years there have been great advances in our knowledge of the system serving the recognition and transmission of β-adrenergic signals (for a review see Stiles et al.[178]). In these investigations, results obtained on red cells, particularly on mammalian reticulocytes, avian, and frog erythrocytes, have contributed a great deal. For a better understanding of the maturational changes of the system, a short exposition on its basic structure is useful.

The transmission system consists of three main components: (1) β-adrenergic receptors, (2) GTP-regulatory units, which combine GTP-binding and -splitting properties, and (3) adenylate cyclase[179] (Figure 13).

One has to distinguish between two types of receptors: stimulatory receptors, which bind, among others, β-adrenergic agents, and inhibitory receptors, with affinity for α-adrenergic, muscarinic, cholinergic agents, and opiates. Correspondingly, there are two types of guanine nucleotide regulatory proteins called N (or G). These regulatory proteins are heterotrimers of the structure αβγ. They differ apparently only with respect to their α-units, whereas their β- and γ-units, with molecular masses of 35 kD and 5 to 10 kD, respectively, appear to be identical. The stimulatory α-unit has a somewhat higher molecular mass of 45 kD as compared with 41 kD for the inhibitory α-unit.[180] The two α-units differ also in their susceptibility to ATP-ribosylation, the stimulatory N-protein being ADP-ribosylated by cholera toxin, whereas the pertussis toxin does the same to the inhibitory N-protein. In each case it is the α-unit which is modified. By this modification, GTPase activity is inhibited and thereby

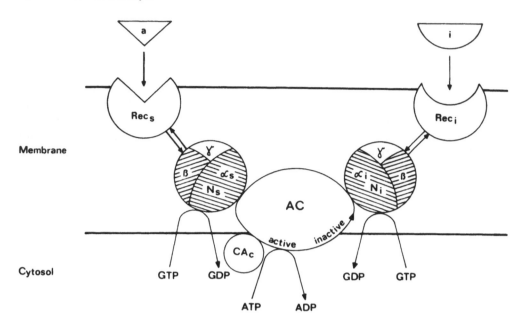

FIGURE 13.   Scheme of interactions between adrenergic receptors and adenylate cyclase. a, Agonist; Rec,, stimulatory receptor; Rec₁, inhibitory receptor; i, inhibitor; AC, adenylate cyclase; N₁, N-protein inhibitory; N₃, N-protein stimulatory; CA_c, cytosolic activator.

the regulatory unit is irreversibly activated. The catalytic unit, adenylate cyclase, the substrate of which is $Mg^{2+}ATP$, requires the presence of the stimulatory regulatory unit for its activity. The receptors are located in the outer layer of the cell membrane. Adenylate cyclase, the structure of which has not been clarified as yet, is confined to the inner layer, whereas the guanine nucleotide regulatory proteins appear to be transmembranous.

Each of the units of the system, i.e., the receptors, the N-proteins, and the catalytic moiety, can exist in active and inactive states through which they alternate with the N-proteins linking the receptor and the cyclase cycles. A crucial intermediate is the formation of a complex between the adrenergic agent receptor and the N-protein, which then binds GTP with high affinity and thereupon undergoes conformational changes and dissociation. The GTP-linked α-subunit of the N-protein forms a complex with adenylate cyclase, thereby activating it. The innate GTPase of the N-protein, in cleaving GTP to GDP, destabilizes this complex so that the cyclase reverts to an inactive state. Nonhydrolyzable guanine nucleotides and fluoride, which inhibit GTPase, cause permanent activation. The fluoride effect depends on the presence of aluminum.[181]

Pharmacological studies have clearly indicated that there are subtypes of the β-adrenergic receptors, designated β₁ and β₂, which differ in their response to a variety of agonists and antagonists. The red cells of different animal species, while containing a homogeneous receptor population, differ in their subtypes. Frog erythrocytes and rat reticulocytes contain exclusively β₂-adrenergic receptors and turkey erythrocytes contain β₁-adrenergic receptors.[182,183]

Purification of the adrenergic receptors has been achieved from frog and turkey erythrocytes as well as N-proteins from human erythrocytes.[184-191]

Most of the work on changes of the adenylate cyclase system preceded the elucidation of the structure of the system. The loss of catecholamine-sensitive or fluoride-stimulated adenylate cyclase activity with maturation of rabbit and rat reticulocytes was noted.[192-197] In the study of Tsamaloukas et al.,[192] fluoride-stimulated adenylate cyclase activity was determined in the membranes of density-fractionated red cells of rabbits during the course of a bleeding anemia. The highest activity was found on the 4th day of bleeding, when "Line 2" cells with high RNA content were produced, whereas no activity could be detected in mature erythrocytes. Catecholamine activation of membranes, even of reticulocytes, was very low in contrast to the results on red cells of rats.

Only in the last few years have the molecular mechanisms responsible for maturational changes been subjected to biochemical analysis.[198,199] The authors found alterations at multiple levels, which included decreases to one third in the number of β-receptors, in the amount of N-proteins, and in the activity of the catalytic unit. These changes appear to be proportional, but by themselves are not sufficient to explain the decline of catecholamine-sensitive cyclase by one order of magnitude. Other alterations in the reaction to agonists and to GTP suggested that the N-proteins might have undergone a change, perhaps by partial loss or limited proteolysis of the α-subunit. However, these changes do not explain the practical absence of adenylate cyclase activity observed in mature erythrocytes. An important role may be played by a cytosolic activator of adenylate cyclase, which is present in reticulocytes but absent in erythrocytes of rats. This cytosolic factor is a protein of 105-kD mass, which acts directly on the enzyme but not on the components of the receptor.[200,201] Its association with the cytosolic face of the membrane is suggested. Recently it was found that the activator protein requires the stimulatory guanine nucleotide-binding protein for its actions on adenylate cyclase.[202,203]

Two types of mechanisms for regulation of the number of β-adrenergic receptors appear to be operative: the first one concerns the dynamics of translocation between plasma membrane and cytosolic sites — similar to other types of receptors. This mechanism has been well documented.[204,205] The other mechanism, presumably operative for long-term regulation, consists of changes in transcription and synthesis of receptor proteins.

The decreased number of receptors in various organs in hypothyroidism is probably effected in this manner. Rat reticulocytes and turkey erythrocytes exhibit marked decreases in the number of β-adrenergic receptors.[206] Reduction in the number of receptors was associated with lower adenylate cyclase activity both under basal conditions and after various types of stimulation. The pattern of alterations indicated a lesser amount of N-protein, which was actually determined to be decreased.[206]

It is reasonable to assume that the recycling of receptors is energy-dependent. Indeed, it was found that the number of accessible receptors is affected by cell metabolism, being decreased by metabolic inhibitors or at low temperature.[207,208] It is also of interest that changes in phospholipids, specifically, synthesis of phosphatidyl choline, unmasks cryptic β-adrenergic receptors in rat reticulocytes.[209]

Under the influence of β-adrenergic agents, particularly if the cyclic AMP phosphodiesterase is inhibited, an overshooting massive accumulation of cAMP occurs.[210,211] Under these conditions, cAMP is exported into the blood plasma.[212]

What are the functions of the adenylate cyclase system in the red cell? As has been discussed in Chapter 1, a variety of reports indicate that the system may serve as a second messenger of erythropoietin. Considering that erythropoietin influences various stages of erythroid differentiation beginning with the CFU-E and extending beyond the basophilic erythroblast stage, the adenylate cyclase system might be of more than transient importance. The effects of the cAMP system consist mainly in the activation of cAMP-dependent protein kinases. In a study of the protein kinase activity associated with the plasma membrane of differentiating erythroid cells from the bone marrow of rabbits, it was found that there was a progressive decrease from the early erythroblasts to the reticulocytes. The cAMP dependence was found only in the early stages of differentiation and was completely lost after the final cell division.[213] It is interesting that there was also a coordinate decrease in the activity of cAMP phosphodiesterase.[214]

An unexpected role of the β-adrenergic receptor system in the transport of Mg^{2+} is suggested by a recent investigation.[215] Inhibition of Mg^{2+} uptake was observed under the influence of β-adrenergic agonists and prostaglandin E_1 in lymphoma cells. This effect is not mediated by cyclic AMP and is not exerted on Ca^{2+} transport.

C. Other Receptors

A stimulatory effect of physiological concentrations of thyroid hormone on Ca^{2+}-ATPase activity has been demonstrated on human erythrocytes in vitro.[216,217] It appeared to be dependent on calmodulin. In a further study, the action of thyroid hormones on the red cell stroma prepared from reticulocyte-rich blood of rabbits was studied. Surprisingly enough, it was found that Ca^{2+}-ATPase activity of the red cell stromata from male rabbits was inhibited, whereas that of females was stimulated. These effects appeared to be conditioned by sex hormones since addition of testosterone in vitro to intact cells or stromata of females converted the stimulation by thyroid hormone to an inhibition. The reverse, i.e., the conversion of inhibition to stimulation, was observed in red cells and stromata of male rabbits if estradiol or estrone were added.[218] Erythrocytes studied before induction of anemia showed only stimulation by thyroid hormones regardless of the sex of the animal. The authors unfortunately did not investigate pure reticulocyte populations so that the effects must have been contaminated by the admixture of erythrocytes.

Recently, the presence of coeroplasmin receptors has been reported which may be important for the protection of the plasma membrane from lipid peroxidation.[219]

D. Red Cell Antigens

The carbohydrate groups of glycoproteins and glycolipids represent the vast majority of the determinants of blood group antigens. Owing to the variety of positional and steric possibilities, a multitude of oligosaccharide isomers can arise, which vastly exceed those that can be formed from amino acids. Consequently, a large variety of antigens result.

The multiplicity of isomers is determined by the properties of glycosyl transferases with specificities to be transferred for both the monosaccharide and the acceptor molecule. The expression of the transferases is genetically determined but also changes during ontogenesis and differentiation.

Because many red cell membrane proteins are glycoproteins, including the three major ones discussed, i.e., glycophorin and anion and glucose transporters, they belong to the carbohydrate antigens. It was mentioned that they carry the ABH as well as the MN blood-group antigens.[220] In addition, glycolipids express the same as well as additional antigens. It is therefore appropriate to discuss this group of substances.

Glycolipids account for about 5% of total lipids and encompass a great variety of compounds (for a review see Hakomori[221]). They are all derivatives of ceramide, which is sphingosin, an 18-C amino alcohol acylated by a long-chain fatty acid. The primary alcohol group of ceramide is connected by a glycosidic bond to a chain consisting of a variety of carbohydrate moieties which include glucose, galactose, *N*-acetyl glucose, *N*-acetyl galactose, as well as mostly one or more sialic acids, and occasionally, fucose. The number of carbohydrate units varies greatly and may reach about 60 in the polyglycosyl ceramides.[222-224] The carbohydrate chains may be linear or branched. Free ceramide also has been demonstrated in human erythrocytes in comparatively large amounts constituting 0.6% of total lipids.[225]

From the largely hydrophilic character of glycolipids, one may assume that they are located in the outer layer of the cell membrane.[226,227] Their accessibility to external reagents is to a large extent masked, possibly owing to interactions among themselves or with other components of the cell membrane.[228,229]

There is comparatively little information on the expression of blood group antigens during erythroid differentiation. In early work it was shown that A, B, H, and Rh antigens are found in normoblasts and their number increases during further differentiation.[230-233] In more recent studies of the expression of blood group A antigens in human bone marrow cells, it was established that the A antigen was absent in pronormoblasts. It makes its appearance in basophilic normoblasts, roughly coincident with the onset of hemoglobin synthesis but slightly later than glycophorin A.[234-236] Accordingly, the amounts of polyactosamines immunologically expressed on the cell surface increase with differentiation. In keeping with these results are reports that human K 562 cells carry polyglycosyl oligosaccharides including the fetal type *i* antigen but not the ABO antigens.[237,238] In consonance with these results are clinical reports that the aplastic anemia occurring in the course of Rh-rh incompatibility of newborns is characterized by the persistence of erythroblasts in the bone marrow, whereas all later stages of differentiation are absent.[239] One may therefore assume that both the ABO- and the Rh-systems make their appearance at the basophilic erythroblast stage.

In a comparative study of the expression of cell surface glycoproteins during differentiation of human erythroid cells in which fetal, neonatal, and adult cells were investigated, it was found that with progressive cell maturation, the expression of band 3 and 4.5, as well as polygalactosamino-conjugates, increased to reach the maximal antigenic expression in the mature erythrocyte.[240]

Both I- and *i*-antigens are cryptic, i.e., masked to a large extent by one further peripheral glycosylation, by which the characteristics of blood group antigens of the ABH system arise.[241] They differ among each other mainly by one feature — the *i*-antigen represents a linear chain of *N*-acetyllactosamines, while the I-antigen is branched. It was found that both erythroblasts and erythrocytes of adult men carry the I-antigen, while fetal and newborn red cells express the *i*-antigen.[70] It would therefore appear that the additional step needed to make the I-antigen, i.e., the introduction of branches, catalyzed by an as yet undefined enzyme system, is ontogenetically determined but not a feature of cell differentiation.

Specific differences in the antigen profile between mature and immature red cells of mice have been utilized to isolate erythropoietin-responsive cells which were not hemolyzed by an anti-erythrocyte antibody.[242]

A specific erythroid developmental agglutinin, which is a protein lectin, has been described in rabbits, the function of which may be to mediate specific cell-cell adhesion between differentiating rabbit erythroblasts. It is assumed that this lectin is important for the coherence of erythropoietic islands.[243]

REFERENCES

1. **Steck, T.,** The organization of proteins in the human red cell membrane, *J. Cell Biol.,* 62, 1, 1974.
2. **Marchesi, V. and Furthmair, H.,** The red cell membrane, *Annu. Rev. Biochem.,* 45, 667, 1976.
3. **Marchesi, V. T.,** Functional proteins of the human red blood cell membrane, *Semin. Hematol.,* 16, 3, 1979.
4. **Goodman, S. R. and Schiffer, K.,** The spectrin membrane skeleton of normal and abnormal human erythrocytes, *Am. J. Physiol. Cell. Physiol.,* 244, C121, 1983.
5. **Cohen, C. M.,** The molecular organization of the red cell membrane skeleton, *Semin. Hematol.,* 20, 141, 1983.
6. **Van Deenen, L. L. M. and de Gier, J.,** Lipids of the red cell membrane, in *The Red Blood Cell,* Vol. 1, 2nd ed., Surgenor, D. M. N., Ed., Academic Press, New York, 1974, 147.
7. **Furthmair, H.,** Glycophorin A, B and C: a family of sialoglycoproteins. Isolation and preliminary characterization of trypsin derived peptides, *J. Supramol. Struct.,* 9, 79, 1978.
8. **Furthmair, H.,** Structural comparison of glycophorins and immunochemical analysis of genetic variants, *Nature,* 271, 519, 1978.
9. **Tomita, M., Furthmair, H., and Marchesi, V. T.,** Primary structure of human erythrocyte glycophorin A. Isolation and characterization of peptides and complete amino acid sequence, *Biochemistry,* 17, 4756, 1978.

10. **Wasniowska, K., Drzeniek, Z., and Lisowska, E.,** The amino acids of M and N blood group glycopeptides are different, *Biochem. Biophys. Res. Commun.,* 76, 385, 1977.

11. **Steck, T. L.,** The band 3 protein of the human erythrocyte membrane: a review, *J. Supramol. Struct.,* 8, 311, 1978.

12. **Jennings, M. L.,** Oligomeric structure and the anion transport function of human erythrocyte band 3 protein, *J. Membr. Biol.,* 80, 105, 1984.

13. **Tomita, M. and Marchesi, V. T.,** Amino acid sequence and oligosaccharide attachment sites of human erythrocyte glycophorin, *Proc. Natl. Acad. Sci. U.S.A.,* 72, 2964, 1975.

14. **Kaul, R. K., Murthy, S. N. P., Reddy, A. G., Steck, T. I., and Kohler, H.,** Amino acid sequence of the N^α-terminal 201 residues of human erythrocyte membrane band 3, *J. Biol. Chem.,* 258, 7981, 1983.

15. **Drickamer, L. K.,** Orientation of the band 3 polypeptide from human erythrocyte membranes, *J. Biol. Chem.,* 253, 7242, 1978.

16. **Markowitz, S. and Marchesi, V. T.,** The carboxyl-terminal domain of human erythrocyte band 3, *J. Biol. Chem.,* 256, 6463, 1981.

17. **Brock, C. J., Tanner, M. J. A., and Kempf, C.,** The human erythrocyte anion-transport protein, *Biochem. J.,* 213, 577, 1983.

18. **Ramjeesingh, M., Gaarn, A., and Rothstein, A.,** The localization of the three cysteine residues in the primary structure of the intrinsic segments of band 3 protein, and the implications concerning the arrangement of band 3 protein in the bilayer, *Biochim. Biophys. Acta,* 729, 150, 1983.

19. **Jennings, M. L. and Nicknish, J. S.,** Erythrocyte band 3 protein: evidence for multiple membrane-crossing segments in the 17 000 Dalton chymotryptic fragment, *Biochemistry,* 23, 6432, 1984.

20. **Jennings, M. L., Adams-Lackey, M., and Denney, G. H.,** Peptides of human erythrocyte band 3 protein produced by extracellular papain cleavage, *J. Biol. Chem.,* 259, 4652, 1984.

21. **Baldwin, S. A. and Lienhard, G. E.,** Immunological identification of the human erythrocyte monosaccharide transporter, *Biochem. Biophys. Res. Commun.,* 94, 1401, 1980.

22. **Allard, W. J. and Lienhard, G. E.,** Monoclonal antibodies to the glucose transporter from human erythrocytes — identification of the transporter as a M_r = 55,000 protein, *J. Biol. Chem.,* 260, 8668, 1985.

23. **Baldwin, S. A., Baldwin, J. A., and Lienhard, G. E.,** Monosaccharide transporter of the human erythrocyte. Characterization of an improved preparation, *Biochemistry,* 21, 3836, 1982.

24. **Sase, S., Takata, K., Hirano, H., and Kasahara, M.,** Characterization and identification of the glucose transporter of human erythrocytes, *Biochim. Biophys. Acta,* 693, 253, 1982.

25. **Shanahan, M. F. and D'Artel-Ellis, J.,** Orientation of the glucose transporter in the human erythrocyte membrane, *J. Biol. Chem.,* 259, 13878, 1984.

26. **Young, J. D., Jarvis, S. M., Robins, M. J., and Paterson, A. R. P.,** Photoaffinity labeling of the human erythrocyte nucleoside transporter by N^6-(p-azidobenzyl) adenosine and nitrobenzylthioinosine, *J. Biol. Chem.,* 258, 2202, 1983.

27. **Wu, J.-S., Kwong, F. Y. P., Jarvis, S. M., and Young, J. D.,** Identification of the erythrocyte nucleoside transporter as a band 4.5 polypeptide, *J. Biol. Chem.,* 258, 13745, 1983.

28. **Shalton, D., Burke, B., and Branton, D.,** The molecular structure of human erythrocyte spectrin, *J. Mol. Biol.,* 131, 303, 1979.

29. **Morrow, J. S., Speicher, D. W., Knowles, W. J., Hsu, C. J., and Marchesi, V. T.,** Identification of functional domains of human erythrocyte spectrin, *Proc. Natl. Acad. Sci. U.S.A.,* 77, 6592, 1980.

30. **Speicher, D. W., Morrow, J. S., Knowles, W. J., and Marchesi, V. T.,** A structure model of human erythrocyte spectrin, *J. Biol. Chem.,* 257, 9093, 1982.

31. **Speicher, D. W., Davin, G., and Marchesi, V. T.,** Structure of human erythrocyte spectrin, *J. Biol. Chem.,* 258, 14938, 1983.

32. **Speicher, D. W. and Marchesi, V. T.,** Erythrocyte spectrin is comprised of many homologous helical segments, *Nature,* 311, 177, 1984.

33. **Tyler, J., Hargreaves, W., and Branton, D.,** Purification of two spectrin-binding proteins: biochemical and electron microscopic evidence for site-specific reassociation between spectrin and bands 2.1 and 4.1, *Proc. Natl. Acad. Sci. U.S.A.,* 76, 5192, 1979.

34. **Tyler, J., Reinhardt, B., and Branton, D.,** Associations of erythrocyte membrane proteins. Binding of purified bands 2.1 and 4.1 to spectrin, *J. Biol. Chem.,* 255, 7034, 1980.

35. **Chanian, V., Wolfe, L. C., John, K. M., Pinder, J. C., Lux, S. E., and Gratzer, W. B.,** Analysis of the ternary interaction of the red cell membrane skeletal proteins spectrin, actin, and 4.1, *Biochemistry,* 23, 4416, 1984.

36. **Liv, S. C. and Paler, J.,** Spectrin tetramer-dimer equilibrium and the stability of erythrocyte membrane skeletons, *Nature,* 285, 586, 1980.

37. **Pinder, J., Ungewickell, E., and Bray, D.,** The spectrin-actin complex and erythrocyte shape, *J. Supramol. Struct.,* 8, 439, 1978.

38. **Nakashima, K. and Beutler, E.**, Comparison of structure and function of human erythrocyte and human muscle actin, *Proc. Natl. Acad. Sci. U.S.A.*, 76, 935, 1979.

39. **Tilley, L. and Ralston, G.**, Purification and kinetic characterisation of human erythrocyte actin, *Biochim. Biophys. Acta*, 790, 46, 1984.

40. **Brenner, S. and Korn, E.**, Spectrin/actin complex isolated from sheep erythrocytes accelerates actin polymerization by simple nucleation, *J. Biol. Chem.*, 255, 1670, 1980.

41. **Pinder, J. C. and Gratzer, W. B.**, Structural and dynamic states of actin in the erythrocyte, *J. Cell Biol.*, 96, 768, 1983.

42. **Bennet, V. and Stenbuck, P. J.**, Human erythrocyte ankyrin. Purification and properties, *J. Biol. Chem.*, 255, 2540, 1980.

43. **Weaver, D. C., Pasternack, G. R., and Marchesi, V. T.**, The structural basis of ankyrin function, *J. Biol. Chem.*, 259, 6170, 1984.

44. **Wallin, R., Culp, E. N., Coleman, D. B., and Goodman, S. R.**, A structural model of human erythrocyte band 2.1: alignment of chemical and functional domains, *Proc. Natl. Acad. Sci. U.S.A.*, 81, 4095, 1984.

45. **Goodman, S. R., Shiffer, K., Coleman, D. B., and Whitfield, C. F.**, Red cell membranes, erythrocyte membrane skeletal protein-4.1-a, *Progr. Clin. Biol. Res.*, 165, 415, 1984.

46. **Goodman, S. R., Yu, J., Whitfield, C. F., Culp, E. N., and Posnak, E. J.**, Erythrocyte membrane skeletal protein bands 4.1 a and b are sequence-related phosphoproteins, *J. Biol. Chem.*, 257, 4564, 1982.

47. **Leto, T. L. and Marchesi, V. T.**, A structural model of human erythrocyte protein 4.1, *J. Biol. Chem.*, 259, 4603, 1984.

48. **Ohanian, V., Wolfe, L. C., John, K. M., Pinder, J. C., Lux, S. E., and Gratzer, W. B.**, Analysis of the ternary interaction of the red cell membrane skeletal proteins, spectrin, actin, and 4.1, *Biochemistry*, 23, 4416, 1984.

49. **Shiffer, K. A. and Goodman, S. R.**, Protein 4.1: its association with the human erythrocyte membrane, *Proc. Natl. Acad. Sci. U.S.A.*, 81, 4404, 1984.

50. **Anderson, R. A. and Lovrien, R. E.**, Glycophorin is linked by band 4.1 protein to the human erythrocyte membrane skeleton, *Nature*, 307, 655, 1984.

51. **Granler, B. T. and Lazarides, E.**, Membrane skeletal protein 4.1 of avian erythrocytes is composed of multiple variants that exhibit tissue-specific expression, *Cell*, 37, 595, 1984.

52. **Ohanian, V. and Gratzer, W.**, Preparation of red-cell-membrane cytoskeletal constitutents and characterisation of protein 4.1, *Eur. J. Biochem.*, 144, 375, 1984.

53. **Siegel, D. and Branton, D.**, Human erythrocyte band 4.9, *J. Cell Biol.*, 95, 265a, 1982.

54. **Fowler, V. M. and Bennett, V.**, Erythrocyte membrane tropomyosin. Purification and properties, *J. Biol. Chem.*, 259, 5978, 1984.

55. **Whitfield, C. F., Mylin, L. M., and Goodman, S. R.**, Species-dependent variations in erythrocyte membrane skeletal proteins, *Blood*, 61, 500, 1983.

56. **Harris, J. R.**, Release of a macromolecular protein component from human erythrocyte ghosts, *Biochim. Biophys. Acta*, 150, 534, 1968.

57. **Harris, J. R. and Naeem, I.**, The subunit composition of two high molecular weight extrinsic proteins from human erythrocyte membranes, *Biochim. Biophys. Acta*, 537, 495, 1978.

58. **White, M. D. and Ralstone, G. B.**, The hollow cylinder protein of erythrocyte membranes, *Biochim. Biophys. Acta*, 554, 469, 1979.

59. **Malech, H. and Marchesi, V. T.**, Hollow cylinder protein in the cytoplasm of human erythrocytes, *Biochim. Biophys. Acta*, 670, 385, 1981.

60. **Lande, W. M., Thiemann, P. V. W., Fisher, K. A., and Mentzer, W. C.**, Two-dimensional electrophoretic analysis of human erythrocyte cylindrin, *Biochim. Biophys. Acta*, 778, 105, 1984.

61. **Branton, D., Cohen, C. M., and Tyler, J.**, Interaction of cytoskeletal proteins in the human erythrocyte membrane, *Cell*, 24, 24, 1981.

62. **Goodman, S. R. and Branton, D.**, Spectrin binding and the control of membrane protein mobility, *J. Supramol. Struct.*, 8, 455, 1978.

63. **Haest, C., Plasa, G., Kamp, D., and Deuticke, B.**, Spectrin as a stabilizer of the phospholipid asymmetry in the human erythrocyte membrane, *Biochim. Biophys. Acta*, 509, 21, 1978.

64. **Williamson, P., Bateman, J., Kozarsky, K., and Mattocks, K.**, Involvement of spectrin in the maintenance of phase-state asymmetry in the erythrocyte membrane, *Cell*, 30, 725, 1982.

65. **Mohandas, N., Wyatt, J., Mel, S. F., Rossi, M. E., and Shohet, S. B.**, Lipid translocation across the human erythrocyte membrane, *J. Biol. Chem.*, 257, 6537, 1982.

66. **Lange, Y., Hadesman, R., and Steck, T.**, Role of the reticulum in the stability and shape of the isolated human erythrocyte membrane, *J. Cell Biol.*, 92, 714, 1982.

67. **Lange, Y., Cutler, H., and Steck, T.**, The effect of cholesterol and other intercalated amphipaths on the contour and stability of the isolated red cell membrane, *J. Biol. Chem.*, 255, 9331, 1980.

68. **Lange, Y., Gough, A., and Steck, T. L.**, Role of the bilayer in the shape of the isolated erythrocyte membrane, *J. Membr. Biol.*, 69, 113, 1982.

82 *The Reticulocyte*

69. **Chang, H., Langer, P. J., and Lodish, H. F.,** Asynchronous synthesis of erythrocyte membrane proteins, *Proc. Natl. Acad. Sci. U.S.A.,* 73, 3206, 1976.
70. **Chan, L.,** Changes in the composition of plasma membrane proteins during differentiation of embryonic chick erythroid cells, *Proc. Natl. Acad. Sci. U.S.A.,* 74, 1062, 1977.
71. **Eisen, H., Bach, R., and Emery, R.,** Induction of spectrin in erythroleukemic cells transformed by Friend virus, *Proc. Natl. Acad. Sci. U.S.A.,* 74, 3898, 1977.
72. **Weise, M. and Chan, L.,** Membrane protein synthesis in embryonic chick erythroid cells, *J. Biol. Chem.,* 253, 1892, 1978.
73. **Yurchenco, P., Lemay, A., and Marchesi, V.,** Expression of red cell membrane proteins in erythroid precursor cells, *J. Cell Sci.,* 34, 91, 1978.
74. **Rossi, G., Aducci, P., Gambari, R., Minetti, M., and Vernale, P.,** Presence of spectrin in untreated Friend erythroleukemic cells. Its accumulation upon treatment of the cells with dimethyl sulfoxide, *J. Cell. Physiol.,* 97, 293, 1978.
75. **Hasthorge, S.,** Quantification of spectrin-containing erythroid precursor cells in normal and perturbed erythropoiesis, *Exp. Hematol.,* 8, 1001, 1980.
76. **Tong, B. D. and Goldwasser, E.,** The formation of erythrocyte membrane proteins during erythropoietin-induced differentiation, *J. Biol. Chem.,* 256, 12666, 1981.
77. **Pfeffer, S. and Redman, C.,** Biosynthesis of mouse erythrocyte membrane proteins by Friend erythroleukemia cells, *Biochim. Biophys. Acta,* 641, 254, 1981.
78. **Koury, M. J., Bondurant, M. C., Duncan, D. T., Krantz, S. B., and Hankins, W. D.,** Specific differentiation events induced by erythropoietin in cells infected in vitro with the anemia strain of Friend virus, *Proc. Natl. Acad. Sci. U.S.A.,* 79, 635, 1982.
79. **Light, N. D. and Tanner, M. J. A.,** Erythrocyte membrane proteins. Sequential accumulation in the membrane during reticulocyte maturation, *Biochim. Biophys. Acta,* 508, 571, 1978.
80. **Harrison, P. R.,** Molecular analysis of erythropoesis. A current appraisal, *Exp. Cell Res.,* 155, 321, 1984.
81. **Lazarides, E. and Moon, R. J.,** Assembly and topogenesis of the spectrin-based membrane skeleton in erythroid development, *Cell,* 37, 354, 1984.
82. **Moon, R. J. and Lazarides, E.,** β-Spectrin limits α-spectrin assembly on membranes following synthesis in a chicken erythroid cell lysate, *Nature,* 305, 62, 1983.
83. **Woods, C. M. and Lazarides, C. M.,** Degradation of unassembled α- and β-spectrin by distinct intracellular pathways: regulation of spectrin topogenesis by β-spectrin degradation, *Cell,* 40, 959, 1985.
84. **Bodine, D. M., Birkenmeier, C. S., and Barker, J. E.,** Spectrin deficient inherited hemolytic anemias in the mouse: characterization by spectrin synthesis and mRNA activity in reticulocytes, *Cell,* 37, 721, 1984.
85. **Geiduschek, J. and Singer, S.,** Molecular changes in the membranes of mouse erythroid cells accompanying differentiating, *Cell,* 16, 149, 1979.
86. **Zweig, S., Tokuyasu, K., and Singer, S.,** Membrane-associated changes during erythropoiesis. On the mechanism of maturation of reticulocytes to erythrocytes, *J. Supramol. Struct.,* 17, 163, 1981.
87. **Pan, B. T. and Johnstone, R. M.,** Fate of the transferrin receptor during maturation of sheep reticulocytes in vitro: selective externalization of the receptor, *Cell,* 33, 967, 1983.
88. **Zweig, S. and Singer, S.,** Concanavalin A-induced endocytosis in rabbit reticulocytes and its decrease with reticulocyte maturation, *J. Cell Biol.,* 80, 487, 1979.
89. **Schekman, R. and Singer, S. J.,** Clustering and endocytosis of membrane receptors can be induced in mature erythrocytes of neonatal but not adult humans, *Proc. Natl. Acad. Sci. U.S.A.,* 73, 4075, 1976.
90. **Tokuyasu, K., Schekman, R., and Singer, S.,** Domains of receptor mobility in the membranes of neonatal human erythrocytes and reticulocyte are deficient in spectrin, *J. Cell Biol.,* 80, 481, 1979.
91. **Jokinen, M., Gahmberg, C. G., and Andersson, L. C.,** Biosynthesis of the major human red cell sialoglycoprotein, glycophorin A, in a continuous cell line, *Nature,* 279, 604, 1979.
92. **Gahmberg, C. G., Jokinen, M., and Andersson, L. C.,** Expression of the major red cell sialoglycoprotein, glycophorin A, in the human leukemic cell line K 562, *J. Biol. Chem.,* 254, 7442, 1979.
93. **Gahmberg, C. G., Jokinen, M., Karhi, K. K., and Andersson, L. C.,** Effect of tunicamycin on the biosynthesis of the major human red cell sialoglycoprotein, glycophorin A, in the leukemia cell line K 562, *J. Biol. Chem.,* 255, 2169, 1980.
94. **Gahmberg, C. G., Jokinen, M., and Andersson, L. C.,** Expression of the major sialoglycoprotein (glycophorin) on erythroid cells in human bone marrow, *Blood,* 52, 379, 1978.
95. **Robinson, J., Sieff, C., Delia, D., Edwards, P. A. W., and Greaves, M.,** Expression of cell-surface HLA-DR, HLA-ABC and glycophorin during erythroid differentiation, *Nature,* 289, 68, 1981.
96. **Gahmberg, C. G., Ekblom, M., and Andersson, L. C.,** Differentiation of human erythroid cells is associated with increased O-glycosylation of the major sialoglycoprotein, glycophorin A, *Proc. Natl. Acad. Sci. U.S.A.,* 81, 6752, 1984.
97. **Braell, W. H. and Lodish, H. F.,** Biosynthesis of the erythrocyte anion transport protein, *J. Biol. Chem.,* 256, 11337, 1981.

98. **Braell, W. H. and Lodish, H. F.,** The erythrocyte anion transport protein is cotranslationally inserted into microsomes, *Cell,* 28, 23, 1982.

99. **Dellagi, K., Vainchenker, W., Vinci, G., Paulin, D., and Brouet, J. C.,** Alteration of vimentin intermediate filament expression during differentiation of human hemopoietic cells, *EMBO J.,* 2, 1509, 1983.

100. **Granger, B. L. and Lazarides, E.,** Structural associations of synemin and vimentin filaments in avian erythrocytes revealed by immunoelectron microscopy, *Cell,* 30, 263, 1982.

101. **Barrett, L. A. and Dawson, R. P.,** Avian erythrocyte development; microtubules and the formation of disc shape, *Dev. Biol.,* 36, 72, 1974.

102. **Cohen, W. D., Bartelt, D., Jaeger, R., Langford, G., and Nemhauser, I.,** The cytoskeletal system of nucleated erythrocytes. I. Composition and function of major elements, *J. Cell Biol.,* 93, 828, 1982.

103. **Joseph-Silverstein, J. and Cohen, W. D.,** The cytoskeletal system of nucleated erythrocytes. III. Marginal band function in mature cells, *J. Cell Biol.,* 98, 2118, 1984.

104. **Goniakowska-Witalinski, L. and Witalinski, W.,** Occurrence of microtubules during erythropoiesis in llama, *Llama glama, J. Zool.,* 181, 309, 1977.

105. **Wong, J. A., Kiehart, D. P., and Pollard, Th. D.,** Myosin from human erythrocytes, *J. Biol. Chem.,* 260, 46, 1985.

106. **Fowler, V. M., Davis, V. Q., and Bennett, V.,** Human erythrocyte myosin: identification and purification, *J. Cell Biol.,* 100, 47, 1985.

107. **Widdas, W. F.,** Hexose permeability of fetal erythrocytes, *J. Physiol.,* 127, 318, 1955.

108. **Augustin, H. W., Häcker, M. R., and Hofmann, E.,** Aufnahme und Phosphorylierung von 2-Desoxy-D-glucose in Kaninchenerythrozyten und -retikulozyten, *Hoppe-Seyler's Z. Physiol. Chem.,* 339, 42, 1964.

109. **Kim, H. D. and Luthra, M. G.,** Pig reticulocytes. III. Glucose permeability in naturally occurring reticulocytes and red cells from newborn piglets, *J. Gen. Physiol.,* 70, 171, 1977.

110. **Kim, H. D. and Luthra, M. G.,** Pig reticulocytes. I. Transitory glucose permeability and metabolism, *Am. J. Physiol.,* 230, 1668, 1976.

111. **Zeidler, R. B., Lee, P., and Kim, H. D.,** Kinetics of 3-O-methyl glucose transport in red blood cells of newborn pigs, *J. Gen. Physiol.,* 67, 67, 1976.

112. **Zeidler, R. B. and Kim, M. D.,** Pig reticulocytes. IV. In vitro maturation of naturally occurring reticulocytes with permeability loss to glucose, *J. Cell. Physiol.,* 112, 360, 1982.

113. **Augustin, H. W., Rohden, L. V., and Häcker, M. R.,** Some properties of the monosaccharide transport system in erythrocytes of newborn and adult rabbits, *Acta Biol. Med. Ger.,* 19, 723, 1967.

114. **Mooney, N. A. and Young, J. D.,** Nucleoside and glucose transport in erythrocytes from new-born lambs, *J. Physiol.,* 284, 229, 1978.

115. **Young, J. D.,** Nucleoside transport in sheep erythrocytes: genetically controlled transport variation and its influence on erythrocyte ATP concentrations, *J. Physiol.,* 277, 325, 1978.

116. **Javis, S. M. and Young, J. D.,** Nucleoside translocation in sheep reticulocytes and fetal erythrocytes. A proposed model for the nucleoside transporter, *J. Physiol.,* 324, 47, 1982.

117. **Syllm-Rapoport, I., Daniel, A., and Rapoport, S.,** Creatine transport into red blood cells, *Acta Biol. Med. Ger.,* 39, 771, 1980.

118. **Ku, Ch.-P. and Passow, H.,** Creatine and creatinine transport in old and young human red blood cells, *Biochim. Biophys. Acta,* 600, 212, 1980.

119. **Holzhütter, H.-G., Syllm-Rapoport, I., and Daniel, A.,** A mathematical model for the cell age-dependent decline of creatine in human red blood cells, *Biomed. Biochim. Acta,* 43, 153, 1984.

120. **Christensen, H. N.,** On the development of amino acid transport systems, *Fed. Proc. Fed. Am. Soc. Exp. Biol.,* 32, 19, 1973.

121. **Young, J. D. and Ellory, J. C.,** Red cell amino acid transport, in *Membrane Transport in Red Cells,* Ellory, J. C. and Lew, V. L., Eds., Academic Press, London, 1977, 301.

122. **Guidotti, G., Borghetti, A., and Gazzola, G.,** The regulation of amino acid transport in animal cells, *Biochim. Biophys. Acta,* 515, 329, 1978.

123. **Riggs, T. R., Christensen, H. N., and Palastine, I. M.,** Concentrating activity of reticulocytes for glycine, *J. Biol. Chem.,* 194, 53, 1952.

124. **Winters, C. G. and Christensen, H. N.,** Contrasts in neutral amino acid transport by rabbit erythrocytes and reticulocytes, *J. Biol. Chem.,* 240, 3594, 1965.

125. **Wheeler, K. P. and Christensen, H. N.,** Role of Na$^+$ in transport of amino acids in rabbit red cells, *J. Biol. Chem.,* 242, 1450, 1967.

126. **Imler, J. R. and Vidaver, G. A.,** Anion effects on glycine entry into pigeon red blood cells, *Biochim. Biophys. Acta,* 288, 153, 1972.

127. **Ellory, J. C., Johnes, S. E. M., and Young, J. D.,** Glycine transport in human erythrocytes, *J. Physiol.,* 320, 403, 1981.

128. **Al-Saleh, E. A. S. and Wheeler, K. P.,** Transport of neutral amino acids by human erythrocytes, *Biochim. Biophys. Acta,* 684, 157, 1982.

129. **Fincham, D. A., Willis, J. S., and Young, J. D.,** Red cell amino acid transport. Evidence for the presence of system gly in guinea pig reticulocytes, *Biochim. Biophys. Acta*, 777, 147, 1984.

130. **Wheeler, K. P. and Christensen, H. N.,** Interdependent fluxes of amino acids and sodium ion in the pigeon red blood cell, *J. Biol. Chem.*, 242, 3782, 1967.

131. **Christensen, H. N. and Antionelli, J. A.,** Cationic amino acid transport in the rabbit reticulocyte, *J. Biol. Chem.*, 244, 1497, 1969.

132. **Ellory, J. C. and Wolowyk, M. W.,** A new density gradient technique for age separation of human erythrocytes and reticulocytes, *J. Physiol.*, 295, 9, 1979.

133. **Young, J. D., Wolowyk, M. W., Johnes, S. E. M., and Ellory, J. C.,** Red-cell amino acid transport. Evidence for the presence of system ASC in mature human red blood cells, *Biochem. J.*, 216, 349, 1983.

134. **Vidaver, G. A., Romain, L. F., and Haurowitz, F.,** Some studies on the specificity of amino acid entry routes in pigeon erythrocytes, *Arch. Biochem. Biophys.*, 107, 82, 1964.

135. **Antionelli, J. A. and Christensen, H. N.,** Differences in schedules of repression of transport systems during reticulocyte maturation, *J. Biol. Chem.*, 244, 1505, 1969.

136. **Friend, C., Scher, W., Holland, J. G., and Sato, T.,** Hemoglobin synthesis in murine virus-induced leukemic cells in vitro: stimulation of erythroid differentiation by dimethyl sulfoxide, *Proc. Natl. Acad. Sci. U.S.A.*, 68, 378, 1976.

137. **Hempling, H. G. and Wise, W. C.,** Maturation of membrane function: the permeability of the rat erythroblastic leukemic cell to water and to non-electrolytes, *J. Cell. Physiol.*, 85, 195, 1975.

138. **Germinario, R., Kleiman, L., Peters, S., and Oliveira, M.,** Decreased deoxy-D-glucose transport in Friend cells during exposure to inducers of erythroid differentiation, *Exp. Cell Res.*, 110, 375, 1977.

139. **Mager, D. and Bernstein, A.,** Early transport changes during erythroid differentiation of Friend leukemic cells, *J. Cell. Physiol.*, 94, 275, 1978.

140. **Gordon, P. B. and Rubin, M. S.,** Membrane transport during erythroid differentiation, *J. Membr. Biol.*, 64, 11, 1982.

141. **Miles, P. and Lee, P.,** Sodium and potassium content and membrane transport properties in red blood cells from newborn puppies, *J. Cell. Comp. Physiol.*, 79, 367, 1972.

142. **Israel, Y., MacDonald, A., Bernstein, J., and Rosenmann, E.,** Changes from high potassium (HK) to low potassium (LK) in bovine red cells, *J. Gen. Physiol.*, 59, 270, 1972.

143. **Lee, P., Woo, A., and Tosteson, D. C.,** Cytodifferentiation and membrane transport properties in LK sheep red cells, *J. Gen. Physiol.*, 50, 379, 1966.

144. **Dunham, P. B. and Blostein, R.,** Active potassium transport in reticulocytes of high-K^+ and low-K^+ sheep, *Biochim. Biophys. Acta*, 455, 749, 1976.

145. **Lauf, P. K. and Dunham, N. C.,** Thiol dependent passive potassium-chloride transport in sheep red cells. II. Loss of chloride and *N*-ethylmaleimide sensitivity in maturing high potassium cells, *J. Membr. Biol.*, 73, 247, 1983.

146. **Blostein, R., Drapeau, P., Benderoff, S., and Weigensberg, A. M.,** Changes in Na^+-ATPase and Na, K-pump during maturation of sheep reticulocytes, *Can. J. Biochem. Cell Biol.*, 61, 23, 1983.

147. **Furukawa, H., Bilezikian, J. P., and Loeb, J. N.,** Potassium fluxes in the rat reticulocyte: ouabain sensitivity and changes during maturation, *Biochim. Biophys. Acta*, 649, 625, 1981.

148. **Maede, Y. and Inaba, M.,** (Na,K)-ATPase and ouabain binding in reticulocytes from dogs with high K and low K erythrocytes and their changes during maturation, *J. Biol. Chem.*, 260, 3337, 1985.

149. **Panet, R. and Atlan, H.,** Characterization of a potassium carrier in rabbit reticulocyte cell membrane, *J. Membr. Biol.*, 52, 273, 1980.

150. **Wiley, I. S. and Shaller, C. C.,** Selective loss of calcium permeability on maturation of reticulocytes, *J. Clin. Invest.*, 59, 1113, 1977.

151. **Shimoda, M., Migashima, K., and Yawata, Y.,** Increased calcium uptake in the red cell of unsplenectomized patients with hereditary spherocytosis: significant contribution of reticulocytosis, *Clin. Chim. Acta*, 142, 183, 1984.

152. **Siems, W., Dubiel, W., Dumdey, R., Müller, M., and Rapoport, S. M.,** Accounting for the ATP-consuming processes in rabbit reticulocytes, *Eur. J. Biochem.*, 139, 101, 1984.

153. **Luthra, M. G. and Sears, D. A.,** Increased Ca^{2+}, Mg^{2+}, Na^+, K^+ ATPase activities in erythrocytes of sickle cell anemia, *Blood*, 58, 1332, 1982.

154. **Monson, C. M., Pennisten, J. T., Fairbanks, V. F., and Burgert, E. O.,** Erythrocyte calmodulin correlates with red cell age, *Br. J. Haematol.*, 51, 261, 1982.

155. **Gambhir, K. K., Archer, J. A., and Bradley, C. J.,** Characteristics of human erythrocyte insulin receptors, *Diabetes*, 27, 701, 1978.

156. **Dons, R. F., Corash, L. M., and Gordon, P.,** The insulin receptor is an age-dependent integral component of the human erythrocyte membrane, *J. Biol. Chem.*, 256, 2982, 1981.

157. **Herzberg, V., Bongter, J. M., Carlisle, S., and Hill, D. E.,** Evidence for two insulin receptor populations of human erythrocytes, *Nature*, 286, 279, 1980.

158. **Leibush, B. N., Kolychev, A. P., and Bondareva, V. M.,** Erythrocyte insulin receptors in the antogeny of chickens, *Ontogenez*, 15, 290, 1984.

159. **Fehlmann, M., Carpentier, J.-L., van Obberghen, E., Freychet, P., Thamm, P., Saunders, S. D., Brandenburg, D., and Orci, L.,** Internalized insulin receptors are recycled to cell surface in rat hepatocytes, *Proc. Natl. Acad. Sci. U.S.A.*, 79, 5921, 1982.

160. **Kasuga, M., Zick, J., Blith, D. I., Karlsson, F. A., Häring, H. U., and Kahn, C. R.,** Insulin stimulation of phosphorylation of the β subunit of the insulin receptor, *J. Biol. Chem.*, 257, 9891, 1982.

161. **Roth, R. A. and Cassell, D. I.,** Insulin receptor: evidence that it is a protein kinase, *Science*, 219, 299, 1983.

162. **Grigorescu, F., White, M. F., and Kahn, C. R.,** Insulin binding and insulin-dependent phosphorylation of the insulin receptor solubilized from human erythrocytes, *J. Biol. Chem.*, 258, 13708, 1983.

163. **Grigorescu, F., Flier, J. S., and Kahn, C. R.,** Defect in insulin receptor phosphorylation in erythrocytes and fibroblasts associated with severe insulin resistance, *J. Biol. Chem.*, 259, 15003, 1984.

164. **Im, J. H., Meezan, E., Rackley, Ch. E., and Kim, H. D.,** Isolation and characterization of human erythrocyte insulin receptors, *J. Biol. Chem.*, 258, 5021, 1982.

165. **Im, J. H., Zeidler, R. B., Rackley, C. E., and Kim, H. D.,** Developmental changes in insulin receptors of pig red blood cells, *Arch. Biochem. Biophys.*, 232, 26, 1984.

166. **Ward, G. M., Clark, S., and Harrison, L. C.,** Absence of covalent binding by insulin to erythrocyte and reticulocyte insulin receptors, *Biochem. Biophys. Res. Commun.*, 123, 849, 1984.

167. **Ginsberg, B. H. and Brown, T. J.,** Regulation of insulin receptors in erythroid cells, *Metabolism*, 31, 728, 1982.

168. **Thomopoulos, P., Berthellier, M., and Laudat, M. H.,** Loss of insulin receptors on maturation of reticulocytes, *Biochem. Biophys. Res. Commun.*, 85, 1460, 1978.

169. **Thomopoulos, P., Testa, U., Flamier, A., and Berthelier, M.,** Insulin receptors and protein synthesis in bone marrow cells and reticulocytes, *Diabetes*, 29, 820, 1980.

170. **Eng, J., Lee, L., and Yalow, R. S.,** Influence of the age of erythrocytes on their insulin receptors, *Diabetes*, 29, 164, 1980.

171. **Peterson, S. W.,** Insulin receptor down regulation in human erythrocytes, *J. Biol. Chem.*, 258, 9605, 1983.

172. **Dustin, M. L., Jacobson, G. R., and Peterson, S. W.,** Effects of insulin receptor down-regulation on hexose transport in human erythrocytes, *J. Biol. Chem.*, 259, 13660, 1984.

173. **Albert, S. G.,** The effects of insulin on glucose transport in rabbit erythrocytes and reticulocytes, *Life Sci.*, 31, 265, 1982.

174. **Thomopoulos, P., Postel-Vinay, M. C., Testa, U., Guyda, H. J., and Posner, B. I.,** Receptors for insulin-like peptides (ILAs) in rat reticulocytes and erythrocytes, *Endocrinology*, 108, 1087, 1981.

175. **Ivarsson, S. A. and Thorell, J. I.,** Reticulocytes and insulin binding to erythrocytes, *Acta Med. Scand.*, Suppl., 565, 23, 1981.

176. **Knip, M., Punkka, R., Lantala, P., Leppilampi, M., and Punkka, M.,** Basal insulin secretion and erythrocyte insulin binding in preterm and term newborn infants, *Biol. Neonate*, 43, 172, 1983.

177. **Sunyer, T., Codina, J., and Birnbaumer, L.,** GTP hydrolysis by pure N, the inhibitory regulatory component of adenyl cyclase, *J. Biol. Chem.*, 259, 1544, 1984.

178. **Stiles, G. L., Caron, M. G., and Lefkowitz, R. J.,** β-Adrenergic receptors: biochemical mechanisms of physiological regulation, *Physiol. Rev.*, 64, 661, 1984.

179. **Gilman, A. G.,** G Proteins and dual control of adenylate cyclase, *Cell*, 36, 577, 1984.

180. **Bokoch, G. M.,** Identification of the predominant substrate for ADP-ribosylation by islet activating protein, *J. Biol. Chem.*, 258, 2072, 1983.

181. **Sternweis, P. C. and Gilman, A. G.,** Aluminium: a requirement for activation of the regulatory component of adenylate cyclase by fluoride, *Proc. Natl. Acad. Sci. U.S.A.*, 79, 4888, 1982.

182. **Dickinson, K., Richardson, A., and Nakashi, S. R.,** Homogeneity of beta₂-receptors on rat erythrocytes and reticulocytes. A comparison with heterogeneous rat lung beta-receptors, *Mol. Pharmacol.*, 19, 194, 1981.

183. **Rashidbaigi, A. and Rucho, A. E.,** Photoaffinity labelling of β-adrenergic receptors: identification of the β-receptor binding site(s) from turkey, pigeon and frog erythrocytes, *Biochem. Biophys. Res. Commun.*, 106, 139, 1982.

184. **Hanski, E., Sternweis, P. C., Northup, J. K., Dromerick, A. W., and Gilman, A. G.,** The regulatory component of adenylate cyclase, *J. Biol. Chem.*, 256, 12911, 1981.

185. **Sternweis, P. C., Northup, J. K., Smigel, M. D., and Gilman, A. G.,** The regulatory component of adenylate cyclase, purification and properties, *J. Biol. Chem.*, 256, 11517, 1981.

186. **Shorr, R., Lefkowitz, R. J., and Caron, M. G.,** Purification of the β-adrenergic receptor: identification of the hormone-binding subunit, *J. Biol. Chem.*, 256, 5820, 1981.

187. **Shorr, R. G. L., Strohsacker, M. W., Lavin, T. N., Lefkowitz, R. J., and Caron, M. G.,** The β-adrenergic receptor of the turkey erythrocyte: molecular heterogeneity revealed by purification and photoaffinity labelling, *J. Biol. Chem.,* 257, 12341, 1982.
188. **Northup, J. K., Smigel, M. D., Sternweis, P. C., and Gilman, A. G.,** The subunits of the stimulatory regulatory component of adenylate cyclase, *J. Biol. Chem.,* 258, 11369, 1983.
189. **Codina, J., Hildebrandt, J., Iyengar, R., Birnbaumer, L., Sekura, R. D., and Manclark, C. R.,** Pertussis toxin substrate, the putative N_i component of adenylyl cyclases, is an αβ heterodimer regulated by guanine nucleotide and magnesium, *Proc. Natl. Acad. Sci. U.S.A.,* 80, 4276, 1983.
190. **Hildebrandt, J., Codina, J., Risinger, R., and Birnbaumer, L.,** Identification of a γ-subunit associated with the adenylyl cyclase regulatory proteins N_s and N_i, *J. Biol. Chem.,* 259, 2039, 1984.
191. **Codina, J., Hildebrandt, J. D., Sekuras, R. D., Birnbaumer, M., Bryan, J., Manclark, Ch. R., Iyengar, R., and Birnbaumer, L.,** N_s and N_i, the stimulatory and inhibitory regulatory components of adenylyl cyclase, *J. Biol. Chem.,* 259, 5871, 1984.
192. **Tsamaloukas, A. G., Maretzki, D., Setchenska, M., and Rapoport, S.,** Maturation dependence of fluoride-sensitive adenylate cyclase in red blood cells, *Acta Biol. Med. Ger.,* 35, 523, 1976.
193. **Gauger, D., Palm, D., Kaiser, C., and Quiring, K.,** Adenyl cyclase activities in rat erythrocytes during stress erythropoiesis: localization of the enzyme in the reticulocytes, *Life Sci.,* 13, 31, 1973.
194. **Charness, M. E., Bylund, D. B., Beckman, B. S., Hollenberg, M. D., and Snyder, S. H.,** Independent variation of β-adrenergic receptor binding and catecholamine stimulated adenylate cyclase activity in rat erythrocytes, *Life Sci.,* 19, 243, 1976.
195. **Bilezikian, J. P., Spiegel, A. M., Brown, E. M., and Aurbach, G. D.,** Identification and persistence of beta adrenergic receptors during maturation of the rat reticulocyte, *Mol. Pharmacol.,* 13, 775, 1977.
196. **Bilezikian, J. P. and Gammon, D. E.,** The effect of age on beta adrenergic receptors and adenylate cyclase activity in rat erythrocytes, *Life Sci.,* 23, 253, 1978.
197. **Beckman, B. S. and Hollenberg, M. D.,** Beta-adrenergic receptors and adenylate cyclase activity in rat reticulocytes and mature erythrocytes, *Biochem. Pharmacol.,* 28, 239, 1979.
198. **Limbird, L. E., Gill, D. M., Stadel, J. M., Hickey, A. R., and Lefkowitz, R. J.,** Loss of β-adrenergic receptor-guanine nucleotide regulatory protein interactions accompanies decline in catecholamine-responsiveness of adenylate cyclase in maturing rat erythrocytes, *J. Biol. Chem.,* 255, 1854, 1980.
199. **Larner, A. C. and Ross, E. M.,** Alteration in the protein components of catecholamine-sensitive adenylate cyclase during maturation of rat reticulocytes, *J. Biol. Chem.,* 256, 9551, 1981.
200. **Shane, E., Gammon, D. C., and Bilezikian, J. P.,** A cellular activator of catecholamine-sensitive adenylate cyclase in rat reticulocytes and erythrocytes: changes during reticulocyte development and effects in the β receptor, *Arch. Biochem. Biophys.,* 208, 418, 1981.
201. **Sahyoon, N., Levine, H., III, Steinbruck, P., and Cuatrecasas, P.,** Cytosolic activator of adenylate cyclase: reconstitution, characterization, and mechanism of action, *Proc. Natl. Acad. Sci. U.S.A.,* 80, 3646, 1983.
202. **Shane, E., Yeh, M., Feigin, A. S., Owens, J. M., and Bilezikian, J. P.,** Cytosol activator protein from rat reticulocytes requires the stimulatory guanine nucleotide binding protein for its actions on adenylate cyclase, *Endocrinology,* 117, 255, 1985.
203. **Shane, E., Yeh, M., Feigin, A. S., and Bilezikian, J. P.,** Reticulocyte cytosol activator protein: Its actions upon the beta-adrenergic receptor and the N-proteins of adenylate cyclase, *J. Recept. Res.,* 4, 475, 1984.
204. **Stadel, J. M., Strulovici, B., Nambi, P., Lavin, T. N., Briggs, M. M., Caron, M. G., and Lefkowitz, R. J.,** Desensitization of the beta-adrenergic receptor of frog erythrocytes: recovery and characterization of the down regulated receptors in sequestered vesicles, *J. Biol. Chem.,* 258, 3032, 1983.
205. **Strulovici, B., Stadel, J. M., and Lefkowitz, R. J.,** Structural integrity of desensitized beta-adrenergic receptors: internalized receptors reconstitute catecholamine-stimulated adenylate cyclase activity, *J. Biol. Chem.,* 258, 6410, 1983.
206. **Stiles, G. L., Stadel, J. M., De Lean, A., and Lefkowitz, R. J.,** Hypothyroidism modulates beta adrenergic receptor-adenylate cyclase interactions in rat reticulocytes, *J. Clin. Invest.,* 68, 1450, 1981.
207. **Beer, M. and Porzig, H.,** Cell metabolism effects the density of β-adrenergic receptors in intact rat reticulocytes, *FEBS Lett.,* 111, 205, 1980.
208. **Montandon, J. B. and Porzig, H.,** Changes in beta-adrenoceptor binding properties and receptor-cyclase coupling during in vitro maturation of rat reticulocytes, *J. Recept. Res.,* 4, 91, 1984.
209. **Strittmatter, W. J., Hirata, F., and Axelrod, J.,** Phospholipid methylation unmasks cryptic β-adrenergic receptors in rat reticulocytes, *Science,* 204, 1205, 1979.
210. **Wiemer, G., Hellwich, U., Wellstein, A., Dietz, J., Hellwich, M., and Palm, D.,** Energy-dependent extension of cyclic 3′,5′-adenosine-monophosphate. A drug-sensitive regulatory mechanism for the intracellular nucleotide concentration of rat erythrocytes, *Naunyn Schmiedebergs Arch. Pharmakol.,* 321, 239, 1982.

211. **Kostič, M. M., Müller, M., Krause, E. G., and Rapoport, S.,** Metabolic effects of (−) isoprenalin stimulation of adenylate cyclase in reticulocytes, *Biomed. Biochim. Acta,* in press.

212. **Brunton, L. L. and Buss, I. G.,** Export of cyclic AMP by mammalian reticulocytes, *J. Cycl. Nucl. Res.,* 6, 369, 1980.

213. **Setchenska, M. S., Vasileva-Popova, J. G., and Arnstein, H. R. V.,** Plasma membrane-associated proteinkinase activity of differentiating rabbit bone marrow erythroid cells, *Int. J. Biochem.,* 11, 393, 1980.

214. **Setchenska, M. S., Arnstein, H. R. V., and Vasileva-Popova, J. G.,** Cyclic AMP phosphodiesterase activity during differentiation of rabbit erythroid bone marrow cells, *Biochem. J.,* 196, 887, 1981.

215. **Maguire, M. E. and Erdos, I. I.,** Inhibition of magnesium uptake by β-adrenergic agonists and prostaglandin E_1 is not mediated by cyclic AMP, *J. Biol. Chem.,* 255, 1030, 1980.

216. **Davis, P. J. and Blas, S. D.,** In vitro stimulation of human red blood cell Ca^{2+}-ATPase by thyroid hormone, *Biochem. Biophys. Res. Commun.,* 99, 1073, 1981.

217. **Davis, F. B., Cody, V., Davis, P. J., Borzynski, L. J., and Blas, S. D.,** Stimulation by thyroid hormone analogues of red blood cell Ca^{2+}-ATPase activity in vitro. Correlations between hormone structure and biological activity in a human cell system, *J. Biol. Chem.,* 258, 12373, 1983.

218. **Lawrence, W. D., Davis, P. J., Blas, S. D., and Schoene, M.,** Interaction of thyroid hormone and sex steroids at the rabbit reticulocyte membrane in vitro: control by 17β-estradiol and testosterone of thyroid hormone-responsive Ca^{2+}-ATPase activity, *Arch. Biochem. Biophys.,* 235, 78, 1984.

219. **Barnes, G. and Frieden, E.,** Ceruloplasmin receptors of erythrocytes, *Biochem. Biophys. Res. Commun.,* 125, 157, 1984.

220. **Finne, J.,** Identification of the blood-group ABH-active glycoprotein components of human erythrocyte membrane, *Eur. J. Biochem.,* 104, 181, 1980.

221. **Hakomori, S.,** Glycosphingolipids in cellular interaction, differentiation and oncogenesis, *Annu. Rev. Biochem.,* 50, 733, 1981.

222. **Koscielak, J., Miller-Podraza, H., Kranze, R., and Piasek, A.,** Isolation and characterization of poly (glycosyl) ceramides (megaloglycolipids) with A, H and I blood-group activities, *Eur. J. Biochem.,* 71, 9, 1976.

223. **Koscielak, J.,** Immunochemistry of Ic-active glycosphingo lipids, *Eur. J. Biochem.,* 96, 331, 1979.

224. **Zdebska, E. and Koscielak, J.,** Studies on the structure and I-blood-group activity of poly (glycosyl) ceramides, *Eur. J. Biochem.,* 91, 517, 1978.

225. **Bouhours, J.-F. and Bouhours, D.,** Identification of free ceramide in human erythrocyte membrane, *J. Lipid Res.,* 25, 613, 1984.

226. **Gahmberg, C. G. and Hakomori, S.-I.,** External labeling of cell surface galactose and galactosamine in glycolipid and glycoprotein of human erythrocytes, *J. Biol. Chem.,* 248, 4311, 1973.

227. **Steck, T. and Dawson, G.,** Topographical distribution of complex carbohydrates in the erythrocyte membrane, *J. Biol. Chem.,* 249, 2135, 1974.

228. **Hakomori, S., Teather, C., and Andrews, H.,** Organizational difference of cell surface "hematoside" in normal and virally transformed cells, *Biochem. Biophys. Res. Commun.,* 33, 563, 1968.

229. **Lampio, A., Finne, J., Homer, D., and Gahmberg, C. G.,** Exposure of the major human red-cell glycolipid, globoside, to galactose oxidase, *Eur. J. Biochem.,* 145, 77, 1984.

230. **Yunis, J. J. and Yunis, E.,** Cell antigen and cell specialisation. I. A study of blood group antigens on normoblasts, *Blood,* 22, 53, 1963.

231. **Mazumdar, P. M. H.,** Agglutination of normoblasts with anti-D, *Vox Sang.,* 11, 90, 1966.

232. **Furusawa, M. and Adachi, H.,** Immunological analysis of the structural molecules of erythrocyte membrane in mice. II. Staining of erythroid cells with labelled antibody, *Exp. Cell Res.,* 50, 497, 1968.

233. **Minio, F., Howe, C., Hsu, K. C., and Rifkind, R. A.,** Antigen density on differentiating erythroid cells, *Nature,* 237, 187, 1972.

234. **Karki, K. K., Andersson, L. C., Unopso, P., and Gahmberg, C. G.,** Expression of blood group A antigens in human bone marrow cells, *Blood,* 57, 147, 1981.

235. **Robinson, J., Sieff, C., Delia, D., Edwards, P. A. W., and Greaves, M.,** Expression of cell-surface HLA-DR, HLA-ABC and glycophorin during erythroid differentiation, *Nature,* 289, 68, 1981.

236. **Fitchen, J. H., Foon, K., and Cline, M. J.,** The antigenic characteristics of hematopoetic stem cells, *N. Engl. J. Med.,* 305, 17, 1981.

237. **Turco, S. J., Rush, J. S., and Laine, R. A.,** Presence of erythroglycan on human K-562 chronic myelogenous leukemia-derived cells, *J. Biol. Chem.,* 255, 3266, 1980.

238. **Fukuda, M.,** K 562 human leukaemic cells express fetal type (i) antigen on different glycoproteins from circulating erythrocytes, *Nature,* 285, 405, 1980.

239. **Reitzig, Ch., Gmyrek, D., and Syllm-Rapoport, I.,** Zur Klinik der Rh-Inkompatibilität. I. Passagere Erythropoeseschädigung infolge persistierender Rh-Antikörper nach Austauschtransfusion, *Dtsch. Gesundheitswes.,* 26, 497, 1971.

240. **Fukuda, M., Fukuda, M. N., Papayamopoulou, Th., and Hakomori, S.-J.**, Membrane differentiation in human erythroid cells: unique profiles of cell surface glycoproteins expressed in erythroblasts in vitro from three ontogenic stages, *Proc. Natl. Acad. Sci. U.S.A.*, 77, 3474, 1980.

241. **Feizi, T. and Childs, R. A.**, Carbohydrate structures of glycoproteins and glycolipids as differentiation antigens, tumour-associated antigens and components of receptor systems, *Trends Biochem. Sci.*, 10, 24, 1985.

242. **Cantor, L. N., Morris, A. J., Marks, P. A., and Rifkind, R. A.**, Purification of erythropoietin-responsive cells by immune hemolysis, *Proc. Natl. Acad. Sci. U.S.A.*, 69, 1337, 1972.

243. **Harrison, E. L. and Chesterton, C. J.**, Erythroid developmental agglutinin is a protein lectin mediating specific cell-cell adhesion between differentiating rabbit erythroblasts, *Nature*, 286, 502, 1980.

244. **Heller, M. and Hanahan, D. J.**, Human erythrocyte membrane bound enzyme acetylcholinesterase, *Biochim. Biophys. Acta*, 255, 251, 1972.

245. **Niday, E., Wang, C. S., and Alaupovic, P.**, Studies on the characterization of human erythrocyte acetylcholine-esterase and its interaction with antibodies, *Biochim. Biophys. Acta*, 469, 180, 1977.

246. **Wenkstern, T. W. and Engelhardt, W. A.**, Die an der Oberfläche der kernhaltigen roten Blutkörperchen lokalisierte Adenosinpolyphosphatase, *Folia Haematol.*, 76, 422, 1959.

247. **Boninsegna, A., Deana, R., and Silirandi, D.**, The localization of enzymatic activities involved in uridine nucleotides reactions in human erythrocytes, *J. Cell. Physiol.*, 83, 53, 1974.

248. **Hofmann, E. and Noll, F.**, Verteilung von DPN- and TPN-spezifischen Nukleosidasen in Erythrozyten verschiedener Tierarten, *Acta Biol. Med. Ger.*, 6, 1, 1961.

249. **Hofmann, E. C. G., Karadschova, M., and Klecker, J.**, Über vergleichende Untersuchungen DPN- und TPN-spezifischer Nukleosidasen in Erythrozyten und Reticulocyten, *Folia Haematol.*, 76, 372, 1959.

250. **Pekala, P. H. and Anderson, B. M.**, Studies of bovine erythrocyte NAD glycohydrolase, *J. Biol. Chem.*, 253, 7453, 1978.

251. **Berry, D. H. and Hochstein, P.**, Membrane acid phosphatase in rabbit erythrocytes, *Arch. Biochem. Biophys.*, 131, 170, 1969.

252. **Bosmann, H. B.**, Glycosidase activities of human erythrocyte plasma membranes, *J. Membr. Biol.*, 4, 113, 1971.

253. **Yatziv, S., Kahane, I., Abeliuk, P., Cividalli, G., and Rachmilewitz, E. A.**, "Lysosomal" enzyme activities in red blood cells of normal individuals and patients with homozygous β-thalassaemia, *Clin. Chim. Acta*, 96, 67, 1979.

254. **Mulder, E. and Van Deenen, L. L. M.**, Metabolism of red cell lipids. I. Incorporation in vitro of fatty acids into phospholipids from mature erythrocytes, *Biochim. Biophys. Acta*, 106, 106, 1965.

255. **Hirata, F. and Axelrod, J.**, Enzymatic synthesis and rapid translocation of phosphatidylcholine by two methyltransferases in erythrocyte membranes, *Proc. Natl. Acad. Sci. U.S.A.*, 75, 2348, 1978.

256. **Hokin, L. E. and Hokin, M. R.**, Diglyceride kinase and other pathways for phosphatidic acid synthesis in the erythrocyte membrane, *Biochim. Biophys. Acta*, 67, 470, 1963.

257. **Mack, S. E. and Palmer, F. B. St. C.**, Evidence for a specific phosphatidylinositol 4-phosphate phosphatase in human erythrocyte membrane, *J. Lipid Res.*, 25, 75, 1984.

258. **Koutousov, S. and Marche, P.**, The Mg^{2+}-activated phosphatidylinositol 4,5-bisphosphate-specific phosphomonoesterase of erythrocyte membrane, *FEBS Lett.*, 144, 16, 1982.

259. **Allan, D. and Michell, R. H.**, A calcium-activated polyphosphoinositide phosphodiesterase in the plasma membrane of human and rabbit erythrocytes, *Biochim. Biophys. Acta*, 508, 277, 1978.

260. **Downes, C. P. and Michell, R. H.**, The polyphosphoinositide phosphodiesterase of erythrocyte membranes, *Biochem. J.*, 198, 133, 1981.

261. **Downes, C. P., Mussat, M. C., and Michell, R. H.**, The inositol triphosphate phosphomonoesterase of the human erythrocyte membrane, *Biochem. J.*, 203, 169, 1982.

262. **Kaya, K., Miura, T., and Kubotka, K.**, Different incorporation rates of arachidonic acid into alkenylacyl-, alkylacyl- and diacylphosphatidylethanolamine of rat erythrocytes, *Biochim. Biophys. Acta*, 796, 304, 1984.

263. **Zamudio, I., Cellino, M., and Canessa-Fischer, M.**, The relation between membrane structure and NADH: (acceptor) oxido-reductase activity of erythrocyte ghosts, *Arch. Biochem. Biophys.*, 129, 336, 1969.

264. **Beutler, E. and Kuhl, W.**, Guanosine triphosphatase activity in human erythrocyte membranes, *Biochim. Biophys. Acta*, 601, 372, 1980.

265. **Ferber, E., Munder, P. G., Kohlschütter, A., and Fischer, H.**, Lysolecithin-Stoffwechsel in Erythrocytenmembranen. Lysolecithin-Acylierung und Lysophospholipase in alternden Erythrocyten, *Eur. J. Biochem.*, 5, 395, 1968.

266. **Cartron, J.-P., Andreu, G. A., Cartron, J., Bird, G. W. G., Salmon, C., and Gerbal, A.**, Demonstration of T-transferase deficiency in Tn-polyagglutinable blood samples, *Eur. J. Biochem.*, 92, 111, 1978.

267. **Berger, E. G. and Kozdrowski, I.**, Permanent mixed-field polyagglutinable erythrocytes lack galactosyltransferase activity, *FEBS Lett.*, 93, 105, 1978.

268. **Kim, Y. S., Perdomo, J., Bella, A., Jr., and Nordberg, J.,** *N*-acetyl-D-galactosaminyl transferase in human serum and erythrocyte membranes, *Proc. Natl. Acad. Sci. U.S.A.,* 68, 1753, 1971.

269. **Bosmann, H. B.,** Red cell hydrolases. III. Neuraminidase activity in isolated human erythrocyte plasma membranes, *Vox Sang.,* 26, 497, 1974.

270. **Martin-Barrientos, J. and Parodi, A. J.,** Synthesis of dolichol derivatives in human erythrocyte membranes, *Mol. Cell. Biochem.,* 16, 111, 1977.

271. **Parodi, A. J. and Martin-Barrientos, J.,** Glycosylation of endogenous proteins through dolichol derivatives in reticulocyte plasma membranes, *Biochim. Biophys. Acta,* 500, 80, 1977.

272. **Delbauffe, D., Paysant, M., and Polonovski, J.,** Phosphatidyl-glycerolphospholipase A des globules rouges de rat. I. Influence des effecteurs, *Bull. Soc. Chim. Biol.,* 50, 1431, 1968.

273. **Garcia-Parra, M., Schewe, T., and Rapoport, S. M.,** On the presence of a calcium-stimulated phospholipase A in the stroma-free supernatant fluid of rabbit reticulocytes, *Acta Biol. Med. Ger.,* 34, 1075, 1975.

274. **Roach, P. D. and Palmer, F. B. St. C.,** Human erythrocyte cytosol phosphatidyl-inositol-bisphosphate phosphatase, *Biochim. Biophys. Acta,* 661, 323, 1981.

275. **Siefring, G. E., Jr., Apostol, A. B., Velasco, P. T., and Lorand, L.,** Enzymatic basis for the Ca^{2+}-induced cross-linking of membrane proteins in intact human erythrocytes, *Biochemistry,* 17, 2598, 1978.

276. **Maretzki, D., Groth, J., Tsamaloukas, A. G., Gründel, M., Krüger, S., and Rapoport, S.,** The membrane association and dissociation of human GAPD under various conditions of hemolysis. Immunochemical evidence for the lack of binding under cellular conditions, *FEBS Lett.,* 39, 83, 1974.

277. **Brindle, K. M., Campbell, I. D., and Simpson, R. I.,** A nmr study of the kinetic properties expressed by glyceraldehyde phosphate dehydrogenase in the intact human erythrocyte, *Biochem. J.,* 108, 583, 1983.

278. **Maretzki, D.,** unpublished data.

297. **Földes-Papp, Z., Tsamaloukas, A. G., and Maretzki, D.,** Reevaluation of membrane binding of erythrocyte glutathione reductase, *Acta Biol. Med. Ger.,* 40, 1129, 1981.

280. **Földes-Papp, Z. and Maretzki, D.,** Enzymatic *t*-butyl-hydroperoxide reduction on human erythrocyte membranes. NADPH- and GSH-dependent activities, *Biochim. Biophys. Acta,* 43, 271, 1984.

281. **Agutter, P. S. and Suckling, K. E.,** Models of the interactions between membranes and intracellular protein structures, *Biochem. Soc. Trans.,* 12, 713, 1984.

282. **Rich, G. T., Dawson, A. P., and Pryor, J. S.,** Glyceraldehyde-3-phosphate dehydrogenase release from erythrocytes during haemolysis. No evidence for substantial binding of the enzyme to the membrane in the intact cell, *Biochem. J.,* 221, 197, 1984.

Chapter 5

INDUCTION OF RETICULOCYTOSIS AND METHODS OF SEPARATING RED CELLS

I. INDUCTION OF RETICULOCYTOSIS

Stress reticulocytosis in experimental animals is best produced by a schedule of bleeding. An effective procedure is herein described.[1] Young lean rabbits of 3 to 4 kg body weight are most suitable. Reticulocytosis is provoked by withdrawal of 17 mℓ of blood per kilogram of body weight from the ear vein daily. The hematocrit is checked every day. If the hematocrit decreases below 16%, the amount of blood withdrawn is reduced. The regeneration of blood in the animals can be improved, especially in winter, by daily oral administration to each animal of a mixture containing 15 g of soybean grits, 5 g of egg albumin, 150 mg of iron, e.g., as powdered ferrous salt, ascorbate tablets, 3 g of partially hydrolyzed chlorophyll, and 0.5 g of sodium chloride; subcutaneous administration of 10 μg of vitamin B_{12} and 3 mg of folic acid is also given. After an interval of 2 days, i.e., beginning Day 6, a high reticulocytosis with large amounts of lipoxygenase is achieved. The rationale for adding iron is apparent from data on ferrokinetics,[2] according to which rabbits subjected to bleeding have only half the normal plasma iron concentration and low transferrin saturation.

Blood is obtained by nicking one of the ear veins as far as possible away from the base of the ear to allow further incisions more proximately. The vein is distended by means of gentle heating with a light bulb. Chemical irritants such as xylene should be avoided, since they lead to severe inflammatory and proliferative tissue reactions. Similar schedules of bleeding may be employed for other species. A bleeding schedule for sheep was proposed, according to which sheep weighing 50 to 70 kg are bled twice daily from the jugular vein for 3 days;[3] about 500 mℓ is usually removed each time. Should the animal become restless, less blood is taken. In this manner the hematocrit is reduced to about 18% in 3 days. Peak reticulocyte response is observed 3 to 6 days after the last bleeding.

Phenylhydrazine hydrochloride is frequently used to induce reticulocytosis. A common protocol consists of the subcutaneous application of neutralized phenylhydrazine hydrochloride in a dosage of 0.3 mℓ of a 2.5% solution in 0.9% NaCl, i.e., 75 mg/kg body weight for 3 to 5 days. Each application of phenylhydrazine is followed immediately by the hemolytic destruction of circulating red cells; 2 days later a massive influx of reticulocytes is observed. Reticulocytosis may reach more than 80%. It is advisable to wait at least 2 days after cessation of phenylhydrazine application before blood is taken for further studies.

II. THE EFFECTS OF PHENYLHYDRAZINE

A brief discussion of the effects of phenylhydrazine is appropriate since this compound and some of its derivatives, particularly acetylphenylhydrazine, have been commonly used to induce reticulocytosis in various experimental animals.

Phenylhydrazine attacks heme proteins in a complex reaction sequence. Hemoglobin is denatured, forming aggregates of oxidized and denatured products,[4] that is, Heinz bodies (for older literature see Beutler[5] and Jandl et al.[6]). The heme moiety is transformed to N-arylprotoporphyrins by reaction of C-radicals stemming from phenylhydrazine.[7,8] The reactions go hand in hand with the formation of HO_2^-[9,10] as well as of superoxide.[11-13] There is also evidence of various membrane changes, partly caused by peroxidation of its lipids and thereby causing fluorescent products to arise.[14,15] The membrane proteins undergo a multitude of changes. Decreases in spectrin polypeptides, the virtual absence of normal glycoproteins, and the appearance of protein aggregates are demonstrable.[15,16] These alterations are related to losses of membrane fluidity and deformability.[17,18] The activities of various enzymes are impaired. Among these are acetylcholinesterase, and particularly, adenylate cyclase.[15,19] The membranes of phenylhydrazine-induced reticulocytes exhibit only $1/10$ to $1/30$ the activity of those induced by bleeding, while the stimulatory effect of prostaglandin E_1 is even more impaired and the stimulatory adenine receptor is practically abolished.

Phenylhydrazine decreases electron transport as well as cytochrome content of mitochondria.[20] The amount of lipoxygenase is usually lower in phenylhydrazine-induced reticulocytes than in those induced by bleeding.[21] With respect to proteolysis, phenylhydrazine-induced reticulocytes exhibit several characteristics which are absent in the bleeding-induced cells.[22] These include (1) the existence of proteolysis in the cytosol. Here one may distinguish both an ATP-dependent and ATP-independent process; (2) the presence of sizable antimycin A-resistant proteolysis; (3) the existence of a large salicylhydroxamate (SHAM)-resistant proteolysis. From these deleterious effects of phenylhydrazine on the various components of the cell several consequences follow. First, there is the formation of denatured and aggregated proteins in the cytosol. These are the substrates of both ATP-independent and ATP-dependent protease systems which have been demonstrated by several authors (see Chapter 13). Second, ATP-independent proteolysis would not be inhibited by either antimycin A or salicylhydroxamate. The mitochondria damage may be a subsidiary reason for the lack of a salicylhydroxamate effect since it would make the attack of lipoxygenase superfluous.

In addition there may be a decreased amount of lipoxygenase. From the results and considerations presented, one may conclude that in phenylhydrazine-induced reticulocytes there is an overlap between the basic biological process of maturation-dependent proteolysis of mitochondria and the breakdown of denatured and aggregated "abnormal" proteins, which is an artificial process. Furthermore, there are probably several different proteolytic processes, both ATP-dependent and ATP-independent.

Based on the evidence that lipid peroxidation products occur predominantly in reticulocytes but are absent in erythrocytes (see Jain and Hochstein[15]), it has been argued[23] that hemolysis is closely related to hemoglobin denaturation rather than to lipid peroxidation. On the other hand, it has been proposed that the membranes of hemoglobin-poor, immature erythroid cells are less protected against the deleterious effects of phenylhydrazine and therefore more prone to alterations. It is also worth noting that lipid peroxidation products persist up to 6 days after cessation of phenylhydrazine treatment. Thus the conclusion emerges that despite the convenience of producing reticulocytosis by means of phenylhydrazine administration, there are grave reasons to caution against its use as a tool for the study of normal processes in the red cell.

III. SEPARATION TECHNIQUES FOR ERYTHROID CELLS

The interest in erythroid differentiation and maturation has led to the development of various methods of separating red blood cells according to age. Here a cursory review is given of the principles of the procedures. The aims of the various separation procedures can be classified as follows: (1) provision of bone marrow cells at different stages of development; (2) separation of reticulocytes from erythrocytes; within this framework methods have been devised to subfractionate the reticulocytes according to maturity; (3) fractionation of erythrocytes according to age, particularly to identify the senescent cell population.

Various principles have been applied to achieve such separations. The density difference of cells is most commonly used as an age marker. It is based on the condition that up to and including the reticulocyte, hemoglobin content of cells increases during differentiation and maturation owing to ongoing hemoglobin synthesis. The further increase in density of the erythrocytes during their aging process is caused by loss of water, which to some extent occurs by vesiculation. Based on density differences, various procedures have been employed.

Before separation techniques are applied, there are several possible ways of modifying the starting conditions in experimental studies. By inducing anemia, the number of reticulocytes as well as the percentage of erythroid cells in the bone marrow may be increased. The reverse can be achieved by overtransfusion. Radiation may be used both to deplete and to stimulate the bone marrow. The proliferation of red cells may be stopped by the administration of actinomycin. Contamination by leukocytes may be a serious problem, particularly for such enzymes where the activity is much higher in white cells compared to red cells, e.g., most glycolytic enzymes and proteases. Effective procedures have been devised for the removal of white cells and platelets by means of separation on columns with α-cellulose or cotton.[24,25] Some components and activities occur in reticulocytes but are low or absent in erythrocytes. For such constituents one may dispense with the separation of reticulocytes from erythrocytes. The contribution of erythrocytes to total content or activity can be computed if a separate determination of either a pure sample of erythrocytes or reticulocytes is performed.

A. Sedimentation at 1 g

This method has been applied to large-scale preparations of bone marrow cells. It is based on both differences in cell size and density.[26-29] Sucrose, Ficoll, or Percoll were used for the preparation of gradients. These methods require long separation periods as well as large amounts of gradient solutions and are beset by problems of cell aggregation. Nevertheless, they always yield clear-cut fractions of immature bone marrow reticulocytes that are nearly uncontaminated by nucleated cells.

B. Differential Centrifugation

For the preparation of reticulocytes in the peripheral blood, centrifugation procedures of red blood cells in plasma have been described.[30-32] Murphy's modification has become most popular,[33] involving high-speed centrifugation of red blood cells in plasma with a hematocrit of 0.85 at 3000 g for 1 hr at 30°C. The use of a fixed-angle rotor, which favors internal circulation in the tube, and the high temperature employed, which reduces viscosity, makes this procedure highly effective. The method is suitable for separation on a large scale; however, its resolution is limited.

A simple centrifugation of reticulocyte-rich suspensions in 0.9% NaCl was quite successful with anemic rabbit blood.[34] On this basis a three-step fractionation procedure was developed for the subfractionation of reticulocytes according to their maturity.[56] It is carried out as follows: a rabbit on the 6th to 8th day of anemia is bled, whereby more than 100 mℓ of blood containing about 20 mℓ of cells can usually be obtained. The cells are separated from the plasma and after a thorough mixing are suspended in an equal volume of 0.9% NaCl. Portions of 10 mℓ of the suspension are centrifuged in narrow tubes for 30 min at 2000 g at 4°C. The upper 10% of the cells are carefully collected and constitute the fraction of young reticulocytes (Fraction I). The next 30% of the cells are taken from each centrifuge tube and after suspension in an equal volume of 0.9% NaCl and pooling are subjected to a second centrifugation. The upper half of the cell mass is removed. This constitutes the fraction of reticulocytes of intermediate maturity (Fraction II). The remainder of the cell mass of the first and second centrifugation yields after resuspension and a third centrifugation in its uppermost 15% the fraction of mature reticulocytes (Fraction III). Starting with a reticulocyte count of about 30%, Fraction I always has more than 80%, Fraction II >50%, and Fraction III >35% of reticulocytes, whereas the remainder has less than 10%.

C. Density Gradient Centrifugation

The majority of studies make use of density gradient separation procedures. A variety of compounds have been used since the demands on a satisfactory gradient medium are manyfold and high.

The media used for density gradient separations should meet several conditions:

1. The compounds should be inert or at least nontoxic.
2. The physical chemical properties of the solution should be known and it should be possible to use one or more property to determine the precise concentration.
3. The gradient medium should not interfere with the monitoring of the fractionated cells in the gradient.
4. The medium should provide iso-osmotic conditions and be adjustable to a physiological pH over the whole gradient.
5. The compound should not permeate the cell membrane.
6. The gradient media should not cause cell aggregation.
7. The viscosity should be low.
8. For clinical purposes the media should withstand sterilization.
9. The medium should not be expensive.

A survey on the theoretical background, sources of experimental error, and older work on several gradient media has been provided by Harwood.[35]

Most of the procedures utilize isopycnic continuous density gradients since they usually, but not always, produce a higher degree of separation than discontinuous gradients or other methods of centrifugation.[36]

Among the first media to be used for the separation of reticulocytes was bovine serum albumin.[37-42] It was also applied for the separation of erythroid cells in the bone marrow.[43,44] However, factors such as batch-to-batch variation, viscosity, osmolarity, preparation time for the gradient, and cost have stimulated the search for other gradient materials. Among these, polymeric carbohydrates have been favored. Dextran has been recommended by several authors.[45-54] This material is cheap and may be sterilized without loss of its separative qualities; in fact, there is some decrease in the tendency to aggregate red cells. This tendency is not disturbing with human red cells but much more so with those of rabbits. Nevertheless, good separations were achieved during the course of bleeding anemia of rabbits.

Several authors have used Ficoll, which is a synthetic high polymer prepared by copolymerization of sucrose and epichlorhydrine.[26,55] Among its advantages are stability, high solubility, low osmotic activity, low viscosity, as well as nontoxicity. Its solutions can be autoclaved.

Stractan, a polysaccharide consisting of arabinose and galactose with a molecular mass of 30 kD, is biologically inert. Its solutions can be easily adjusted to suitable osmolarities and pH values and has given excellent results in the hands of several workers.[57-59] Stractan requires extensive purification before use. It has been used mainly for the fractionation of erythrocytes.

Percoll, a new colloidal silica sol coated with polyvinylpyrrolidone, has the advantage of low viscosity, thus permitting short centrifugation times.[60-68] It can be used with self-generated continuous density gradients. The gradient generated is highly dependent on the centrifugation conditions used, such as volume of gradient material, speed and duration of centrifugation, as well as type and angle of rotors used. The preformed gradient is stable for extended periods of time even at room temperature. A procedure to adjust the osmolarity of Percoll solutions has been reported recently.[69] Occasionally nonionic iodinated density gradient media have been used.[70]

D. Nonaqueous Separations

Differential floating of red cells by centrifugation in a mixture of phthalate esters has been proposed.[71,72] Mixtures of *n*-dibutyl phosphate (sp gr 1.0416) and methyl phthalate (sp gr 1,189) can be so adjusted as to cover the range of cell densities to be expected. This method can be applied both on a small scale and on a large scale. According to our experience, however, the cell fractions thus prepared are not suitable for studies of metabolism such as glycolysis, which appears to be inhibited.

An interesting method, which only serves to obtain reticulocytes, however, is centrifugation through a layer of silicon oil which can even be used in a micro-modification for the assessment of the percentage of reticulocytes in a blood sample.[74] The reticulocytes obtained by this procedure are well suited for metabolic studies. Silicon oil has also been used for the separation of erythroid cells.[75,76]

E. Separations Based on Properties of the Cell Surface

Various properties of the cell surface have been utilized for the separation of red cells. Among these is the surface charge, which has been thought to be higher in young erythrocytes on account of a larger amount of sialic acid.[77] A procedure involving a special apparatus has been reported for the preparative electrophoresis of red cells.[78,79] However, there is some disagreement concerning the age dependence of the surface charge of red cells.

The use of antibodies to age-dependent surface antigens appears to be very promising. An affinity chromatography method was used for the preparation of human reticulocytes.[80,81] The reticulocytes were collected by means of a transferrin-coupled Sepharose column. This method is based on the fact that reticulocytes exhibit a large number of transferrin receptors (see Chapter 10) which are eliminated during maturation.

Mature red cells have receptors for lectins, among others for the lectin of *Helix pomatia*, while reticulocytes have not yet exposed them. A preliminary report on the enrichment of reticulocytes was given.[82] A technique to select red blood cells on nylon fibers is not suitable for large-scale preparation, whereas another procedure in which the lectin receptors are occupied by fluorescein-conjugated lectin has been utilized for computerized cell sorting of bone marrow cells.[83,84] A procedure to remove reticulocytes and mature erythroblasts by a combination of an antierythrocyte antibody and albumin density gradient centrifugation was reported by Borsook et al.[85]

Methods for large-scale separation of young red cells which are mainly based on size differences and use counterflow centrifugation have been described recently.[86-88] In early work the greater resistance of young red cells to osmotic hemolysis was utilized for their enrichment.[89] This method has obvious limitations.

IV. INDEXES OF SEPARATION EFFECTIVENESS

The quality of the age-dependent separation of red cells may be tested in various ways. For the establishment of the methods, cohort labeling, usually with Fe^{59}, is employed. The location of the labeled cells in the gradient is monitored after different periods of time. The band of labeled cells moves with time from the top to the lower fractions. For this method Fe^{59} has to be administered to the experimental subject or animal in a single dose, so that only one cohort is labeled.

A multitude of indexes show differences during maturation and/or senescence. They include constituents, enzymes, receptors, and transport activities and are listed in Table 8.

The best marker with the highest discriminatory power both for the separation of reticulocytes from erythrocytes and for aging red cells appears to be creatine.[66,115-121] Polyamines are also supposed to have a high discriminating quality. However, their application requires the rigorous removal of leukocytes, which contain them in a manyfold higher concentration than do red cells. For enzymes, the best among those most commonly determined appear to be hexokinase, pyruvate kinase,[60,66] and aspartate amino transferase.[52] The frequently used indexes, glucose-6-phosphate dehydrogenase and acetylcholinesterase, show a lesser dependence on age. A survey of the methods discussed is tabulated in Table 9.

Table 8
AGE-DEPENDENT MARKERS FOR SEPARATION OF RETICULOCYTES FROM ERYTHROCYTES

Marker	Remarks	Ref.
MCHC	Also suitable as indicator of red cell aging	56, 57, 49, 90—92
Reticulocyte count		
RNA		55, 92
O_2-consumption and respiratory enzymes	See also Chapter 6	59, 92, 93
Uncoupling of respiratory chain	See also Chapter 6; useful to subdivide reticulocytes	94
Hexokinase and other enzymes of glycolysis	See also Chapter 6; some enzymes suitable to indicate aging of red cells	60, 66
5-ALA synthase and other enzymes of heme synthesis	See Chapter 9	
Aspartate aminotransferase and other enzymes of mitochondrial matrix	See Chapter 8	52
Protein synthesis	See Chapter 11	
Protein kinases	See also Chapter 11	58, 95
Lipoxygenase	See Chapter 12; excellent for subdivision of reticulocytes	
Proteases	See Chapter 13	
Inorganic pyrophosphatase		96—99
Adenylate cyclase	See also Chapter 4	58, 100—109
Guanase	Only applicable to red cells of rats	110
Monoamine oxidase		111
Catechol-D-methyl transferase		112
Guanylating enzyme		113
Superoxide dismutase		114
Na-dependent ALA-transport	See Chapter 4	61
Transferrin and other receptors	See Chapter 4; useful for preparative fractionation	80—84
Creatine	Best marker both for separation of reticulocytes and for aging of red cells	66, 115—121
Polyamines	Requires rigorous removal of leukocytes	122, 123
ATP		124, 125
Guanine		126
K^+		125, 127
Mg^{2+}		92—94, 127—129

Table 9
SELECTED METHODS OF SEPARATING RED CELLS

Cell type	Principle	Method	Ref.
Bone marrow			
erythroblasts	Density and cell size	Sedimentation 1 g; Ficoll gradient	26, 29
reticulocytes		Sucrose gradient	28
		Percoll gradient	27
Blood			
reticulocytes	Osmotic resistance	Osmotic hemolysis	89
young erythrocytes	Surface charge immunological properties	Electrophoresis	78
	Transferrin receptors	Affinity chromatography	80
	Lectin receptors	Agglutination	82
		Adsorption	83
		Cell sorting	84
	Anti-erythrocyte antibodies	Density gradient centrifugation	85
	Density	Differential centrifugation	30—34
		Differential centrifugation, silicon oil	74
		Density gradient centrifugation: albumin	37—44
		Dextran	45—54
		Ficoll	26, 55, 56
		Stractan	57—59
		Percoll	60—69
		Iodinated medium	70
		Nonaqueous, phthalate	39—41, 71—73
		Nonaqueous silicon oil	75, 76

REFERENCES

1. **Schewe, T., Wiesner, R., and Rapoport, S.,** Lipoxygenase from rabbit reticulocytes, *Methods Enzymol.,* 71 (Part C), 430, 1981.
2. **Celada, A., Stray, S., Sivarajan, M., and Finch, C.,** Iron supply for erythropoiesis in the rabbit, *J. Clin. Invest.,* 74, 161, 1984.
3. **Tucker, E. M. and Young, J. D.,** Reticulocyte maturation in culture, in *Red Cell Membranes — A Methodological Approach,* Ellory, J. C. and Young, J. D., Eds., Academic Press, London, 1982, 31.
4. **Winterbourne, C. C. and Carrell, R. W.,** Absence of lipid peroxidation in human red cells exposed to acetylphenylhydrazine, *Br. J. Haematol.,* 23, 499, 1972.
5. **Beutler, E.,** Drug-induced hemolytic anemia, *Pharmacol. Res.,* 21, 73, 1969.
6. **Jandl, J. H., Engle, L. K., and Allen, D. W.,** Oxidative hemolysis and precipitation of hemoglobin. I. Heinz body anemias as an acceleration of red cell aging, *J. Clin. Invest.,* 39, 1818, 1960.
7. **Augusto, O., Kunze, K. L., and Ortiz de Montellano, P. R.,** N-Phenylprotoporphyrin. IX. Formation in the hemoglobin-phenylhydrazine reaction, *J. Biol. Chem.,* 257, 6231, 1982.
8. **Saito, S. and Itano, H. A.,** β-Meso-phenylbiliverdin α and N-phenylprotoporphyrin. XI. Products of the reaction of phenylhydrazine with oxyhemoproteins, *Proc. Natl. Acad. Sci. U.S.A.,* 78, 5508, 1981.
9. **Cohen, G. and Hochstein, P.,** Generation of hydrogen peroxide in erythrocytes by hemolytic agents, *Biochemistry,* 3, 895, 1964.
10. **Goldberg, B. and Stern, A.,** The mechanism of oxidative hemolysis produced by phenylhydrazine, *Molec. Pharmacol.,* 13, 832, 1977.
11. **Goldberg, B. and Stern, A.,** The generation of O_2^- by the interaction of the hemolytic agent, phenylhydrazine, with human hemoglobin, *J. Biol. Chem.,* 250, 2401, 1975.
12. **Goldberg, B., Stern, A., and Peisach, J.,** The mechanism of superoxide anion generation by the interaction of phenylhydrazine with hemoglobin, *J. Biol. Chem.,* 251, 3045, 1976.

13. **Misra, H. and Fridovich, I.,** The oxidation of phenylhydrazine: superoxide and mechanism, *Biochemistry,* 15, 681, 1976.

14. **Jain, S. K. and Subramanyam, D.,** On the mechanism of phenylhydrazine-induced hemolytic anemia, *Biochem. Biophys. Res. Commun.,* 82, 1320, 1978.

15. **Jain, S. K. and Hochstein, P.,** Membrane alterations in phenylhydrazine-induced reticulocytes, *Arch. Biochem. Biophys.,* 201, 683, 1980.

16. **Tsao, D., Colton, D. G., Chang, J. S., Buck, R. L., Hudson, B. G., and Carraway, K. L.,** Alteration of red cell membranes from phenylhydrazine-treated rabbits, *Biochim. Biophys. Acta,* 469, 61, 1977.

17. **Rice-Evans, C. and Hochstein, P.,** Alteration in erythrocyte membrane fluidity by phenylhydrazine-induced peroxidation of lipids, *Biochem. Biophys. Res. Commun.,* 100, 1537, 1981.

18. **Nagasawa, T.,** Deformability and osmotic fragility of phenylhydrazine-injected rat erythrocytes fractionated by Percoll density-gradients, *Jpn. J. Physiol.,* 32, 161, 1982.

19. **Cooper, D. M. F. and Jagus, R.,** Impaired adenylate cyclase activity of phenylhydrazine-induced reticulocytes, *J. Biol. Chem.,* 257, 4684, 1982.

20. **Asami, K.,** Mode of inhibition of the electron transport system by phenylhydrazine, *J. Biochem.,* 63, 425, 1968.

21. **Rapoport, S. and Gerischer-Mothes, W.,** Biochemische Vorgänge bei der Reticulocyten-Reifung: Auftreten und Abklingen des Reticulocyten-Hemmstoffes, *Hoppe-Seyler's Z. Physiol. Chem.,* 304, 213, 1956.

22. **Rapoport, S. M. and Dubiel, W.,** The effect of phenylhydrazine on protein breakdown in rabbit reticulocytes, *Biomed. Biochim. Acta,* 43, 23, 1984.

23. **Bates, D. A. and Winterbourne, C. C.,** Hemoglobin denaturation, lipid peroxidation and haemolysis in phenylhydrazine-induced anaemia, *Biochim. Biophys. Acta,* 798, 84, 1984.

24. **Nakao, M., Nakayama, T., and Kankura, T.,** A new method for separation of human blood components, *Nature (London),* 246, 94, 1973.

25. **Beutler, E., West, C., and Blume, K.-G.,** The removal of leucocytes and platelets from whole blood, *J. Lab. Clin. Med.,* 88, 328, 1976.

26. **Denton, M. J. and Arnstein, H. R. V.,** Characterization of developing adult mammalian erythroid cells separated by velocity sedimentation, *Br. J. Haematol.,* 24, 7, 1973.

27. **Clissold, P. M., Arnstein, H. R. V., and Chesterton, G. J.,** Quantitation of globin mRNA levels during erythroid development in the rabbit and discovery of a new β-related species in immature erythroblasts, *Cell,* 11, 353, 1977.

28. **Peterson, E. R. and Evans, W. H.,** Separation of bone marrow cells by sedimentation at unit gravity, *Nature (London),* 214, 824, 1967.

29. **Harrison, F. L., Beswick, T. M., and Chesterton, C. J.,** Separation of haemopoietic cells for biochemical investigation, *Biochem. J.,* 194, 789, 1981.

30. **Chalfin, D.,** Differences between young and mature rabbit erythrocytes, *J. Cell. Comp. Physiol.,* 47, 215, 1956.

31. **Rigas, D. A. and Koler, R. D.,** Ultracentrifugal fractionation of human erythrocytes on the basis of cell age, *J. Lab. Clin. Med.,* 58, 242, 1961.

32. **Prentice, T. C. and Bishop, C.,** Separation of rabbit red cells by density methods and characteristics of separated layers, *J. Cell. Comp. Physiol.,* 65, 113, 1965.

33. **Murphy, J.,** Influence of temperature and method of centrifugation on the separation of erythrocytes, *J. Lab. Clin. Med.,* 82, 334, 1973.

34. **Kahrig, C. and Rapoport, S.,** Das Verhalten der Atmung und der Glukose-6-phosphat-Dehydrogenase roter Blutkörperchen verschiedenen Alters während einer Entblutungsanämie bei Kaninchen, *Acta Biol. Med. Ger.,* 6, 238, 1961.

35. **Harwood, R.,** Cell separation by gradient centrifugation, *Int. Rev. Cytol.,* 38, 369, 1974.

36. **Piomelli, S., Larinsky, G., and Wasserman, L. R.,** The mechanism of red cell aging. I. Relationship between cell age and specific gravity evaluated by ultracentrifugation in a discontinuous density gradient, *J. Lab. Clin. Med.,* 69, 659, 1967.

37. **Borsook, H., Lingrel, J. B., Scaro, J. L., and Millette, R. L.,** Synthesis of haemoglobin in relation to the maturation of erythroid cells, *Nature (London),* 196, 347, 1962.

38. **Leif, R. C. and Vinograd, J.,** The distribution of buoyant density of human erythrocytes in bovine serum albumin solution, *Proc. Natl. Acad. Sci. U.S.A.,* 51, 520, 1964.

39. **Millette, R. L. and Glowacki, E. R.,** In vivo maturation of immature reticulocytes transfused into a normal rabbit, *Nature (London),* 204, 1207, 1964.

40. **Glowacki, E. R. and Millette, R. L.,** Polyribosomes and the loss of hemoglobin synthesis in the maturing reticulocyte, *J. Molec. Biol.,* 11, 116, 1965.

41. **Rowley, P. T.,** Protein synthesis in reticulocytes maturing in vivo, *Nature (London),* 208, 244, 1965.

42. **Bishop, C. and Prentice, T. C.,** Separation of red cells by density in a bovine serum albumin gradient and correlation of red cell density with cell age after in vivo labeling with Fe[59], *J. Cell. Physiol.,* 67, 197, 1966.

43. **Shortman, K. and Seligman, K.,** The separation of different cell classes from lymphoid organs. III. The purification of erythroid cells by pH-induced density changes, *J. Cell Biol.,* 42, 783, 1969.

44. **Zucker, R. M.,** Fetal erythroid cell development: density gradients and size distributions, *J. Cell. Physiol.,* 75, 241, 1970.

45. **Schulman, H. M.,** The fractionation of rabbit reticulocytes in dextran gradients, *Biochim. Biophys. Acta,* 148, 251, 1967.

46. **Gross, J., Rosenthal, S., and Syllm-Rapoport, I.,** Dichte-gradientenzentrifugation von Erythrozyten im Dextranmedium, *Acta Biol. Med. Ger.,* 26, 643, 1971.

47. **Gross, J., Hartwig, A., Botscharowa, L., Syllm-Rapoport, I., and Rosenthal, S.,** Dextran als Medium für die Dichtegradiententrennung von roten Blutzellen, *Acta Biol. Med. Ger.,* 29, 765, 1972.

48. **Gross, J., Coutelle, Ch., Schulz, W., Rosenthal, S., Syllm-Rapoport, I., and Rapoport, S.,** Dichte-gradientenzentrifugation neonataler Erythrozyten im Dextran-Medium, *Acta Biol. Med. Ger.,* 28, 615, 1972.

49. **Gross, J., Papies, B., Grauel, E. L., Hartwig, A., and Rosenthal, S.,** Leitkriterien der Retikulozyten-reifung. I. Dichte von roten Blutzellen während einer Entblutungsanämie des Kaninchens im Dichtegra-dienten, *Acta Biol. Med. Ger.,* 30, 617, 1973.

50. **Coutelle, Ch., Rosenthal, S., Gross, J., David, H., and Uerlings, I.,** Leitkriterien der Reticulocyten-reifung. VI. Verhalten der RNS- und Reticulocytenwerte sowie des Ribosomengehaltes in peripheren er-ythroiden Zellpopulationen, *Acta Biol. Med. Ger.,* 31, 781, 1973.

51. **Schmidt, G., Gross, J., Moller, R., and Staak, R.,** Trennung von roten Blutzellen im isopyknischen Dextran- und Albumingradienten, *Acta Biol. Med. Ger.,* 34, 1621, 1975.

52. **Gross, J., Rapoport, S. M., Rosenthal, S., and Syllm-Rapoport, I.,** Criteria of maturity and age of red blood cells, *Acta Biol. Med. Ger.,* 40, 665, 1981.

53. **Fitzgibbons, J. F., Koler, R. D., and Jones, R. T.,** Red cell age-related changes of hemoglobins A_{Ia+b} and A_{Ic} in normal and diabetic subjects, *J. Clin. Invest.,* 58, 820, 1976.

54. **Gross, J., Schmalisch, G., and Syllm-Rapoport, I.,** A simple method for the approximation of the behaviour of marker constituents in density-fractionated red blood cells by means of normalized cumulative distribution functions, *Biomed. Biochim. Acta,* 42, 1103, 1983.

55. **Boyd, E. M., Thomas, D. R., Horton, B. F., and Huisman, T. H. J.,** The quantities of various minor hemoglobin components in old and young human red blood cells, *Clin. Chim. Acta,* 16, 333, 1967.

56. **Piomelli, S., Seaman, C., Reibman, J., Tytun, A., Graziano, J., Tabachnik, N., and Corash, L.,** Separation of younger red cells with improved survival in vivo: An approach to chronic transfusion therapy, *Proc. Natl. Acad. Sci. USA,* 75, 3474, 1978.

57. **Corash, L. M., Piomelli, S., Cheu, H. C., Seaman, C., and Gross, E.,** Separation of erythrocytes according to age on a simplified density gradient, *J. Lab. Clin. Med.,* 84, 147, 1974.

58. **Pfeffer, S. R. and Swislocki, N. I.,** Age-related decline in the activities of erythrocyte membrane adenylate cyclase and protein kinase, *Arch. Biochem. Biophys.,* 177, 117, 1976.

59. **Piomelli, S., Seaman, C., Reibman, J., Tytun, A., Graziano, J., Tabachnik, N., and Corash, L.,** Separation of younger red cells with improved survival in viyo: an approach to chronic transfusion therapy, *Proc. Natl. Acad. Sci. U.S.A.,* 75, 3474, 1978.

60. **Rennie, C. M., Thompson, S., Parker, A. C., and Maddy, A.,** Human erythrocyte-fractionation in "Percoll" density gradients, *Clin. Chim. Acta,* 98, 119, 1979.

61. **Ellory, J. C. and Wolowyk, M. W.,** A new density gradient technique; age-separation of human eryth-rocytes and reticulocytes, *J. Physiol.,* 295, 9, 1979.

62. **Olofsson, T., Gärtner, J., and Olsson, J.,** Separation of human marrow cells in density gradients of polyvinylpyrrolidone coated silica gel (Percoll), *Scand. J. Haematol.,* 24, 254, 1980.

63. **Pertoft, H. and Laurent, T. C.,** Isopycnic separation of cells and cell organelles by centrifugation in modified colloidal silica gradients, in *Methods of Cell Separation,* Vol. 1, Catsimpoolas, N., Ed., Plenum Press, New York, 1977, chap. 2.

64. **Spooner, R. J., Percy, R. A., and Rimley, A. G.,** The effect of erythrocyte ageing on some vitamin and mineral dependent enzymes, *Clin. Biochem.,* 12, 289, 1979.

65. **Alderman, E. M., Fudenberg, H. H., and Lorius, R. E.,** Binding of immunoglobin classes to subpo-pulations of human red blood cells separated by density-gradient centrifugation, *Blood,* 55, 817, 1980.

66. **Salvo, G., Caprari, P., Samoggia, P., Mariani, G., and Salvati, A. M.,** Human erythrocyte separation according to age on a discontinuous "Percoll" density gradient, *Clin. Chim. Acta,* 122, 293, 1982.

67. **Graziano, J. H., Piomelli, S., Seaman, C., Wang, Th., Cohen, A. R., Kelleher, J. F., and Schwartz, E.,** A simple technique for preparation of young red cells for transfusion from ordinary blood units, *Blood,* 59, 865, 1982.

68. **Lombarts, A. J. P. F., Jagdewsing, J. K., Bot, A. G. M., and Leijnse, B.,** Sterile, medium scale age fractionation of human red blood cells, *Haematologia,* 17, 419, 1984.

69. **Vincent, R. and Nadeau, D.,** Adjustment of the osmolality of Percoll for the isopycnic separation of cell and cell organelles, *Anal. Biochem.,* 141, 322, 1984.

70. **Ogunmola, G. B., Dada, O. A., and Ejike, E. N.,** Isopycnic separation of human pigeon and guinea pig erythrocytes, *Acta Haematol.,* 63, 312, 1980.

71. **Danon, D. and Marikovsky, Y.,** Determination of density distribution of red cell population, *J. Lab. Clin. Med.,* 64, 668, 1964.

72. **Marikovsky, Y., Danon, D., and Katchalsky, A.,** Agglutination by polylysine of young and old red cells, *Biochim. Biophys. Acta,* 124, 154, 1966.

73. **Rapoport, S., Dubiel, W., and Müller, M.,** Proteolysis of mitochondria in reticulocytes during maturation is ubiquitin-dependent and is accompanied by a high rate of ATP hydrolysis, *FEBS Lett.,* 180, 249, 1985.

74. **Gellerich, F. N., Sprengler, V., and Augustin, W.,** A simple procedure for the enrichment of reticulocytes by isodense differential centrifugation on silicon oil mixture, *Acta Biol. Med. Ger.,* 40, 611, 1981.

75. **Kovach, J. S., Marks, P. A., Russel, E. S., and Epler, H.,** Erythroid cell development in fetal mice: ultrastructural characteristics and hemoglobin synthesis, *J. Molec. Biol.,* 25, 131, 1967.

76. **Fantoni, A., DeLaChapelle, A., Rifkind, R. A., and Marks, P. A.,** Erythroid cell development in fetal mice: synthetic capacity for different proteins, *J. Molec. Biol.,* 33, 79, 1968.

77. **Bangham, A. D., Flemans, R., Heard, D. H., and Seaman, G. V. F.,** An apparatus for microelectrophoresis of small particles, *Nature (London),* 182, 642, 1958.

78. **Gear, A. R. L.,** Age dependent separation of erythrocytes by preparative electrophoresis, *J. Lab. Clin. Med.,* 90, 744, 1977.

79. **Bocci, V.,** Determinants of erythrocyte ageing: a reappraisal, *Br. J. Haematol.,* 48, 515, 1981.

80. **Light, N. D. and Tanner, M. J. A.,** An affinity chromatography method for the preparation of human reticulocytes, *Anal. Biochem.,* 87, 263, 1978.

81. **Seligman, F. A., Allen, R. H., Kirchanski, S. J., and Natale, G. P.,** Automated analysis of reticulocytes using fluorescent staining with both acridine orange and an immuno-fluorescent technique, *Am. J. Hematol.,* 14, 57, 1983.

82. **Tannert, Ch. and Tsamaloukas, A. G.,** Enrichment of reticulocytes by lectin-mediated agglutination, *Acta Biol. Med. Ger.,* 40, 969, 1981.

83. **Phillips, S. G., Kabat, A., and Miller, O. J.,** Nylon fiber affinity selection of red blood cells and tissue culture cells on the basis of cell surface determinants, *Exp. Cell Res.,* 127, 361, 1980.

84. **Nicola, N. A., Morstyn, G., and Metcalf, D.,** Lectin receptors on human blood and bone marrow cells and their use in cell separation, *Blood Cells,* 6, 563, 1980.

85. **Borsook, H., Ratner, K., and Tattrie, B.,** Studies on erythropoiesis. II. A method of segregating immature from mature adult rabbit erythroblasts, *Blood,* 34, 32, 1969.

86. **Thompson, C. B., Galli, R. L., Melaragno, A. J., and Valeri, C. R.,** A method for the separation of erythrocytes on the basis of size using counterflow centrifugation, *J. Hematol.,* 17, 177, 1984.

87. **Pisciotto, P., Kiraly, T., Rosen, P., Paradis, L., Kakaiya, M., and Morse, E. E.,** Preparation of young red cells for transfusion using the Fenwal CS 300 cell separator, *J. Hematol.,* 17, 185, 1984.

88. **Walter, H.,** Partition of cells in two-polymer aqueous phases: a surface affinity method for cell separation, in *Methods of Cell Separation,* Vol. 1, Catsimpoolas, N., Ed., Plenum Press, New York, 1977, chap. 8.

89. **Marks, P. A. and Johnson, A. B.,** Relationship between the age of human erythrocytes and their osmotic resistance: a basis for separating young and old erythrocytes, *J. Clin. Invest.,* 37, 1542, 1958.

90. **Lowenstein, L. M.,** The mammalian reticulocyte, *Int. Rev. Cytol.,* 8, 135, 1959.

91. **Fornaini, G.,** Biochemical modifications during the life span of the erythrocyte, *Ital. J. Biochem.,* 16, 257, 1967.

92. **Rosenthal, S., Gross, J., Grauel, E. L., Papies, B., Schulz, W., Belkner, J., Botscharowa, L., Coutelle, C., Hawemann, M., Nieradt-Hiebsch, C., Müller, M., Opitz, M., Prehn, S., Schultze, M., Staak, R., and Wiesner, R.,** Leitkriterien der Retikulocytenreifung, in *VIth Internationales Symposium über Struktur und Funktion der Erythrocyten,* Rapoport, S. and Jung, F., Eds., Akademie Verlag, Berlin, 1972, 513.

93. **Wiesner, R., Rosenthal, S., and Hiebsch, C.,** Leitkriterien der Retikulocytenreifung. II. Das Verhalten von Zytochromoxydase und Hemmstoff F der Atmungskette bei der Retikulocytenreifung, *Acta Biol. Med. Ger.,* 30, 631, 1973.

94. **Thilo, Ch., Schewe, T., Belkner, J., and Rapoport, S.,** In vitro-Reifund von Kaninchenretikulozyten: Verhalten des Sauerstoffverbrauchs, *Acta Biol. Med. Ger.,* 38, 1431, 1979.

95. **Clari, G., Michielin, E., Zen, F., and Moret, V.,** Maturation-related change of the protein kinase activity in the rabbit red blood cells, *Biochim. Biophys. Acta,* 677, 403, 1981.

96. **Malkin, A. and Denstedt, O. F.,** The metabolism of the erythrocyte. X. The inorganic pyrophosphatase of the erythrocyte, *Can. J. Biochem. Physiol.,* 34, 121, 1956.

97. **Scheuch, D. and Rapoport, S.,** Biologische Dynamik der anorganischen PP-ase im Verlauf einer Entblutungsanämie, *Acta Biol. Med. Ger.,* 6, 23, 1961.

98. **Mai, A., Sandring, D., Belkner, J., Prehn, S., and Rapoport, S.,** In vitro-Reifung von Retikulozyten. Verhalten von RNS und anorganischer Pyrophosphatase, *Acta Biol. Med. Ger.,* 39, 217, 1980.

99. **Fisher, R. A., Turner, B. M., Dorkin, H., and Harris, H.,** An investigation of some factors affecting levels of erythrocyte inorganic pyrophosphatase activity, *Clin. Chim. Acta,* 61, 27, 1975.

100. **Gauger, D., Palm, D., Kaiser, G., and Quiring, K.,** Adenylate cyclase activities in rat erythrocytes during stress erythropoiesis: localization of the enzyme in the rat reticulocyte, *Life Sci.,* 13, 31, 1973.

101. **Babu, C. R., Azhar, S., and Murti, C. R. K.,** Loss of epinephrine stimulated synthesis of cyclic adenosine 3':5' monophosphate during maturation of rabbit and human reticulocytes, *Med. Biol.,* 53, 148, 1975.

102. **Tsamaloukas, A. G., Maretzki, D., Setchenska, M., and Rapoport, S.,** Maturation dependence of fluoride-sensitive adenylate cyclase in red blood cells, *Acta Biol. Med. Ger.,* 35, 523, 1976.

103. **Pfeffer, S. R. and Swislocki, N. I.,** Age-related decline in the activities of erythrocyte membrane adenylate cyclase and protein kinase, *Arch. Biochem. Biophys.,* 177, 117, 1976.

104. **Bilezikian, J. P.,** Dissociation of beta-adrenergic receptors from hormone responsiveness during maturation of the rat reticulocyte, *Biochim. Biophys. Acta,* 542, 263, 1978.

105. **Piau, J. P., Fischer, S., and Delauney, J. G.,** Maturation dependent decline of adenylate cyclase in rabbit red blood cell membranes, *Biochim. Biophys. Acta,* 544, 482, 1978.

106. **Piau, J. P., Delauney, J., Fischer, S., Tortolero, M., and Schapira, G.,** Human red cell membrane adenylate cyclase in normal subjects and patients with hereditary spherocytosis sickle cell disease and unidentified hemolytic animals, *Blood,* 56, 963, 1980.

107. **Limbird, L. E., Gill, D. M., Stadel, J. M., Hickey, A. R., and Lefkowitz, R. J.,** Loss of β-adrenergic receptor-guanine nucleotide regulatory protein interactions accompanies decline in catecholamine responsiveness of adenylate cyclase in maturing rat erythrocytes, *J. Biol. Chem.,* 225, 1854, 1980.

108. **Carner, A. C. and Ross, E. M.,** Alteration in the protein components of catecholamine sensitive adenylate cyclase during maturation of rat reticulocytes, *J. Biol. Chem.,* 256, 9551, 1981.

109. **Nakagawa, M., Willner, J., Cerri, C., and Reydel, P.,** The effect of membrane preparation and cellular maturation on human erythrocyte adenylate cyclase, *Biochim. Biophys. Acta,* 770, 122, 1984.

110. **Hindman, H. B. and Knox, W. E.,** Guanase in reticulocytes and other rat tissues, *Enzyme,* 23, 395, 1978.

111. **Quiring, K., Kaiser, G., and Gauger, D.,** Monoamine oxidase activity in rat erythrocytes: evidence for its localization in reticulocyte mitochondria, *Experientia,* 32, 1132, 1976.

112. **Quiring, K., Kaiser, G., and Gauger, D.,** Activities of membrane-bound and soluble catechol-O-methyltransferase in premature and mature erythrocytes from rats, *Experientia,* 31, 1011, 1975.

113. **Dubrul, E. F. and Farkas, W. R.,** Partial purification and properties of the reticulocyte guanylating enzyme, *Biochim. Biophys. Acta,* 442, 379, 1976.

114. **Kobayashi, Y., Yoshimitsu, Y., Okahata, S., and Usui, T.,** Superoxide dismutase activity in rabbit reticulocytes, *Experientia,* 39, 69, 1983.

115. **Griffiths, W. J. and Fitzpatrick, M.,** The effect of age on the creatine in red cells, *Br. J. Haematol.,* 13, 175, 1967.

116. **Griffiths, W. J. and Lothian, E. J.,** Erythropoiesis, red-cell creatine and plasma aldolase activity in anemia in the rabbit and man, *Br. J. Haematol.,* 17, 477, 1969.

117. **Fehr, J. and Knob, M.,** Comparison of red cell creatine level and reticulocyte count in appraising the severity of hemolytic processes, *Blood,* 53, 966, 1979.

118. **Syllm-Rapoport, I., Daniel, A., Starck, H., Götze, W., Hartwig, A., Gross, J., and Rapoport, S.,** Creatine in red cells: transport and erythropoietic dynamics, *Acta Biol. Med. Ger.,* 40, 653, 1981.

119. **Syllm-Rapoport, I., Daniel, A., Starck, H., Hartwig, A., and Gross, J.,** Creatine in density fractionated red cells, a useful indicator of erythrocytic dynamics and of hypoxia past and present, *Acta Haematol.,* 66, 86, 1981.

120. **Brewster, M. A. and Berry, D. H.,** Detection of G6PD and pyruvate kinase deficiencies in reticulocytosis by reference to erythrocyte creatine, *Clin. Biochem.,* 14, 132, 1981.

121. **Smith, B. J., Mohler, D. N., Wells, M. R., and Savory, J.,** Erythrocyte creatine levels in anemia, *Am. Clin. Lab. Sci.,* 12, 439, 1982.

122. **Cooper, K. D., Shukla, J. B., and Rennert, O. M.,** Polyamine distribution in cellular compartments of blood and in aging erythrocytes, *Clin. Chim. Acta,* 73, 71, 1976.

123. **Tokunaya, A.,** Determination of amino acids and polyamines in human erythrocytes. I. Fundamental studies, *Hiroshima J. Med. Sci.,* 32, 113, 1983.

124. **Rapoport, S., Guest, G. M., and Wing, M.,** Size, hemoglobin content, and acid-soluble phosphorus of erythrocytes of rabbits with phenylhydrazine-induced reticulocytosis, *Proc. Soc. Exp. Biol. Med.,* 57, 344, 1944.

125. **Cohen, N. S., Ekholm, J. E., Luthra, M. G., and Hanahan, D. J.,** Biochemical characterization of density separated human erythrocytes, *Biochim. Biophys. Acta,* 419, 229, 1976.

126. **Farkas, W. R., Hankins, W. D., and Singh, R. D.,** The guanylation of transfer RNA: an enzymatic reaction, *Biochim. Biophys. Acta,* 294, 94, 1973.

127. **Vulberg, L. S., Card, R. T., Paulson, E. J., and Szirek, J.,** Alterations in cellular sodium, potassium, calcium, magnesium, copper and zinc levels during the development and maturation of erythrocytes in the rabbit, *Br. J. Haematol.*, 13, 115, 1967.
128. **Ginsburg, S., Smith, J. G., Ginsburg, G. M., Reardon, J. Z., and Aikawa, J. K.,** Magnesium metabolism of human and rabbit erythrocytes, *Blood*, 20, 722, 1962.
129. **Guest, G. M. and Rapoport, S.,** Organic acid-soluble phosphorus compounds of the blood, *Physiol. Rev.*, 21, 410, 1941.

Metabolism of Reticulocytes

Chapter 6

ENERGY AND CARBOHYDRATE METABOLISM

I. INTRODUCTION

Corresponding to its intermediate position in the differentiation program, the reticulocyte no longer exhibits the full range of metabolic pathways of proliferating cells, since some pathways are lost with the disappearance of the nucleus, endoplasmic reticulum, and Golgi apparatus. On the other hand, it is still equipped with a set of metabolic pathways, most of which are lost during its transition to the mature erythrocyte. A survey of the metabolic pathways in the reticulocyte, as compared to those in the erythrocyte, is presented in Table 10.

The following chapter will deal with the various metabolism pathways in some detail. In addition, the balance of ATP-producing and -consuming processes will be discussed. Some aspects which are germane to a discussion of metabolism, the transport of metabolites and cations, have been presented in Chapter 4.

The reticulocyte is characterized by the existence of an extensive supply of endogenous substrates for its oxidative metabolism, and for the synthesis of protein, primarily hemoglobin. The source of the endogenous substrates is mainly mitochondria, but also ribosomes which are degraded during maturation. Thus, a restructuring of the cell occurs, the "Umbau" of classic cell biology. At the same time external substrates are utilized. Under in vivo conditions, i.e., in the peripheral blood, two main substrates are to be considered: glucose and glutamine, the concentration of which is 5 mM and 0.4 to 0.7 mM, respectively, in the blood plasma.[1,2] Alanine, with a plasma concentration of 0.2 to 0.3 mM, may also participate significantly in the external substrate mixture.[3] In Figure 14, a schematic diagram illustrates the sources and fates of the metabolic substrates of the reticulocyte.

II. OXYGEN UPTAKE AND RESPIRATION

The oxygen uptake of reticulocytes varies according to stage of maturation and during the course of anemia. Furthermore, there are some differences between species, the most striking being the reticulocytes in man which usually have low oxygen consumption.[4] This situation is related to the low activity of cytochrome oxidase (see Table 1 in Chapter 3). A contributory factor is their usually high degree of maturity.

Oxygen consumption of rabbit reticulocytes ranges from 15 to 50 μM O_2 per mℓ cells per hour.[5-8] Similar values have been observed in rats and sheep.[9,10] Even chicken reticulocytes correspond to mammalian species[11,12] in O_2 consumption. Thus, the conclusion that the reticulocytes are generally characterized by a specific range of oxygen consumption in many species appears to be justified. In contrast, there is much greater variation in the oxygen consumption of mature erythrocytes. If reticulocytes, platelets, and white blood cells are scrupulously excluded, their oxygen consumption amounts to less than 1 μM per mℓ cells per hour. Nucleated mature erythrocytes have higher values ranging from 5 to 10 μM O_2 per mℓ cells per hour.[11]

About 80% of the oxygen consumption of reticulocytes is accounted for by mitochondrial respiration.[7,13] The remainder represents antimycin A-resistant O_2 consumption, which is the result of several processes. About one fourth is accounted for by lipoxygenase-mediated oxygenation of polyenic fatty acids.[14] However, the amount of O_2 consumed is severalfold greater than that calculated for the hydroperoxide formation; secondary O_2-consuming reactions of the primary products of lipoxygenase have to be assumed, which may be both

Table 10

COMPARISON OF MAIN CONSTITUENTS AND METABOLIC
PATHWAYS BETWEEN RETICULOCYTES AND ERYTHROCYTES IN
RABBITS

Pathway	Reticulocyte	Erythrocyte	Remarks
RNA synthesis	−	−	
Protein synthesis	+	−	
Heme synthesis	+	−	Some enzymes remain
Purine synthesis	+	−	Some enzymes remain
Pyrimidine synthesis	+	−	Some enzymes remain
Glycogen synthesis	−	−	Some enzymes remain
Glycogenolysis	−	−	Some enzymes remain
Lipid synthesis	(+)	−	Only phosphatidic acid synthesized in erythrocytes
Respiration	+	−	
Citrate cycle	+	−	Some enzymes remain
Glycolysis	+ +	+	
Gluconeogenesis	−	−	
OPP pathway	+	+	
Fatty acid breakdown	+	−	
Amino acid catabolism	+	−	Some enzymes remain
Active transport amino acids	+ +	(+)	Some carriers remain
Active transport cations	+ +	+	
Receptors	+ +	(+)	Some remain
Adenyl cyclase	+	−	

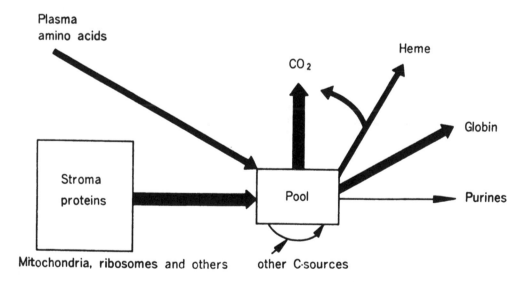

FIGURE 14.   Influx and efflux of the reticulocyte amino acid pool.

enzymatic and nonenzymatic.[15] Another one fourth of antimycin A-resistant O_2 consumption is inhibited by 3-amino-1,2,4-triazole. The point of attack of this inhibitor has not been defined. Further O_2-consuming processes could possibly be extra-mitochondrial flavoproteins, such as NADH-cytochrome b_5 reductase, NADPH-oxidases, and nonenzymatic processes such as autoxidation of hemoglobin and glutathione. Antimycin A-resistant O_2 uptake is accompanied by CO_2 formation. A survey of various labeled substrates was carried out.

It was found that the $^{14}CO_2$ production from fatty acids was completely inhibited, whereas $^{14}CO_2$ formation from substances such as [1-^{14}C]-glucose, ^{14}C-formate, and [3-^{14}C]-serine, which can be oxidized by NADP-dependent enzymes, was not inhibited at all. An unexpected and highly interesting result was that metabolites of the citrate cycle such as [5-^{14}C]-oxo-glutarate, [1,4-^{14}C]-succinate, and [1-^{14}C]-pyruvate contribute significantly to CO_2 formation in the presence of antimycin A. For an explanation of these results it must be assumed that reducing equivalents must be transferred from the mitochondrial matrix to the cytosol in order to be oxidized there.[16] Antimycin A-resistant O_2 uptake represents a constant percentage of total O_2 consumption under all circumstances and is thus strictly correlated with the reticulocyte count.

Oxygen consumption of the reticulocyte is extraordinarily stable and continues for many hours even without any external substrate, decreasing slowly with a half-life of about 10 hr.[17] The addition of glucose causes a small decrease of about 10%, the so-called "Crabtree effect".[17-20] Other substrates also produce minor effects. With oxaloacetate, an increase in O_2 consumption by about 10% was observed.[18]

The degree of respiration uncoupling may be determined using the effect of oligomycin, a specific inhibitor of coupled respiration.[21] It amounts to about 15% on the 6th day of a bleeding anemia; at that time the reticulocyte population represents cohorts of differing stages of maturity. In experiments on maturity-separated reticulocytes, the degree of uncoupling was higher in the most mature cohort.[22]

During in vitro maturation, both O_2 consumption and the degree of uncoupling change drastically. Respiration decreases to one half within 12 hr, while the degree of uncoupling increases to about 50%.[21] There was also a decrease in antimycin A-resistant O_2 consumption. The presence of elevated concentrations of inorganic phosphate does not affect the oxygen consumption of reticulocytes, which indicates that its concentration does not contribute significantly to the control of respiration.[23] However, in vitro maturation in a phosphate-rich medium — conditions under which the reticulocyte count decreases more rapidly than in a medium with a physiological phosphate concentration — results in accelerated decline of respiration, which decreases to one third in 12 hr and in a higher degree of uncoupling which reaches 70% during the same period.[21]

Addition of uncouplers such as dinitrophenol, CCCP (carbonyl cyanide *m*-chlorophenyl-hydrazone), and FCCP (carbonyl cyanide *p*-trifluoromethoxyphenylhydrazone) produce a large increase in O_2 consumption of more than twofold.[8,17,21,24,25] By titrations with inhibitors of specific steps of the respiratory chain and oxidative phosphorylation, it is possible to assess their contribution to the control of the respiration pathway. This contribution can be quantified on the basis of the theories of Heinrich and Rapoport[26-28] and Kacser and Burns.[29]

These investigators defined the control coefficient as a measure of the importance of a given enzyme for the control of the flux through the system. The control coefficient is defined as the differential relative change of the metabolic flux resulting from a relative change in the activity of an enzyme. The sum of the control coefficients adds up to 1, regardless of the value of any single control coefficient. With this methodology it was found that by far the highest control coefficient, 0.53, is exhibited by the proton leak of the inner mitochondrial membrane, which is the point of attack of the uncouplers according to their mode of action as ionophores.[25] On the other hand, the mitochondrial ATPase system contributes rather little to the control of respiratory flux with a control coefficient of 0.14. The contribution of other components of the system, approximately one third, remains to be determined. The estimates are subject to various corrections, but still give an approximate picture of the control of respiration in the rabbit reticulocyte. With inhibition of ATPase by oligomycin nearly all of the control is exerted by the proton leak with a control coefficient of 0.97. Under these experimental conditions the proton gradient is the decisive parameter. During in vitro maturation, the effect of the protonophores disappears, as is to be expected from the increased degree of uncoupling.[21]

Addition of respiratory substrates, in particular, oxaloacetate, in the presence of dinitrophenol leads to a further increase in O_2 consumption which remains elevated for a much longer period of time.[19-21] Thus, one may conclude that under normal coupled conditions only about 40% of the respiratory chain capacity is utilized and protonophores alone fall short of producing maximal O_2 consumption. Therefore, under conditions of uncoupling the supply of metabolic substrate contributes a large proportion to the control of respiration.

Respiratory properties of isolated mitochondria from reticulocytes have been delineated.[30-32] An inventory of cytochromes and enzymes was discussed in Chapter 3. Their functional properties correspond to the expectations derived from studies on intact cells. The isolated mitochondria are well coupled with a respiratory control quotient of about 4. Their "control characteristics", i.e., the dependence of oxidative phosphorylation on the extramitochondrial ATP/ADP ratio, corresponds in principle to that observed for mitochondria of other cells.[8,33,34] The presence of mitochondrially bound hexokinase modifies these characteristics somewhat, since the enzyme exhibits an apparent preference for the ATP formed by the mitochondria, thereby shifting the control range to higher ATP/ADP ratios.[8] There is, however, some discrepancy between the calculated rate of respiration derived from the content of cytochrome oxidase and the measured overall ATP/ADP ratio in intact reticulocytes, on the one hand, and the actual values of respiration measured, on the other. One possible reason may be that part of the ADP measured is sequestered, for instance, by binding to proteins.

III. CARBOHYDRATE METABOLISM

Glucose is almost the only physiological substrate of the metabolism of mature erythrocytes, except that in its absence, ribose from the breakdown of adenine nucleotides, 2,3-DPG and other intermediates of glycolysis, may serve as sources of ATP and lactate until depleted.[35] At physiological pH values, about 90% of glucose is utilized by the glycolytic pathway, with the 2,3-biphosphoglycerate bypass contributing about 25%.[36,37] The oxidative pentose pathway amounts to 10%.[38-41] Its share is pH-dependent in an inverse manner and its magnitude is primarily limited by the NADP concentration.[41,42] In the reticulocyte, glucose is utilized as both a respiratory substrate and in glycolysis; in addition, it furnishes the carbon skeleton for the formation of serine and glycine (see Chapter 8). Oxidation of glucose via the oxidative pentose pathway amounts to only 2%. Previous high estimates were shown to be erroneous as isotopic dilution of glucose breakdown intermediates in the Embden-Meyerhof pathway, which form large pools, was neglected.[6] The oxidative pentose pathway can be greatly stimulated by agents which increase the concentration of $NADP^+$, e.g., methylene blue; these cause a higher state of oxidation of cytosolic pyridine nucleotides without any uncoupling of oxidative phosphorylation.[43] Under the influence of the dye, $^{14}CO_2$ formation from glucose via the oxidative pentose pathway may increase 30-fold without any effect on the Embden-Meyerhof pathway.

The glycolytic capacity of the reticulocyte is about 40 to 50 times higher than that of the erythrocyte, but is only expressed if respiratory ATP production is suppressed, be it by inhibitors or protonophores.[25,38]

The extent of aerobic glycolysis depends on the maturity of reticulocytes and stage of anemia. On the 4th or 5th day of bleeding anemia of rabbits, at which time immature reticulocytes with high respiration predominate, aerobic glycolysis may be absent, while in late stages of anemia as much as 30 μmol of lactate per milliliter of cells per hour may be produced — almost equal to the anaerobic glycolytic rate of the same samples.[20,44] From the 6th to 8th day of bleeding anemia, an average of 10 μM lactate per milliliter of reticulocytes · hour is produced. Similar relations have been found in rat reticulocytes with an aerobic lactate production of 3 μmol/mℓ cells · hour, which rises tenfold if oxidative phosphorylation is inhibited.[45]

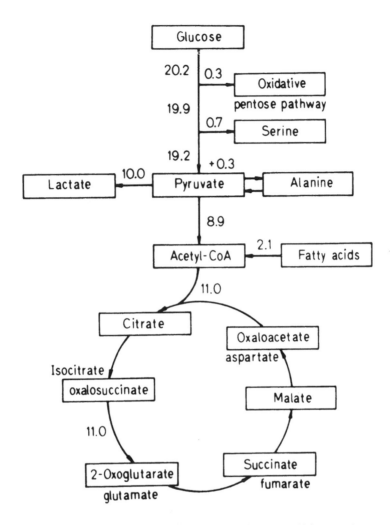

FIGURE 15. Balance sheet of glucose utilization by rabbit reticulocytes. Values are given as mmol C_3 units $\times$ 1ℓ reticulocyte^{-1} $\times$ hr^{-1}. (From Siems, W. et al., *Eur. J. Biochem.*, 124, 567, 1982. With permission.)

In contrast to the erythrocyte, glucose utilization of the reticulocyte is dependent on glucose concentration over a wide range. This effect is due to the action of glucokinase, which has a high K_m — in the range of 30 mM — for glucose.[46] Glucose, if present as the sole external substrate, constitutes the main substrate of energy metabolism.[6] By means of a new mathematical technique the processes utilizing glucose can be estimated. The overall balance sheet of glucose utilization and ATP generation of the reticulocyte is presented in Figure 15.

About 45% of glucose is catabolized via the citrate cycle, and about the same percentage yields lactate. The formation of serine amounts to about 5% of the glucose utilized. Of the total ATP produced, which amounts to 120 to 170 mmol/ℓ cells · hour, glycolysis yields about 10%, while 90% is derived from respiration. Of this, about five sixths is furnished by glucose oxidation and the remainder by fatty acids. With glucose as the external substrate, amino acids contribute little to energy metabolism (Table 11).

Of course, matters are different if no exogenous substrate is present. Under these conditions the main carbohydrate utilized is ribose phosphate, which originates from two sources: first, from the breakdown of ribonucleic acids, a process which also occurs in the presence of

Table 11

THE FORMATION OF ATP BY RETICULOCYTES[a]

Method	ATP formation $(mmol/\ell^{-1}/hr^{-1})$
Coupled oxygen consumption (22.4 mmol/$1\ell^{-1}$/hr^{-1}) + glycolysis	134.4
	+ 14.1
	148.5
Metabolic fluxes of reactions:	
(1) Glucose → glucose 6-phosphate	− 10.1
(2) Fructose 6-phosphate → fructose 1.6 bisphosphate	− 10.0
(3) 1.3-Bisphosphoglycerate → 3-phosphoglycerate (20% via 2.3 P$_2$-glycerate)	+ 15.9
(4) Phosphoenolpyruvate → pyruvate	+ 19.2
(5) Oxidation of NADH stemming from glycolysis (glyceraldehyde 3-phosphate → 1.3-bisphosphoglycerate → pyruvate → lactate)	+ 19.8
(6) Oxidation of NADH stemming from the reaction pyruvate → CoASAc	+ 17.7
(7) Oxidation of NADH and reduced FAD stemming from the splitting of the fatty acids	+ 7.0
(8) Citrate cycle yielding GTP and H$_2$ oxidized in the respiratory chain	+ 91.7
Total reactions	+ 150.3 (100%)
Contribution of glycolysis (Reactions 1,2,3, and 4	+ 14.1 (9.4%)
Contribution of the fatty acid degradation (Reactions 7 and 8)	+ 24.5 (16.3%)

[a]   ATP formation (mmol × 1ℓ reticulocytes^{-1} × hr^{-1}) was calculated first from oxygen consumption and then from metabolic fluxes.

Modified from Siems, W. et al., *Eur. J. Biochem.*, 124, 567, 1982.

external substrate, and, second, from the degradation of low-molecular nucleotides, mainly ATP. From early data on the ratio of $^{14}CO_2$ formation from (2-^{14}C)-glucose as compared with that from (6-^{14}C)-glucose, a rate of recombination from pentose phosphates to hexose phosphate of about 10% of glucose utilization was estimated.[47] It is very likely that this value is an overestimate; nevertheless, the conclusion may be drawn that ribose may be metabolized, like glucose, both by respiration and glycolysis. Lactate formation amounts to about 4 to 5 μmol/cells · hour. Among the endogenous substrates, amino acids surely play a major role, expecially those related to the citric cycle such as glutamate, aspartate, and alanine.[48] Glycine also contributes via the Shemin cycle (see Chapter 8). According to Borsook,[49] amino acid contribution to the metabolic substrate mixture is increased in the presence of a medium containing amino acids. A particularly good substrate is glutamine.[48]

Glycerol is a substrate closely related to the carbohydrates. It contributes to the formation of both CO_2 and lactate.[45,50] The production of $^{14}CO_2$ from labeled glycerol proceeds in a linear manner as a function of substrate concentration and may furnish a large part of the lactate, even in the presence of glucose. From these data several conclusions follow: first, that glycerol kinase is present and active; in keeping with this inference glycerol-3-phosphate was found if glycerol was offered as a substrate, a compound missing in its absence. The glycerol kinase appears to be a limiting step in the utilization of glycerol. Second, an active dihydroxyacetate phosphate dehydrogenase must be postulated which indeed has been found in reticulocytes.[51]

Neither the reticulocyte nor the erythrocyte has significant glycogen stores. This condition does not appear to be due to the absence of enzymatic equipment that forms or degrades glycogen, but rather to the predominance of the degrading enzymes, i.e., phosphorylase and amylo-1,6-glucosidase over UDPG-glycogen synthase. This enzyme exists only in a strictly glucose-6-phosphate-dependent form with a high K_d for it.[52-55] The incorporation of labeled glucose occurs mostly in the outer branches of the glycogen, which indicates that glucan-(1,4-1,6)-glycosyl transferase has only little activity. The rate of incorporation was considerably higher in the cell fraction containing reticulocytes and young erythrocytes as compared to the average.

As is to be expected in erythrocytes with deficient phosphorylase and amylo-1,6-glucosidase, accumulation of glycogen is found.[53,56,57] Phosphoglucomutase, which mediates an essential step in the pathways of both synthesis and glycogen breakdown, is present in red cells regardless of their maturity.

IV. THE RELATION BETWEEN AEROBIC AND ANAEROBIC METABOLISM OF GLUCOSE: THE PASTEUR EFFECT

Under aerobic conditions, the mechanism by which both glucose consumption and, even more so, the formation of lactate is suppressed has been a central problem of intermediary metabolism for a long time. The weight of evidence indicates that a key role in the control of glucose utilization is played by phosphofructokinase.[58] This enzyme is subject to the action of a variety of effectors, the most important inhibitors being ATP, and in respiring cells, citrate, while AMP and inorganic phosphate are the most influential positive effectors. Phosphofructokinase control on glucose consumption is exerted by product inhibition of hexokinase via glucose-6-phosphate. Since glucose-6-phosphate and fructose-6-phosphate are always near equilibrium (because of high activity of phosphohexoisomerase), the two compounds may be considered as an intermediate-couple common to both hexokinase and phosphofructokinase. In addition, ATP concentration in erythrocytes and reticulocytes is in the hexokinase control range.[33,59]

Inhibition of ATP formation from respiration, be it by protonophores or inhibitors of the respiratory chain, results in a decrease in ATP with a reduced ratio of ATP/ADP which drops from values of 10 or more to values of less than 4, and even to 2.[25,45,60] At the same time, the concentrations of AMP and inorganic phosphate increase considerably. The concentration of ammonium, which is also a positive effector of phosphofructokinase (and arises mostly from deamination of nucleotides) increases under these conditions.[17,61] Thus there is a rise of all positive effectors combined with a decrease of the negative effector ATP. In a study on the effects of inhibitors of respiratory ATP formation, a close inverse relation between the concentration of glucose-6-phosphate and the rate of lactate formation was found, which may be taken as strong evidence for the paramount role of phosphofructokinase in the control of glucose consumption.[62]

The effects of increased pH in the medium support the importance of phosphofructokinase. At pH 8.5, at which the inhibition of phosphofructokinase by ATP is almost completely released, aerobic lactate formation reaches 90% of the anaerobic rate at the same pH. As mentioned previously, oxygen consumption in the presence of glucose is somewhat lower than without substrate with a tendency for higher ATP/ADP ratios, which again is in keeping with the role of phosphofructokinase.

A subsidiary role may be played by the transfer of adenine nucleotides from mitochondria into the cytosol and vice versa.[60] Nevertheless, the concentration of adenine nucleotides is higher in mitochondria than in the cytosol. Anaerobic glycolysis may be considered as a compensatory mechanism for safeguarding ATP production. How efficient is it in the reticulocyte? From various data it is quite clear that glycolytic ATP production can reach aerobic values under special conditions but usually falls short of it with a consequent breakdown of adenine nucleotides.[38,62]

V. BALANCE OF ATP PRODUCTION AND CONSUMPTION

Generaliy, our knowledge about the quantitative aspects of ATP-producing and -consuming

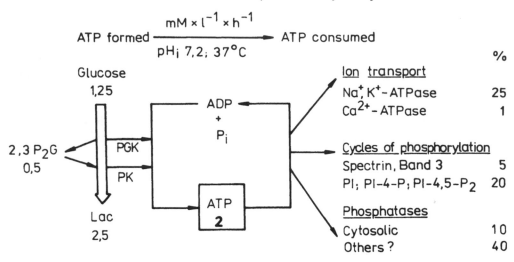

FIGURE 16.   ATP formation and consumption of erythrocytes. The values are given in mmol $\times \ell^{-1} \times hr^{-1}$ at pH_i 7.2 and 37°C.

cell processes is incomplete. The uncertainties exist both in assessment of the shares of various pathways of ATP production, and even more so in ATP consumption. The extent and type of interactions among ATP-consuming processes is an open question as yet. There is a mechanistic, theoretical understanding with experimental verification of the extent and regulation of ATP production for the mature erythrocyte.[27,63-66] Production and consumption of ATP are closely geared to each other. From a functional point of view, consumption governs ATP formation in any type of cell. Therefore it is essential to identify and quantify various ATP-consuming processes of the cell.

Studies on the erythrocyte revealed that ATP consumption in the steady state is mainly restricted to the cell membrane.[66,67] The kinetics of ATP consumption by erythrocytes in a glucose-free medium permitted an assessment of the relation between ATP concentration and rate of ATP breakdown. It was found that rate of overall ATP consumption declines approximately exponentially with time, decreasing to one half at an ATP concentration of 0.8 mM. One may therefore assume that the bulk of ATP-consuming processes has low affinity for ATP. Since most ATP-consuming reactions thus far described, e.g., transport ATPases and phosphokinases, have high affinity for ATP with K_m values of less than 0.1 mM, reactions with high K_m values had to be looked for.

In Figure 16, a balance sheet of ATP formation and consumption in human erythrocytes is presented. It is evident that Na^+K^+-ATPase contributes about 25% to ATP consumption, whereas the Ca^{2+}-ATPase share amounts to only 1% despite its high capacity, brought about by slow permeation of Ca^{2+} through the cell membrane. Phosphorylation and dephosphorylation cycles presumably contribute another one fourth to ATP consumption. A large share is accounted for by phospholipid turnover and is practically limited to the di- and triphosphatidyl-inositols. A study of ATP dependence on the rate of phosphate incorporation in inositol phosphatides indicates that their phosphorylation belongs to the class of processes with low ATP affinity and a K_m value of 0.4 mM ATP.[68] The phosphorylation and dephosphorylation of proteins appear to account only for about 5% of the ATP-consumption, with spectrin being most prominent. About 10% is contributed by a variety of cytosolic phosphatases so that in all, 40% of ATP consumption still remains undetermined and is the object of current research.

As for the reticulocyte, there is a fairly wide variation in both rates of oxygen consumption

Table 12
BALANCE OF ATP PRODUCTION AND CONSUMPTION IN
RETICULOCYTES[a]

Medium	Amino acids + glucose	%	Glucose	%
ATP production	167.2 ± 11.5		134.2 ± 10.3	
ATP consumption				
Globin synthesis	40.5 ± 7.3	24.2 ± 4.4	18.3 ± 3.4	13.6 ± 2.5
Proteolysis	21.6 ± 3.1	12.9 ± 1.9	17.9 ± 4.9	13.3 ± 3.7
Na$^+$K$^+$ATPase	30.0 ± 6.0	17.9 ± 3.6	22.6 ± 6.6	16.8 ± 4.9
Ca^{2+}ATPase	3.6 ± 0.8	2.2 ± 0.4	3.6 ± 0.8	2.7 ± 0.6
Balance consumption	95.7 ± 9.9	57.3 ± 5.6	62.4 ± 8.9	46.6 ± 4.1
Unaccounted for	71.5	42.7	71.8	53.4

[a] The values are given in mmol $\times \ell^{-1} \times$ hr^{-1}. Mean values are of 30 experiments.

and glycolysis, depending on the various factors which were previously mentioned. Thus, in order to draw a valid balance it is imperative to determine both production and consumption of ATP in the same sample of cells. The respiratory ATP production, calculated from O_2 consumption, has to be corrected for both the antimycin A-resistant portion as well as for the degree of uncoupling, while the glycolysis portion requires an adjustment for the 2,3-DPG bypass. Coupled oxygen uptake amounts to about 65% of total O_2 consumption. The balance between ATP production and ATP-consuming processes was determined for two metabolic states of reticulocytes, first in a medium containing glucose only and, second, in an enriched medium containing amino acids and iron salts (see Tables 11 and 12).

Several specific ATP-consuming processes were determined by using, for the most part, two independent methods. These included protein synthesis, ATP-dependent proteolysis, ATP-dependent transport of Na$^+$K$^+$ and Ca^{2+}, and heme synthesis. The indirect method was based on the effect of selective inhibitors of specific processes on coupled respiration, i.e., cycloheximide for protein synthesis, ouabain for Na$^+$K$^+$ATPase, and La^{3+} for Ca^{2+} transport. The estimates obtained were supplemented by direct determination of protein synthesized by incorporating (4,5-^{3}H)-lysine as well as by changing lysine concentration. The rate of proteolysis was also determined in two ways: liberation of lysine from the stroma, and hydrolysis of ATP.[7,69]

Coupled oxygen consumption and therefore ATP synthesis were 25% higher in the enriched medium. Out of a total of 167 mmol ATP produced per liter of reticulocytes per hour, about 40 mmol, ie., 27%, was used for protein synthesis and about half as much was used for ATP-dependent proteolysis. Furthermore, a large contribution of about 30 mmol/ℓ reticulocytes · hour was accounted for by Na$^+$K$^+$-ATPase. Ca^{2+} transport was distinctly higher than in erythrocytes, but was still a minor process, accounting for less than 3% of ATP consumption. An even smaller share is contributed by heme synthesis. Overall, nearly 60% of ATP production could be accounted for by well-defined processes. In the glucose medium, the main difference was much lower protein synthesis which amounted, with 18 mmol ATP per liter reticulocytes per hour, to less than half of that in the amino acid-enriched medium after 1 hr. This difference was progressive with time. Whereas in the enriched medium, the rate of protein synthesis was nearly linear with time, in the absence of amino acids in the medium, protein synthesis fell from 60% after the first 30 min to less than 25% in the second 30 min of incubation. Of the 30-mmol difference in ATP consumption between the two media, three fourths is accounted for by protein synthesis and one fourth by Na$^+$K$^+$-ATPase. This condition is related to Na$^+$ dependence of most amino acid transports. ATP-dependent proteolysis did not differ significantly.

The balance reported thus far is obviously incomplete. The most important omission are cell membrane processes involving both phospholipids and proteins. In addition, the energy expense of cell membrane movements will have to be considered.

A further conclusion can be drawn. Each of the ATP-consuming processes studied appeared to control ATP production in an independent manner. For example, inhibition of protein synthesis led to a corresponding reduction in ATP formation, while stimulation of protein synthesis resulted in an appropriate increase of ATP production. Similarly, inhibitions of ATP-dependent proteolysis or ion transport led to stoichiometric decreases of coupled oxygen consumption. Obviously there is no competition amoung ATP-consuming processes since the elimination or increase of a single ATP-consuming process would result in a compensatory change in the others.

Attempts to test this conclusion under conditions of reduced ATP supply met with complications. Reduction in ATP formation to 60% was achieved by inhibition of the respiratory chain so that only glycolysis served as its source. Both protein synthesis and proteolysis were reduced disproportionately, although ATP concentration decreased only by one third, possibly because of inhibitor effects arising from the breakdown of adenine nucleotides.[70]

There are some indications that reticulocyte respiration and glycolysis respond differentially to changes in ATP demand. Most of the variations in experimentally imposed ATP consumption affected respiration almost exclusively. These results may be explained by the fact that the respiratory function of mitochondria, but not glycolysis, is closely regulated by the ATP/ADP ratio of the cytosol. A predominant response in lactate formation by glycolysis — which may increase up to fivefold without significant change in O_2-uptake — has recently been observed in studies of the metabolic effects of isoprenalin on rat and rabbit reticulocytes.[9] The adrenergic agonist produced massive enhancement of cAMP concentration concomitantly with some decrease in ATP. It may be assumed that the change in the balance of positive and negative effectors of phosphofructokinase was responsible for the specific metabolic response which was limited to the cytosol.

In conclusion, it would appear that the reticulocyte represents an excellent model for studying the interplay of ATP-producing and -consuming reactions, which may help to elucidate this basic problem in cell biology.

Finally one may consider the dramatic changes in energy production and ATP concentration during maturation of the reticulocyte. ATP production in the reticulocytes of many mammalian and avian species probably exceeds $10 \text{ mmol} \cdot \ell^{-1} \cdot hr^{-1}$ and declines by two or even three orders of magnitude during maturation. ATP concentration, on the other hand, reflects these changes to only a small degree. It decreases from about 3 to 1 mM in those mammalian species, the mature erythrocytes of which contain 2,3-P_2G in sizable amounts.

In ruminants ATP concentrations may be lower both in reticulocytes and erythrocytes (see Table 4 in Chapter 3). Low ATP levels in mature erythrocytes go hand in hand with weak or nearly absent Na^+K^+-ATPase activity with a consequent loss of the ionic gradients between red cells and blood plasma. One may conclude that for the mature erythrocyte, the maintenance of the intracellular milieu with its high K^+ concentration is of little importance for its vital functions. A remarkable case of phylogenetic adaptation can be observed in monotremes. The red cells of platypus and echidna exhibit amazingly low ATP concentrations of 0.06 and 0.03 mM, respectively.[71,72] Yet they are able to maintain a high cellular K^+ concentration because of highly active Na^+K^+-ATPase, the amounts of which compensate for low ATP concentration.[73] This ability to utilize glucose at an appreciable rate would also suggest that the hexokinase of these erythrocytes has a lower K_m for glucose than that of other mammlian species.[74] Again one may conclude that wide variations in ATP concentration are compatible with the function of mature erythrocytes.

REFERENCES

1. **Ramadan, M. E. D. and Greenberg, D. M.,** An enzymatic micromethod for determination of glutamine and asparagine in blood, *Anal. Biochem.,* 6, 144, 1963.
2. **Benson, J. V., Gordon, M. J., and Patterson, J. A.,** Accelerated chromatographic analysis of amino acids in physiological fluids containing glutamine and asparagine, *Anal. Biochem.,* 18, 228, 1967.
3. **Stein, H. W. and Moore, S.,** The free amino acids of human blood plasma, *J. Biol. Chem.,* 211, 915, 1954.
4. **Richter-Rapoport, S. K. N., Dumdey, R., Hiebsch, Ch., Thamm, R., Uerlings, I., and Rapoport, S.,** Charakterisierung von Retikulozyten des Menschen: Atmung, Pasteur-Effekt und elektronenmikroskopische Befunde an Mitochondrien, *Acta Biol. Med. Ger.,* 36, 53, 1977.
5. **Kahrig, C. and Rapoport, S.,** Das Verhalten der Atmung und Glukose-6-Phosphat-Dehydrogenase roter Blutkörperchen ver-schiedenen Alters während einer Entblutungsanämie, *Acta Biol. Med. Ger.,* 6, 238, 1961.
6. **Siems, W., Müller, M., Dumdey, R., Holzhütter, H.-G., Rathmann, J., and Rapoport, S. M.,** Quantification of pathways of glucose utilization and balance of energy metabolism of rabbit reticulocytes, *Eur. J. Biochem.,* 124, 567, 1982.
7. **Siems, W., Dubiel, W., Dumdey, R., Müller, M., and Rapoport, S. M.,** Accounting for the ATP-consuming processes in rabbit reticulocytes, *Eur. J. Biochem.,* 139, 101, 1984.
8. **Augustin, W. and Gellerich, F. N.,** Studies on the regulation of mitochondrial ATP formation in rabbit reticulocytes, *Acta Biol. Med. Ger.,* 40, 603, 1981.
9. **Kostić, M. M., Müller, M., Krause, E. G., and Rapoport, S.,** Metabolic effects of (-)isoprenalin stimulation of adenylate cyclase in reticulocytes, *Biomed. Biochem. Acta,* in press.
10. **Jacobasch, G. and Rapoport, S.,** Phosphoglyzerinsäureveränderungen in Retikulozyten und Erythrozyten von Schafen, *Folia Haematol.,* 83, 283, 1965.
11. **Augustin, H. W. and Rapoport, S.,** Über Atmung und Succinat-oxydasesystem bei reifen und jugendlichen Hühnererythrozyten, *Acta Biol. Med. Ger.,* 3, 433, 1959.
12. **Syllm-Rapoport, I., Daniel, A., and Dumdey, R.,** Kreatingehalt in Erythrozyten und Blutplasma des Huhns vor und nach Erythropoese-Stimulation durch Anämie, *Acta Biol. Med. Ger.,* 39, 1015, 1980.
13. **Rapoport, S., Müller, M., Hartwig, A., and Dumdey, R.,** Antimyzin A-resistenter Sauerstoffverbrauch in Kaninchenretikulozyten, *Acta Biol. Med. Ger.,* 34, 1301, 1975.
14. **Salzmann, U., Kühn, H., Schewe, T., and Rapoport, S. M.,** Pentane formation during the anaerobic reactions of reticulocyte lipoxygenase. Comparison with lipoxygenases from soybeans and green pea seeds, *Biochim. Biophys. Acta,* 795, 535, 1984.
15. **Salzmann, U., Ludwig, P., Schewe, T., and Rapoport, S. M.,** The share of lipoxygenase in the antimycin-resistant oxygen uptake of intact rabbit reticulocytes, *Biochim. Biophys. Acta,* 44, 211, 1985.
16. **Rapoport, S.,** Metabolic pathways in the rabbit reticulocyte, *Ergeb. Exp. Med.,* 28, 9, 1978.
17. **Schweiger, H. G. and Rapoport, S.,** Der N-Stoffwechsel bei der Retikulozyten-Reifung. Atmung und Ammoniakbildung, *Acta Biol. Med. Ger.,* 1, 422, 1958.
18. **Ababei, L. and Rapoport, S.,** Über den Einfluß von Glutamin, Glutaminsäure und α-Ketoglutarsäure auf die O_2-Aufnahme und NH_3-Bildung von Kaninchenretikulozyten, *Acta Biol. Med. Ger.,* 5, 636, 1960.
19. **Ababei, L. and Rapoport, S. M.,** Die Wirkung von Glukose und Inosin auf den Oxalazetatabbau unter dem Einfluß von 2,4-Dinitrophenol in Kaninchenretikulozyten, *Acta Biol. Med. Ger.,* 7, 543, 1961.
20. **Rapoport, S. M. and Ababei, L.,** Der Einfluß von 2,4-Dinitrophenol auf die Veratmung von Substraten des Zitronensäurezyklus durch Kaninchenretikulozyten und Hühnererythrozyten, *Acta Biol. Med. Ger.,* 7, 533, 1961.
21. **Thilo, Ch., Schewe, T., Belkner, J., and Rapoport, S.,** In vitro-Reifung von Kaninchenretikulozyten: Verhalten des Sauerstoffverbrauchs, *Acta Biol. Med. Ger.,* 39, 1431, 1979.
22. **Salzmann, U.,** unpublished observation.
23. **Rapoport, S., Hinterberger, U., Ababei, L., v. Jagow, R., and Hofmann, E. C. G.,** Über begrenzende Faktoren von Atmung und Glykolyse in Retikulozyten, *Acta Biol. Med. Ger.,* 7, 528, 1961.
24. **Augustin, W. and Spengler, V.,** Energy balance in rabbit reticulocytes and its control by adenine nucleotides, *Biomed. Biochim. Acta,* 42, 223, 1983.
25. **Spengler, V. and Augustin, W.,** Dissipation of electrochemical ion gradients induced by carbonyl cyanide p-trifluoromethoxyphenylhydrazone and valinomycin in rabbit reticulocytes as loads of energy metabolism, *Biomed. Biochim. Acta,* 44, 403, 1985.
26. **Heinrich, R. and Rapoport, T. A.,** A linear steady-state treatment of enzymatic chains. General properties, control and effector strength, *Eur. J. Biochem.,* 42, 89, 1974.
27. **Rapoport, T. A., Heinrich, R., Jacobasch, G., and Rapoport, S. M.,** A linear steady-state model of glycolysis of human erythrocytes, *Eur. J. Biochem.,* 42, 107, 1974.
28. **Heinrich, R., Rapoport, S. M., and Rapoport, T. A.,** Metabolic regulation and mathematic models, *Prog. Biophys. Molec. Biol.,* 32, 1, 1977.

29. **Kacser, H. and Burns, J. A.**, The control of flux, in *Rate Control of Biological Processes*, Davies, D. D., Ed., Cambridge University Press, London, 1973, 65.

30. **Augustin, H. W. and Kunzendorf, H. J.**, Funktionelle Eigenschaften von Mitochondrien aus Kaninchenretikulozyten, in *VIth Internationales Symposium über Struktur und Funktion der Erythrozyten*, Rapoport, S. and Jung, F., Eds., Abhandlungen der Akademie der Wissenschaften der DDR, Akademie Verlag, Berlin, 1972, 459.

31. **Greksch, G., Wiswedel, I., and Augustin, W.**, Enzymatic characterization of rabbit reticulocyte mitochondria, in *VIIth Internationales Symposium über Struktur und Funktion der Erythrozyten*, Rapoport, S. M. and Jung, F., Eds., Abhandlungen der Akademie der Wissenschaften der DDR, Akademie Verlag, Berlin, 1975, 587.

32. **Gellerich, F. N. and Augustin, H. W.**, Kinetics of hexokinase from reticulocyte mitochondria-release, rebinding and electrophoretical pattern, in *VIIth Internationales Symposium über Struktur und Funktion der Erythrozyten*, Rapoport, S. M. and Jung, F., Abhandlungen der Akademie der Wissenschaften der DDR, Akademie Verlag, Berlin, 1975, 625.

33. **Gellerich, F. N. and Augustin, W.**, Studies on the functional significance of mitochondria-bond hexokinase in rabbit reticulocytes, *Acta Biol. Med. Ger.*, 36, 571, 1977.

34. **Kunz, W., Bohnensack, R., Böhme, G., Küster, U., Letko, G., and Schönfeld, P.**, Relations between extramitochondrial and intramitochondrial adenine nucleotide system, *Arch. Biochem. Biophys.*, 209, 219, 1981.

35. **Rapoport, I., Rapoport, S. M., Maretzki, D., and Elsner, R.**, The break-down of adenine nucleotides in glucose-depleted human red cells, *Acta Biol. Med. Ger.*, 38, 1419, 1979.

36. **Rapoport, I., Berger, H., Elsner, R., and Rapoport, S. M.**, pH-Dependent changes of 2,3-bisphosphoglycerate, *Acta Biol. Med. Ger.*, 36, 515, 1977.

37. **Oxley, S. T., Porteous, R., Brindle, K. M., Boyd, J., and Campbell, I. D.**, A multinuclear NMR study of 2,3 bis-phosphoglycerate metabolism in the human erythrocyte, *Biochim. Biophys. Acta*, 805, 19, 1984.

38. **Hinterberger, U., Rapoport, S., and Gerischer-Mothes, W.**, Der Pasteur-Effekt bei Retikulozyten. I. Die Stoffwechselgrößen der Kaninchen-Retikulozyten und ihre Beziehungen, *Acta Biol. Med. Ger.*, 8, 117, 1962.

39. **Rose, J. S. and O'Connell, E. L.**, The role of glucose-6-phosphate in the regulation of glucose metabolism in human erythrocytes, *J. Biol. Chem.*, 239, 13, 1964.

40. **Brand, K., Arese, P., and Rivera, M.**, Bedeutung und Regulation des Pentosephosphat-Weges in menschlichen Erythrozyten, *Hoppe-Seyler's Z. Physiol. Chem.*, 351, 501, 1970.

41. **Albrecht, V., Roigas, H., Schultze, M., Jacobasch, G., and Rapoport, S.**, The influence of pH and methylene blue on the pathways of glucose utilization and lactate formation in erythrocytes of man, *Eur. J. Biochem.*, 20, 44, 1971.

42. **Roigas, H., Zöllner, E., Jacobasch, G., Schultze, M., and Rapoport, S.**, Regulierende Faktoren der Methylenblaukatalyse in Erythrocyten, *Eur. J. Biochem.*, 12, 24, 1970.

43. **Schultze, M., Rapoport, S., and Lach, A.**, Einfluss von Methylenblau auf die Veratmung von Substraten im Retikulozyten, *Folia Haematol.*, 83, 371, 1965.

44. **Hinterberger, U., Ockel, E., Gerischer-Mothes, W., and Rapoport, S. M.**, Größe und pH-Abhängigkeit der anaeroben Glykolyse und der Hexokinase-Aktivitäten von Erythrozyten und Retikulozyten des Kaninchens, *Acta Biol. Med. Ger.*, 7, 50, 1961.

45. **Ghosh, A. K. and Sloviter, H. A.**, Glycolysis and the Pasteur effect in rat reticulocytes, *J. Biol. Chem.*, 248, 3035, 1973.

46. **Gerber, G., Schröder, K., and Rosenthal, S.**, Leitkriterien der Retikulozytenreifung: Verhalten von Hexokinase und Gluko-kinase roter Blutzellen während einer Entblutungsanämie des Kaninchens, *Acta Biol. Med. Ger.*, 30, 773, 1973.

47. **Schultze, M. and Rapoport, S. M.**, Über die Wege des Glukose-abbaus beim Retikulozyten, *Acta Biol. Med. Ger.*, 13, 310, 1964.

48. **Rapoport, S., Rost, J., and Schultze, M.**, Glutamine and glutamate as respiratory substrates of rabbit reticulocytes, *Eur. J. Biochem.*, 23, 166, 1971.

49. **Borsook, H., Ratner, K., Tattrie, B., Teigler, D., and Lajtha, L. G.**, Erythropoietin and the development of erythrocytes. Effect of erythropoietin in vitro which simulates that of a massive dose in vivo, *Nature (London)*, 217, 1024, 1968.

50. **Rapoport, S. and Schultze, M.**, CO_2-Bildung aus Glyzerin in Kaninchenretikulozyten, *Acta Biol. Med. Ger.*, 20, 553, 1968.

51. **Fessas, P., Anagnou, N. P., and Loukopoulos, D.**, Glycerol-3P-dehydrogenase activity in the red cells of patients with thalassemia, *Blood*, 55, 564, 1980.

52. **Spencer-Peet, J.**, Erythrocyte glycogen synthetase in glycogen storage deficiency resulting from the absence of this enzyme from liver, *Clin. Chim. Acta*, 10, 481, 1964.

53. **Cornblath, M., Steiner, D. F., Bryan, P., and King, J.**, Uridine-diphosphoglucose glucosyltransferase in human erythrocytes, *Clin. Chim. Acta*, 12, 270, 1965.

54. **Moses, S. W., Bashan, N., and Gutman, A.,** Properties of glycogen synthetase in erythrocytes, *Eur. J. Biochem.,* 30, 205, 1972.
55. **Moses, S. W., Bashan, N., and Gutman, A.,** Glycogen metabolism in the normal red blood cell, *Blood,* 40, 836, 1972.
56. **Moses, S. W., Bashan, N., Gutman, A., and Ockerman, P. A.,** Glycogen metabolism in glycogen-rich erythrocytes, *Blood,* 44, 275, 1974.
57. **Sidbury, J. B., Cornblath, M., Fisher, J., and House, E.,** Glycogen in erythrocytes of patients with glycogen storage disease, *Pediatrics,* 27, 103, 1961.
58. **Krebs, H. A.,** The Pasteur effect and the relations between respiration and fermentation, *Essays Biochem.,* 7, 1, 1972.
59. **Gerber, G., Preissler, H., Heinrich, R., and Rapoport, S. M.,** Hexokinase of human erythrocytes: purification, kinetic model and its application to the conditions in the cell, *Eur. J. Biochem.,* 45, 39, 1974.
60. **Jacobasch, G., Matusch, J., Schönian, G., Gerth, C., and Rapoport, S. M.,** Control mechanism of the Pasteur effect, in *VIIth Internationales Symposium über Struktur und Function der Erythrozyten,* Rapoport, S. M. and Jung, F., Eds., Abhandlungen der Akademie der Wissenschaften der DDR, Akademie Verlag, Berlin, 1975, 617.
61. **Schweiger, H. G. and Rapoport, S.,** Der N-Stoffwechsel bei der Erythrocytenreifung: Die N-Bilanz unter endogenen Bedingungen, *Hoppe-Seyler's Z. Physiol. Chem.,* 313, 97, 1958.
62. **Spengler, V.,** Phosphorylierungsgrad der Adeninnukleotide von intakten Kaninchenretikulozyten bei unterschiedlichen Geschwindigkeiten von oxidativer und glykolytischer ATP-Bildung, M.D. Thesis, Magdeburg, 1985.
63. **Heinrich, R. and Rapoport, S. M.,** The utility of mathematical models for the understanding of metabolic systems, *Biochem. Soc. Trans.,* 11, 31, 1983.
64. **Rapoport, T. A., Heinrich, R., and Rapoport, S. M.,** The regulatory principles of glycolysis in erythrocytes in vivo and in vitro, *Biochem. J.,* 154, 449, 1976.
65. **Ataullakhanov, F. I., Vitvitsky, V. M., Zhabotinsky, A. M., Pichugin, A. V., Platonova, O. V., and Kholodenko, B. N.,** The regulation of glycolysis in human erythrocytes, *Eur. J. Biochem.,* 115, 359, 1982.
66. **Maretzki, D., Brenneis, M., Schwarz, Z., Lange, I., and Rapoport, S.,** Glykolyse und ATP-Verbrauch in membranfreien Hämolysaten, *Acta Biol. Med. Ger.,* 36, 625, 1977.
67. **Reimann, B., Klatt, D., Tsamaloukas, G. A., and Maretzki, D.,** Membrane phosphorylation in intact human erythrocytes, *Acta Biol. Med. Ger.,* 40, 487, 1981.
68. **Maretzki, D., Reimann, B., Klatt, D., and Schwarzer, E.,** Involvement of polyphosphoinositides in the ATP turnover of intact human erythrocytes and in the ATPase activity of purified membranes, *Biomed. Biochim. Acta,* 42, 72, 1983.
69. **Rapoport, S., Dubiel, W., Maretzki, D., and Siems, W.,** Balance of ATP-producing and consuming reactions in the red cells, in *Proc. 16th FEBS Meet., Moscow 1984,* Part A, VNU Science Press, Utrecht, 1985, 165.
70. **Freudenberg, H. and Mager, J.,** Studies on the mechanism of the inhibition of protein synthesis induced by intracellular ATP depletion, *Biochim. Biophys. Acta,* 232, 537, 1971.
71. **Kim, H. D., Zeidler, R. B., Sallis, J. D., Nicol, S. C., and Isaacks, R. E.,** Adenosine triphosphate-deficient erythrocytes of the egg-laying mammal, Echidna (Tachyglossus aculeatus), *Science,* 213, 1517, 1981.
72. **Isaacks, R., Nicol, S., Sallis, J., Zeidler, R., Kim, H. D.,** Erythrocyte phosphates and hemoblobin function in monotremes and some marsupials, *J. Physiol.,* 246, R236, 1984.
73. **Kim, H. D., Baird, M., Sallis, J., Nicol, S., and Isaacks, R. E.,** Active cation transport and Na^+K^+Mg ATPase of the monotreme erythrocytes, *Biochem. Biophys. Res. Commun.,* 119, 1161, 1984.
74. **Kim, H. D., Zeidler, R. B., Sallis, J., Nicol, S., and Isaacks, R. E.,** Metabolic properties of low ATP erythrocytes of the monotremes, *FEBS Lett.,* 167, 83, 1984.

Chapter 7

LIPID METABOLISM

I. INTRODUCTION

Lipid metabolism is widely ramified, encompassing the catabolism of fatty acids as well as a variety of synthesis and breakdown pathways, mainly for triacylglycerols, phospholipids, and sphingolipids. Among the catabolic pathways of the reticulocyte, acetone body formation is absent; among the anabolic pathways, only traces of the new formation of fatty acids, as well as their desaturation and elongation, are present.

II. OXIDATION AND *DE NOVO* SYNTHESIS OF FATTY ACIDS

In principle, there are three possible fates for fatty acids in the reticulocyte: (1) incorporation in triacylglycerols and phospholipids; (2) oxygenation by lipoxygenase with the formation of secondary products (see Chapter 12); and (3) β-oxidation.

There is some distinction in the prevalence of pathways among different types of fatty acids. Saturated fatty acids, particularly octanoate, are good substrates of both reticulocytes and their isolated mitochondria.[1] Long-chain fatty acids may originate from intracellular breakdown, and may also originate from external sources since they permeate the cellular membrane freely. β-Oxidation must be preceded by activation, i.e., the formation of acyl-coenzyme-A thioesters at the outer mitochondrial membrane. In order to enter the mitochondrial matrix where the degradation occurs, long-chain fatty acids are first converted into acyl-carnitine compounds at the outer side of the inner mitochondrium membrane by a transferase. They are transported in this form through the inner membrane and are reconverted to acyl-coenzyme-A thioesters by a second transferase located at the matrix side of the inner mitochondrial membrane. Medium-chain fatty acids such as octanoate may enter the mitochondria in a carnitine-independent manner (for a review see Fritz[2]).

The utilization of octanoate and palmitate were compared.[1] It was found that $^{14}CO_2$-formation from palmitate as a function of its external concentration reached a ceiling of about 10% of total CO_2-formation due to the limitation of the capacity of the activating system for long-chain fatty acids or for their translocation through the mitochondrial membrane. On the other hand, octanoate was found to contribute as much as 65% of the respiratory substrate. One may therefore conclude that the steps of fatty acid degradation subsequent to their activation, which are identical for palmitate and octanoate, have large capacity and are mainly limited by the supply of substrate.

Mature erythrocytes do not synthesize long-chain fatty acids from ^{14}C-acetate.[3,4] Their inability for *de novo* synthesis is due to a lack of acetyl-CoA carboxylase activity which is lost during red cell maturation.[4,5] Weak *de novo* synthesis of fatty acids was demonstrated in red blood cells of fowl.[5] Even erythrocytes, however, have acetyl-CoA synthetase and some fatty acid synthetase activity.[4] Purified fatty acid synthases of rabbit reticulocytes and rat and pigeon erythrocytes exhibit only 1% the activity of liver enzymes.[7,8] They also differ in their size, the red cell enzyme being considerably smaller and immunologically nonidentical. Tests of partial enzyme activities revealed that these model reactions were reduced by about one order of magnitude.

Studies on the metabolism of polyunsaturated linoleic and arachidonic acids demonstrated that β-oxidation of exogenous fatty acids amounted to only 5% of total utilization, which is only about one fourth the share of saturated fatty acids;[10,19] 30% or more of polyenoic fatty acids were metabolized via lipoxygenase, mainly to hydroxy-compounds which arise

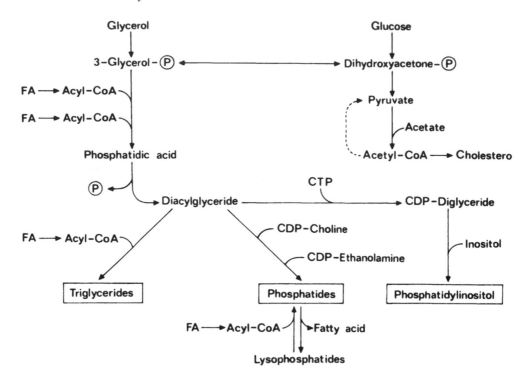

FIGURE 17.   Scheme of lipid synthesis. FA = fatty acids.

by reduction of the primarily formed hydroperoxypolyenoic fatty acids. In addition, some secondary products including volatile hydrocarbons such as pentane arise both by enzymatic and nonenzymatic routes.[11] It was also shown that hydroxypolyenoic acids may be substrates of β-oxidation.

The β-oxidation share is probably determined, at least in part, by competition with the other pathways. This is indicated by the fact that inhibition of lipoxygenase results in increased incorporation of fatty acids into cellular lipids. On the other hand, inhibition of the respiratory chain by antimycin A with consequent interruption of β-oxidation of fatty acids augments the formation of products of the lipoxygenase pathway. The metabolism of polyenoic fatty acids gives rise to oxygen consumption which is antimycin A-resistant, i.e., independent of the respiratory chain. It may be due in part to secondary oxygen-consuming reactions of primary lipoxygenase products. Even in a model system of mitochondrial membranes, oxygen consumption exceeded diene formation from polyenoic fatty acids.

III. LIPID SYNTHESIS*

Metabolism of lipids by mature erythrocytes is limited to two processes which affect the composition of their fatty acids: (1) the exchange of entire phospholipid molecules with plasma.[12-14] and (2) reacylation of endogenous lysophospholipids with exogenous fatty acids.[15-23] This process requires the activation of fatty acids by conversion to acyl-CoA thioesters and has been shown to be dependent on both ATP and CoA.[15-17,22,23] The exchange of phospholipids primarily involves the outer layer of the erythrocyte cell membrane, whereas the incorporation of fatty acids occurs preferentially at the inner layer of the cell membrane.[20,21]

It is well known that red cell phospholipids, particularly phosphatidylethanolamine, contain fatty acid alcohols in the form of alkenyl and alkyl ethers.[23] In the first study where incor-

* See Figure 17.

poration of arachidonic acid in alkenylacyl-, alkylacyl- and diacylphosphatidylethanolamine of rat erythrocytes was differentiated, 90% was found in the 2-position of the diacylphospholipid, although it represented only one third of total phosphatidylethanolamine.[23]

In early work it was found that reticulocytes maintained the ability to turn over lipids in vitro.[3] Some lipid turnover represents only the exchange of preformed components.[24,25] Incorporation of inorganic phosphate into phosphatidic acid occurs in erythrocytes, while the subsequent steps leading to phospholipids are absent. Synthesis of phosphatidic acid is much greater in reticulocytes and *de novo* synthesis of phospholipids is demonstrable.[26] Starting with the precursor ^{14}C-glycerol, it was shown that this compound is incorporated in the total cell lipids of rabbit reticulocytes but not of erythrocytes, mainly in the 3-carbon backbone of phosphatidylcholine and phosphatidylethanolamine. There was, however, little labeling of phosphatidylserine and sphingomylin.[27,28] A fairly large portion was incorporated in diacylglycerols which are the precursors of both triacylglycerols and phospholipids. A further significant share amounting to 10% of the total incorporation was found in cholesterol. These data, combined with decreased incorporation of glycerol by dilution with glucose, pyruvate, or acetate, were interpreted to indicate that glycerol after phosphorylation by glycerolkinase is converted to dihydroxyacetone phosphate and further metabolized to acetyl-CoA.[28] 3-Glycerol phosphate dehydrogenase, which would catalyze the conversion of glycerol phosphate to dihydroxyacetone phosphate, indeed has been demonstrated to occur in reticulocytes.[29]

Starting with ^{14}C-acetate, active incorporation into the lipids of reticulocytes was observed, which again was absent in erythrocytes.[3,24,28] The main portion, about 60%, was incorporated into cholesterol. The remainder was found in the phospholipids. A large portion appeared in the organic phase after partial hydrolysis of the phospholipids. However, its definite identification as a fatty acid was not carried out. A small amount was found in the aqueous phase after hydrolysis and was assumed to represent glycerol. Since a direct carboxylation of acetyl-CoA is unlikely, the labeling may have occurred by way of isotopic dilution in the citrate cycle.

In another study it was demonstrated that exposure of rabbit reticulocytes to agents which presumably damage the cell membrane, such as vincristine, butanol, hydrogen peroxide, and puromycin, stimulated the incorporation of radioactive acetate and glycerol into membrane lipids.[30] Differential effects were observed, with vincristine predominantly increasing the synthesis of cholesterol both from glycerol and acetate, whereas puromycin stimulated the incorporation of glycerol only into phospholipids. The mechanism of these effects was not clarified.

The incorporation of fatty acids into lipids may occur at various stages of triacylglycerol and phospholipid synthesis and also as a reacylation reaction of lysophospholipids which arise by splitting off of one fatty acid from their parent compounds by action of phospholipase A$_2$, an enzyme which occurs in red cells.[31,32] The reacylation of lysophospholipids is mediated by lysophospholipid-acyltransferase and takes place in both reticulocytes and mature erythrocytes.[17,18,33-36] However, acylation activity is considerably higher in reticulocytes than in mature erythrocytes and diminishes with aging of erythrocytes.[34] Reacylation was found in plasma membranes and mitochondria as well as in a postmitochondrial fraction with mitochondria showing the highest activity. The reacylation activity of reticulocyte mitochondria appeared to exceed that of liver mitochondria, possibly due to their relatively larger amount of lysophospholipids. It may be assumed that the lysophospholipid acylation may represent a repair mechanism.[36]

Another process which is active in reticulocytes is glycerol-3-phosphate acylation, which forms lysophosphatidic acid; it occurs almost exclusively in mitochondria,[36] presumably in their outer membrane.[37,38] However, incorporation into phospholipids was not observed.

In earlier work the occurrence of the last steps of phospholipid synthesis, namely, the formation of phosphatidylcholine from diacylglycerol and CDP choline and of phosphatidylinositol from CDP glyceride and inositol, was demonstrated in rabbit reticulocytes. In mature erythrocytes these reactions did not occur.[39] According to the work of Augustin,[36] phosphocholine transferase activity was highest in the postmitochondrial fraction, which should have contained the endoplasmic reticulum. A strict correlation also was found between the activities of transferase and NADPH-cyt c reductase, which is a marker enzyme for the endoplasmic reticulum. Therefore, the conclusion was drawn that transferase activity is located in the still existing remnants of the endoplasmic reticulum. These results correspond to well-established observations in other types of cells. According to these observations, the mitochondria are only able to synthesize lysophosphatidic and phosphatidic acids, and to acylate lysophospholipids, while completion of N-containing phospholipid synthesis proceeds in the endoplasmic reticulum.[36,37]

Recently, the incorporation and fate of palmitic, oleic, linoleic, and arachidonic acids in rabbit reticulocytes was studied.[9,10] All fatty acids were incorporated into reticulocyte lipids by at least one order of magnitude greater than in erythrocytes. The highest rate of incorporation was found with linoleic acid; the lowest was found with oleic acid. However, the percentage incorporation of total fatty acid utilized was considerably lower with arachidonic and linoleic acids, which are lipoxygenase substrates. In keeping with other results, incorporation into phosphatidylcholine predominated, particularly after short periods of incubation, and more so with palmitic acid than with the other fatty acids. In second place was incorporation of fatty acids into phosphatidylethanolamine, whereas phosphatidylserine was much less labeled. About 10% of the radioactivity was found in neutral lipids, mostly di- and triacylglycerols, particularly if incubated with ^{14}C-linoleic acid; cholesterol was labeled only to a small extent, and slowly, presumably on account of the relatively small extent of β-oxidation and the dilution of acetyl-CoA (from which it is formed) by other sources such as amino acids and carbohydrates.

Based on the time course of specific radioactivity, one could estimate a considerable dilution of fatty acids supplied by endogenously liberated fatty acids which presumably arise from the breakdown of mitochondria and perhaps of cellular membranes.

IV. METABOLISM OF PHOSPHATIDYLINOSITIDES*

The phosphatidylinositides differ from other glycerophospholipids in several respects. First, by the polyol head-group which bestows on them a hydrophilic character, and, second, by their negative charge. In addition to phosphatidylinositol (PI), mono- and bisphosphate esters occur, PI-4-P and PI-4,5-P_2, respectively, in the membrane of red cells. This group of substances accounts for less than 10% total phospholipids, and like phosphotidylserine, are located entirely in the cytosolic layer of the cell membrane. It appears likely that the clustered negative charges of the phosphates interact with the membrane skeleton and/or the cytosolic proteins. The phosphatidylinositides participate in a variety of enzymatic transformations (see Figure 15 in Chapter 6). The most rapid reactions are the ATP-dependent phosphorylations by two specific enzymes, which form PI-4-P and PI-4,5-P_2, respectively, on the one hand and the hydrolysis of the compounds by two specific phosphatases in the other.[40-42] Thus, two phosphorylation/dephosphorylation cycles exist between PI and PI-4-P, as well as between PI-4-P and PI-4,5-P_2. Metabolic turnover appears to be high even in erythrocytes, although technical difficulties do not permit an exact assessment as yet.[43] There is some indication for heterogeneity of the phosphatidyl inositide pool. In erythrocytes, synthesis of phosphatidylinositol is absent; it is slow in reticulocytes. Turnover of phosphatidylinositides presupposes the action of a phosphodiesterase, phospholipase C, which would yield 1,2-diacylglycerol and the corresponding water-soluble inositol mono-, bis-, or trisphosphates. This enzyme indeed has been found in the red cells of man and other species

---

*   See Figure 18.

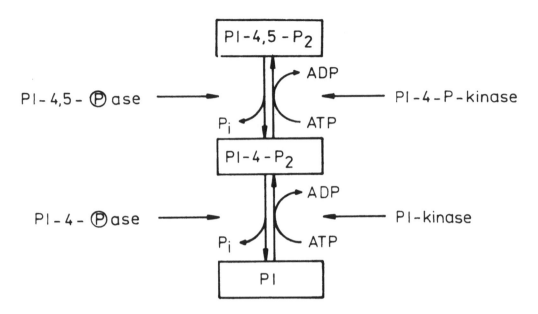

FIGURE 18.   Interconversions of phosphatidylinositides.

(see Table 7 in Chapter 4); and is activated by Ca^{2+} ions.[44,45] Whereas according to one report the enzyme is practically inactive under physiological conditions, i.e., at normal ionic strength and in the presence of 1 mM Mg^{2+},[46] deviating results were obtained in other work on rabbit reticulocyte membranes.[41] Such discrepancies may be due either to species differences or to variations in the preparation of cell membranes. It is obvious that further studies will be required to elucidate the complex properties of the Ca-stimulated phospholipid phosphoesterases. At any rate, it would appear that the irreversible degradation of phosphatidylinositides in mature erythrocytes is a very slow process.

Various functions for phosphatidylinositides have been proposed, such as modulation of ATPase activity, anchoring of acetylcholine esterase,[47] interaction with membrane skeleton, and effects on membrane fluidity, none of which have been established to be functional in mature erythrocytes under physiological conditions. It has been suggested that phosphatidyl inositides may take part in receptor response to hormones. Recent work suggests that diacylglycerols, which are breakdown products of phosphatidylinositols, may have a function in the regulation of erythroid differentiation (see Chapter 1).

V. SPHINGOLIPID SYNTHESIS*

As mentioned in Chapter 4, sphingolipids are components of the cell membrane of red cells and constitute an important part of their antigenic profile. It was also mentioned that there are a multitude representative of this class which belong to various families of compounds, among them sphingomyelins (ceramide phosphorylcholines) and ceramide di-, tri-, and tetrahexosides. These are all derivatives of ceramide, which is a fatty acid amide of sphingosine. Sphingosine is a long-chain 2-amino-1,3-diol. It originates from the condensation of palmitoyl-CoA with serine with the intermediate formation of 3-ketosphingosine and dihydrosphingosine. Pyridoxalphosphate is necessary for the activation of the serine. The formation of ceramide presumably occurs by condensation of sphingosine with fatty acyl-CoA.

---

* For a review, see Kishimoto.[48]

Sphingomyelin can be formed either by transfer of phosphorylcholine from CDP choline to ceramide or from phosphatidylcholine (lecithin). The synthesis of glycosphingolipids proceeds by complicated pathways in which galactose, glucose, acetyl-galactosamine, sialic acid, and fucose are incorporated with different positional and steric specificities, as well as in varied sequence. Some of the glycosphingolipids, as indicated by their names — hematoside and globoside — occur predominantly in red cells. They contribute a large part to the blood group activity of red cell membranes.

Although very few data are available, one may assume that erythroid cells have at their disposal the complete multifarious assembly for total synthesis of sphingolipids. Experiments with palmitic, stearic, and oleic acids in rabbit reticulocytes revealed a significant rate of incorporation into the sphingomyelin fraction.[10] It was nearly one order of magnitude smaller than in phosphatidylcholine and phosphatidylethanolamine but larger than phosphatidylserine. Erythrocytes, on the other hand, exhibited an insignificant rate of fatty acid incorporation into sphingomylin.

REFERENCES

1. **Schultze, M., Rost, J., Augustin, W., Gellerich, F. N., and Rapoport, S. M.,** The oxidation of fatty acids by rabbit reticulocytes and their mitochondria, *Eur. J. Biochem.,* 27, 43, 1972.
2. **Fritz, I. B.,** Carnitin and its role in fatty acid metabolism, *Adv. Lipid Res.,* 1, 285, 1963.
3. **Marks, P. A., Gellhorn, A., and Kidson, C.,** Lipid synthesis in human leukocytes, platelets and erythrocytes, *J. Biol. Chem.,* 235, 2579, 1960.
4. **Pittman, J. G. and Martin, D. B.,** Fatty acid biosynthesis in human erythrocytes. Evidence in mature erythrocytes for an incomplete long chain fatty acid synthesizing system, *J. Clin. Invest.,* 45, 165, 1966.
5. **Weir, G. C. and Martin, D. B.,** Fatty acid biosynthesis in the erythrocyte. The effect of red cell age on the integrity of the pathway, *Diabetes,* 17, 305, 1968.
6. **Webb, J. P. W., Allison, C. C., and James, A. T.,** In vitro lipid synthesis in fowl blood, *Biochim. Biophys. Acta,* 43, 89, 1960.
7. **Jenik, R. A. and Porter, J. W.,** Red blood cell fatty acid synthetase. Nonidentity with the enzyme from liver, *Int. J. Biochem.,* 10, 609, 1979.
8. **Jenik, R. A. and Porter, J. W.,** Fatty acid synthetase from pigeon red blood cells, *Int. J. Biochem.,* 13, 423, 1981.
9. **Reinhardt, U., Kühn, H., Wiesner, R., and Rapoport, S.,** Metabolism of polyunsaturated fatty acids by rabbit reticulocytes, *Eur. J. Biochem.,* 153, 189, 1985.
10. **Siems, W., Mochizuki, S., Ueta, N., and Rapoport, S.,** Fatty acid incorporation by rabbit reticulocytes, in preparation.
11. **Salzmann, U., Kühn, H., Schewe, T., and Rapoport, S. M.,** Pentane formation during the anaerobic reactions of reticulocyte lipoxygenase. Comparison with lipoxygenases from soybeans and green pea seeds, *Biochim. Biophys. Acta,* 795, 535, 1984.
12. **Reed, C. F.,** Phospholipid exchange between plasma and erythrocytes in man and dog, *J. Clin. Invest.,* 47, 749, 1968.
13. **Shohet, S. B.,** Release of phospholipid fatty acids from human erythrocytes, *J. Clin. Invest.,* 49, 1668, 1970.
14. **Shohet, S. B. and Nathan, D. G.,** Incorporation of phosphatide precursors from serum into erythrocytes, *Biochim. Biophys. Acta,* 202, 202, 1970.
15. **Oliveira, M. M. and Vaughan, M.,** Incorporation of fatty acids into phospholipids of erythrocyte membranes, *J. Lipid Res.,* 5, 165, 1964.
16. **Robertson, A. F. and Lands, W. G.,** Metabolism of phospholipids in normal and spherocytic human erythrocytes, *J. Lipid Res.,* 5, 88, 1964.
17. **Mulder, E. and van Deenen, L. L. M.,** Metabolism of red cell lipids. I. Incorporation in vitro of fatty acids into phospholipids from mature erythrocytes, *Biochim. Biophys. Acta,* 106, 106, 1965.
18. **Mulder, E., van den Berg, J. W. O., and van Deenen, L. L. M.,** Metabolism of red-cell lipids. II. Conversions of lysophosphoglycerides, *Biochim. Biophys. Acta,* 106, 118, 1965.
19. **Mulder, E. and van Deenen, L. L. M.,** Metabolism of red cell lipids. III. Pathways for phospholipid renewal, *Biochim. Biophys. Acta,* 106, 348, 1965.

20. **Renooij, W. I., van Golde, L. M. G., Zwaal, R. F. A., Roelofson, B., and van Deenen, L. L. M.,** Preferential incorporation of fatty acids at the inside of human erythrocyte membranes, *Biochim. Biophys. Acta,* 363, 287, 1974.

21. **Renooij, W., van Golde, L. M. G., Zwaal, R. F. A., and van Deenen, L. L. M.,** Topological assymmetry of phospholipid metabolism in rat erythrocyte membranes, *Eur. J. Biochem.,* 61, 53, 1976.

22. **Dise, C. A., Goodman, D. B. P., and Rasmussen, H.,** Definition of the pathway for membrane phospholipid fatty acid turnover in human erythrocytes, *J. Lipid Res.,* 21, 292, 1980.

23. **Kaya, K., Miura, T., and Kubota, K.,** Different incorporation rates of arachidonic acid into alkenyl-acyl-, alkylacyl-, and diacylphosphatidylethanol amine of rat erythrocytes, *Biochim. Biophys. Acta,* 796, 304, 1984.

24. **O'Donnell, V. J., Ottolenghi, P., Malkin, A., Denstedt, O. F., and Heard, R. D.,** The biosynthesis from acetate-1-C^{14} of fatty acids and cholesterol in formed blood elements, *Can. J. Biochem. Physiol.,* 36, 1125, 1958.

25. **Sloviter, H. A. and Bose, R. K.,** Evaluation of synthesis of lipids by rabbit reticulocytes, *Biochim. Biophys. Acta,* 116, 156, 1966.

26. **Raderecht, H. J., Binnewies, S., and Schölzel, E.,** Zum Mechanismus des Einbaus des ^{32}P in Phosphatidfraktionen von Retikulozyten und reifen Erythrozyten, *Acta Biol. Med. Ger.,* 8, 199, 1962.

27. **Sloviter, H. A. and Tanaka, S.,** The biosynthesis of glycerides and glycerophosphatides by rabbit reticulocytes, *Biochim. Biophys. Acta,* 137, 70, 1967.

28. **Ballas, S. K. and Burka, E. R.,** Pathways of de novo phospholipid synthesis in reticulocytes, *Biochim. Biophys. Acta,* 337, 239, 1974.

29. **Fessas, P., Anagnou, N. P., and Loukopoulos, D.,** Glycerol-3P-dehydrogenase activity in the red cells of patients with thalassemia, *Blood,* 55, 564, 1980.

30. **Ballas, S. K. and Burka, E. R.,** Stimulation of lipid synthesis in reticulocytes as a response to membrane damage, *Blood,* 44, 263, 1974.

31. **Delbauffe, D., Paysant, M., and Polonovski, J.,** Phosphatidyl-glycerolphospholipase A des globules rouges de rat. I. Influence des effecteurs, *Bull. Soc. Chim. Biol.,* 50, 1431, 1968.

32. **Garcia Parra, M., Schewe, T., and Rapoport, S. M.,** On the presence of a calcium-stimulated phospholipase A in the stroma-free supernatant fluid of rabbit reticulocytes, *Acta Biol. Med. Ger.,* 34, 1075, 1975.

33. **Warrendorf, E. M. and Rubinstein, D.,** Esterification of linoleate by rabbit reticulocytes, *Can. J. Biochem.,* 49, 919, 1971.

34. **Ferber, E., Munder, P. G., Kohlschütter, A., and Fischer, H.,** Lysolecithin-Stoffwechsel in Erythrocytenmembranen, *Eur. J. Biochem.,* 5, 395, 1968.

35. **Kanoh, H. and Ohno, K.,** Solubilization and purification of rat liver microsomal 1,2-diacylglycerol:CDP-choline-phosphotransferase and 1,2-diacylglcerol:CDP-ethanolamine ethanolaminephosphotransferase, *Eur. J. Biochem.,* 66, 201, 1976.

36. **Augustin, W., Zborowski, J., Baranska, J., Wiswedel, I., and Wojtczak, L.,** Synthesis of phospholipids in mitochondria and other membrane fractions of rabbit reticulocytes, *Biochim. Biophys. Acta,* 489, 298, 1977.

37. **Stoffel, W. and Schiefer, H. G.,** Biosynthesis and composition of phosphatides in outer and inner mitochondrial membranes, *Hoppe-Seyler's Z. Physiol. Chem.,* 349, 1017, 1968.

38. **Zborowski, J. and Wojtczak, L.,** Phospholipid synthesis in rat liver mitochondria, *Biochim. Biophys. Acta,* 187, 73, 1969.

39. **Percy, A. K., Schmell, E., Earles, B. J., and Lennarz, W. J.,** Phospholipid biosynthesis in the membranes of immature and mature red blood cells, *Biochemistry,* 12, 2456, 1973.

40. **Quist, E. E.,** Polyphosphoinositide synthesis in rabbit erythrocyte membranes, *Arch. Biochem. Biophys.,* 219, 58, 1982.

41. **Quist, E.,** Ca^{2+}-stimulated phospholipid phosphoesterase activities in rabbit erythrocyte membranes, *Arch. Biochem. Biophys.,* 236, 140, 1985.

42. **Buckley, J. T.,** Properties of human erythrocyte phosphatidylinositol lipase and inhibition by adenosine, ADP and related compounds, *Biochim. Biophys. Acta,* 498, 1, 1977.

43. **Maretzki, D., Reimann, B., Klatt, D., and Schwarzer, E.,** Involvement of polyphosphoinositides in the ATP turnover of intact human erythrocytes and in the ATPase activity of purified membranes, *Biomed. Biochim. Acta,* 42, 72, 1983.

44. **Allan, D. and Michell, R. H.,** A calcium-activated polyphosphoinositide phosphodiesterase in the plasma membrane of human and rabbit erythrocytes, *Biochim. Biophys. Acta,* 508, 277, 1978.

45. **Downes, C. P. and Michell, R. H.,** The polyphosphoinositide phosphodiesterase of erythrocyte membranes, *Biochem. J.,* 198, 133, 1981.

46. **Downes, C. P. and Michell, R. H.,** The control by Ca^{2+} of the polyphosphoinositide phosphodiesterase and the Ca^{2+}-pump ATPase in human erythrocytes, *Biochem. J.,* 202, 53, 1982.

47. **Low, M. G. and Finean, J. B.,** Non-lytic release of acetylcholinesterase from erythrocytes by a phosphatidylinositol-specific phospholipase, *FEBS Lett.,* 82, 143, 1977.

48. **Kishimoto, Y.,** Sphingolipid formation, in *The Enzymes,* Vol. XVI, 3rd ed., Boyer, P. D., Ed., Academic Press, New York, 1983, 358.

Chapter 8

METABOLISM OF AMINO ACIDS

I. INTRODUCTION

Prolonged oxygen consumption in reticulocytes without substrate brings into question the endogenous substrate in oxidation. Another aspect to be considered is the interaction between endogenous and external substrates. Warburg surmised, from a respiratory quotient of about 1, that proteins constitute the metabolic fuel under endogenous conditions.[1] Later studies showed that amino acids arising from the breakdown of the stroma, mostly mitochondria and ribosomes, constitute a large part of the endogenous substrate.[2,3] The degradation of the stroma also furnishes a sizable proportion of substrate, perhaps as much as one third, to hemoblobin synthesis. The endogenous metabolic substrate in reticulocytes includes besides amino acids, fatty acids and glycerol from the phospholipids, as well as ribose, which originates from ribonucleic acids and low molecular nucleotides, mostly ATP. Amino acids will be discussed in the following section.

II. THE CONCENTRATIONS OF AMINO ACIDS AND THEIR UTILIZATION IN THE RETICULOCYTE

Complete oxidation of a protein yields an RQ of about 0.8. The higher value found by Warburg would lead one to expect that lower aliphatic amino acids would be metabolized preferentially. A survey of $^{14}CO_2$ formation from tracer amounts ($5 \times 10^{-6} M$) of various ^{14}C-labeled amino acids indicated a pronounced preference for glutamine, glutamate, aspartate, proline, serine, and glycine. On the other hand, histidine, phenylalanine, and tyrosine yielded little $^{14}CO_2$.[4] In these studies possible differences in the uptake of amino acids and in their intracellular concentrations were not taken into account.[4] A complete analysis of intracellular concentrations of amino acids in the reticulocyte is not available as yet. The concentrations are strongly dependent on respiratory rate and maturity,[5] presumably as a function of the activity of specific carriers of the cell membrane[6] and the intracellular processes liberating amino acids as well as their rate of utilization for protein synthesis. The concentration of amino acids is also strongly dependent on the time of incubation. Most of the amino acids diminish by one half within 10 min of substrate-free incubation as a result of the cessation of external influx. Thereafter they increase to levels well in excess of initial concentrations. Addition of glucose may suppress this secondary rise, which is probably due to the cessation of hemoglobin synthesis in the absence of glucose as the breakdown of the stroma continues. From these observations one may conclude that the composition of the metabolic mixture undergoes variations both with maturation and during experimental periods under various conditions.

Table 13 contains a compilation of our data on the concentrations of those amino acids that occur in significant amounts in rabbit reticulocytes. It may be seen that glycine predominates with a concentration twice as high as glutamate, three times as high as aspartate and alanine, and four to five times as high as serine, phosphoserine, and threonine. All other amino acids occur in concentrations less that $1/_{50}$ of glycine. The occurrence of phosphoserine in red cells was previously unknown. It is also present in erythrocytes (0.19 ± 0.03 mM).[7]

It is remarkable that the amino acid pool is composed almost exclusively of neutral and acid amino acids with five carbon atoms or less. This is in agreement with the respiratory quotient of the reticulocyte. A second inference may be drawn, namely, that the affinity of other processes utilizing amino acids in protein synthesis must be very high indeed.

Table 13
THE CONCENTRATION OF
SOME AMINO ACIDS IN
RABBIT RETICULOCYTES

Amino acid	Concentration (mM)
Glycine	1.1—1.8
Glutamate	0.6—0.9
Alanine	0.4—0.8
Aspartate	0.4—0.7
Threonine	0.4
Serine	0.2—0.5
Others	0.05

Note: Initial values of reticulocyte-rich suspensions. The lower values correspond to oxygen uptakes of less than 8 μmol of O_2, the higher one to more than 11 μmol of O_2 per mℓ cells per hour.

From Rapoport, S., Müller, M., Dumdey, R., and Rathmann, J., *Eur. J. Biochem.,* 108, 449, 1980. With permission.

Taking intracellular concentrations and the data on $^{14}CO_2$ formation together, it would appear that for the immature reticulocyte, glutamate, aspartate, and alanine furnish the bulk of the metabolic substrate in respiration. With increasing maturity the glycine contribution takes on increasing importance.

Further insights concerning the utilization of amino acids were obtained by studying $^{14}CO_2$ formation as a function of their external concentrations.[8] Under such circumstances it is possible to calculate the share of external substrate in the total CO_2 formed. A useful approximate indicator of the share is the percentage ratio of the $^{14}CO_2$ formed from a given substrate to the total O_2 consumption, i.e., the "relative CO_2 formation". For precise calculations, the RQ of various substrates has to be taken into account. Since oxygen consumption does not vary regardless of which substrate is utilized, an increase in relative CO_2 formation from one amino acid signifies a suppression of the contribution from other substrates. Figure 19 shows some representative examples of the dependence of relative CO_2 formation on substrate concentration.[9] It can be seen that the amino acids exhibit different patterns. Glutamine, glutamate, and aspartate converge to a limit of 80% relative CO_2 formation. Curves similar to those with glutamate were also obtained for proline, lysine, threonine, and leucine, with values of relative CO_2 formation in excess of 30% at high substrate concentrations. On the other hand, glycine, serine, phenylalanine, and tyrosine remain below 10% relative CO_2 formation. The curves for individual amino acids show a strong dependence on cell maturity.[8] This is particularly pronounced for alanine and leucine, which exhibit a decline in relative CO_2 formation to less than one fifth. This means that the capacity for their oxidation declines five times more steeply than for total respiration. On the other hand, relative CO_2 formation from glycine increases two- to threefold, i.e., it declines less than in cellular respiration. The differences between amino acids in the presence and in the absence of external substrate as well as the changes during maturation appear to be from the interplay of three processes, i.e., (1) their transport across the cell membrane, (2) the endogenous supply of amino acids from intracellular breakdown, and (3) the number of key enzymes in the metabolic pathways of the individual amino acids. Each of these

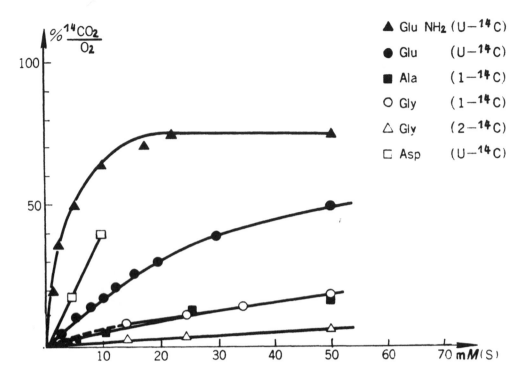

FIGURE 19. "Relative" $^{14}CO_2$ formation from some amino acids as a function of substrate concentration.

processes exhibits a different dependence on maturation. The fact that it is possible to greatly increase relative CO_2 formation for most amino acids by elevation of their external concentration indicates that the decline in oxidation is primarily determined by the first two factors, i.e., the disappearance of carriers and the exhaustion of endogenous sources.

Closer analysis of the metabolic fate of amino acids in the reticulocyte is appropriate at this point. So far, the metabolisms of only two groups of amino acids have been studied in any detail: (1) those metabolized via the citrate cycle, i.e., glutamine, glutamate, aspartate, and alanine, and (2) glycine and serine.

III. METABOLISM OF GLUTAMINE, GLUTAMATE, ASPARTATE, AND ALANINE

These amino acids all have in common that they tend to reach 80% relative CO_2 formation. This fact permits two conclusions: (1) that the enzymatic capacities of their pathways are very large and unsaturated under endogenous conditions as well as in vivo, and (2) that each may monopolize the basic CO_2-forming system, i.e., the citrate cycle, thus suppressing the utilization of other substrates draining into it. It has been demonstrated that there are close interrelations among the amino acids and their deamination products. Glutamate is utilized almost exclusively via aspartate by transamination with oxaloacetate,[10,11] which may yield acetyl-CoA via pyruvate (Figure 20). In this manner, either from glutamate, aspartate, or alanine, decarboxylation of oxaloacetate to pyruvate is functionally reversible. Thus, both compounds necessary for the complete citrate cycle, i.e., acetyl CoA and oxaloacetate, can originate from one source. In this way the high relative CO_2 formation from each single amino acid is accounted for. The equilibrium constant and the steady-state concentrations of ammonium ions, 2-oxoglutarate, and glutamate, as well as direct proof of the low yield of $^{15}NH_3$ from ^{15}N-labeled glutamate all indicate the insignificance of the glutamate dehy-

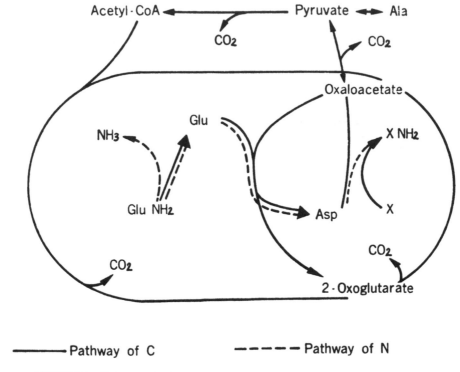

——————— Pathway of C   — — — — — Pathway of N

FIGURE 20.   Pathways of oxidative interconversions of substrate related to the citrate cycle.

drogenase reaction for the catabolism of glutamate.[12] Rather, the enzyme may serve in ammonia fixation. The differences between amino acids, with respect to substrate dependence on CO_2 formation, mainly are caused by their transport properties. The exceptionally steep rise from glutamine, with an apparent K_m of 2 mM, is caused both by its concentrative transport and ease of passive permeation. On the other hand, glutamate, which exhibits an apparent K_m of 40 mM,[9] is known to permeate slowly. Glutamine is split rapidly, mainly by a phosphate-dependent glutaminase located in the mitochondria, and also to a lesser extent by the "glutaminase II" system in the cytosol, which involves transamination and hydrolysis of the oxoglutaramide.[14,15] Thus, permeation is the only significant factor responsible for the difference between glutamine and glutamate. Aspartate, which is next highest in relative CO_2 formation, presumably permeates better than glutamate, while alanine permeation may vary according to the state of maturity of the reticulocyte.

IV. METABOLISM OF SERINE AND GLYCINE

A. Serine and Glycine Synthesis

The carbon skeleton of newly synthesized serine and glycine originates from glucose. Two pathways for the formation of serine appear to be operative in the reticulocyte[7] (Figure 21). Both branch off from the main glycolysis pathway. One starts with 2-phosphoglycerate. In it dephosphorylation precedes dehydrogenation which yields hydroxypyruvate as the immediate precursor of serine. Hydroxypyruvate does not arise in the second pathway, which starts with 3-phosphoglycerate. The phosphate is split off after transamination so that phosphoserine is the immediate precursor of serine. The occurrence of sizable concentrations of phosphoserine apparently is related to the functioning of this pathway.

As acceptors the amino groups serve either hydroxy- or phosphohydroxypyruvate. The synthesis of serine appears to take place in the cytosol with alanine, glutamate, or

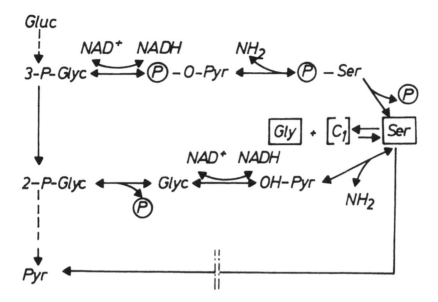

FIGURE 21. Alternative pathways of serine formation. Gly = glycerate.

glutamine as donors of the amino group. The observed dilution of ^{15}N in the serine moiety reflects the sizable intracellular pool of glutamate as well as the exchange of its amino group with that of other amino acids.

B. Fate of Serine and Glycine

The following discussion will be based on the comprehensive scheme in Figure 22. Various data indicate that serine is the main, possibly the exclusive, source of glycine by way of serine hydroxymethyltransferase. Significant glycine formation by transamination with glyoxylate could be excluded.[16] The same goes for *de novo* synthesis of glycine from ammonia and CO_2,[7] which has been described for liver.

Serine is mainly degraded via glycine. Apparently the only serine degradation pathway in the reticulocyte proceeds via a pyridoxalphosphate-dependent serine hydroxymethyltransferase reaction, with the formation of glycine and of N(5), N(10) methylene-FH_4.[17,18] This reaction is located in the mitochondria. Methylene-FH_4 is utilized either for syntheses that occur in the mitochondrium, such as the formation of porphyrins and purines, or is degraded to CO_2. This degradation involves the transfer of methylene-FH_4 from the mitochondrium to the cytosol and is followed by liberation of formate.

Other pathways of serine conversion such as transformation to pyruvate, alanine, or lactate can be excluded.[19]

Glycine is mainly utilized under physiological conditions in porphyrin and globin synthesis and to a small extent in purine synthesis.[19,20] The only pathway of CO_2 formation from glycine is by way of Shemin cycle reactions. All the enzymes in this cycle are located in the mitochondrial matrix.

The first reaction brought about by 5-ALA synthase liberates CO_2 from the glycine carboxyl group. Under physiological or substrate-free conditions, the 5-ALA formed is utilized almost completely for porphyrin synthesis, with CO_2 formation from the 2-carbon of glycine amounting to only 3% of that from the carboxyl carbon. If the substrate concentration of glycine is increased, the subsequent reactions of the Shemin cycle acquire greater prominence, presumably due to an increased concentration of 5-ALA. By transamination, 2-oxoglutaraldehyde is formed, from which two possible pathways have been demonstrated: (1) de-

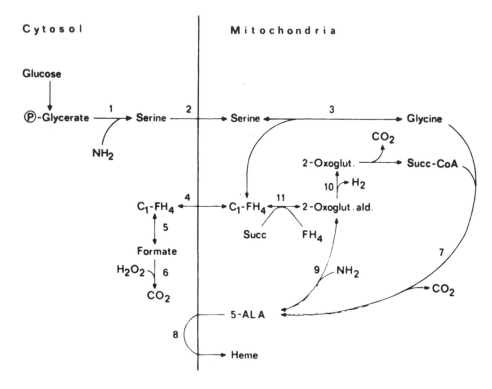

FIGURE 22. Survey of serine and glycine metabolism in reticulocytes. 1. Two pathways of serine formation. 2. Concentrative uptake. 3. Serine hydroxymethyltransferase. 4. Exchange of methylenetetrahydrofolate. 5. N^5,N^{10}-Methylene-FH_4-dehydrogenase, cyclohydrase, FH_4-formylase. 6. Catalase. 7. 5-ALA-synthase. 8. Heme-synthase. 9. 5-ALA-transaminase. 10. 2-Oxoglut.ald.-dehydrogenase. 11. 2-Oxoglut.ald.-hydroxymethyltransferase.

hydrogenation to 2-oxoglutarate, and (2) C_1-transfer to FH_4 and formation of succinyl-CoA. The reactions from 5-ALA via 2-oxoglutaraldehyde to succinate appear to be freely reversible, whereas the dehydrogenation of 2-oxoglutaraldehyde to 2-oxoglutarate is not. At high glycine concentrations the pathway via 2-oxoglutarate, which yields CO_2 by way of the citrate cycle, predominates over that via succinate.

C. Compartmentation of Serine and Glycine in Reticulocytes
   The lower degree of glycine labeling as compared with serine, which was observed both in experiments with (^{14}C)-glucose and (^{15}N)-glutamate, appears to be in contradiction with the reversible nature of the serine hydroxymethyltransferase reaction.
   The explanation was found in the concentrative uptake of serine by mitochondria and the slow permeation of glycine through mitochondrial membranes. In the absence of succinate and pyridine nucleotides no concentrative uptake occurred. The slow, if any, equilibration between serine and glycine indicated by the labeling experiments would suggest that both the efflux and the influx of glycine through the mitochondrial membrane are slow processes. Thus the large pool of cytosolic glycine with its concentration of about 1.5 mmol/ℓ cells is largely insulated from the rapid equilibrium reaction between serine and glycine in the mitochondrial matrix.[7]

D. Balance between Supply and Utilization of Serine and Glycine
   On the basis of the experimental data an overall balance could be given of the quantitative relations between new formation and proteolytic supply of serine and glycine on the one hand and the synthesis of hemoglobin on the other. In Figure 23 the various pathways and

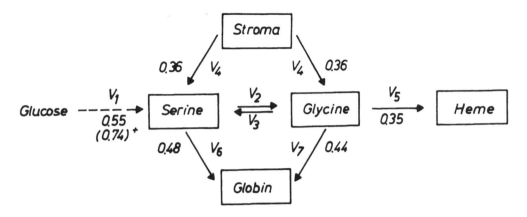

FIGURE 23. Model of the new formation and proteolytic supply of serine and glycine and their utilization for the synthesis of heme and globin. ⁺ Value includes CO_2 formation from C_2 of glycine and purine synthesis. (From Rapoport, S., Müller, M., Dumdey, R., and Rathmann, J., *Eur. J. Biochem.*, 108, 449, 1980. With permission.)

their magnitudes are summarized. The new formation of serine from glucose is designated as v_1 and the interconversion of serine and glycine as v_2 and v_3. The proteolytic breakdown of the stroma, which was determined in an independent manner, is designated as v_4; since the stroma contains equal amounts of serine and glycine a single velocity suffices to characterize the liberation of both amino acids. The rate of formation of heme is represented by v_5 and the amounts of serine and glycine incorporated into globin a v_6 and v_7, respectively. The new formation of serine from glucose may amount to as much as 5% of the total glucose utilized by the reticulocyte and furnishes about one half of the serine and glycine moieties required for the synthesis of heme and globin. The remainder is provided by ATP-dependent proteolysis.[15] Under conditions in vivo, amino acids of the plasma may contribute an as yet undetermined share.

V. ONE-CARBON METABOLISM IN THE RETICULOCYTE

The metabolism of both serine and glycine is closely connected to the one-carbon group. Both the C-3 and the C-2 of serine or glycine, respectively, yield C_1-FH_4 compounds, which are either utilized for synthesis, primarily of purines, or yield formate in a reversible reaction (FH_4-formylase). The formate is catabolized to CO_2 by means of catalase, depending on the availability of H_2O_2. The $H_2O_2^-$ required may be supplied by an NADPH-dependent oxidizing system.[21] The release of CO_2 from methylenetetrahydrofolate via NADP-dependent 10-formyl-FH_4 dehydrogenase described for liver does not occur in the reticulocyte.[19] Utilization of one-carbon units for syntheses predominates, of course, in the immature reticulocyte. The share of one-carbon units catabolized to CO_2 via formate by means of catalase increases with maturity, being practically zero in the most immature and approaching 100% in the most mature reticulocyte, which is the situation in the erythrocyte as well (Figure 24). The reverse is true for CO_2 formation via 2-oxoglutarate and the citrate cycle.

The limitation of CO_2 formation from C-2 of glycine via the catalase pathway appears to be in the reaction leading from 2-oxo-glutaraldehyde to the C_1-FH_4 compound, since increase in H_2O_2 supply (by addition of methylene blue) will not augment the yield of $^{14}CO_2$ from (2-^{14}C)-glycine. CO_2 formation from formate increases fourfold.[22] During maturation the enzymes in the mitochondria disappear completely, while those in the cytosol decline to one tenth, except for catalase, which does not change.

A curious circumstance is the almost complete absence of catalase from mature duck erythrocytes, while there is distinct activity of this enzyme in the bone marrow; this activity

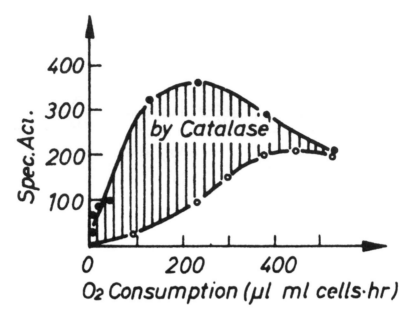

FIGURE 24. $^{14}CO_2$ formation from 3-^{14}C-serine in red cells as a function of maturity (O_2 consumption).

is, however, two orders of magnitude lower than in red cells of man.[23] In the peripheral blood of anemic ducks catalase activity rises up to tenfold and reaches the value achieved in bone marrow. The steep decline in catalase activity during recovery from anemia corresponds to the estimated lifetime of catalase — about 1.5 days.

VI. NITROGEN ECONOMY OF THE RETICULOCYTE: THE FATE OF NITROGEN FROM AMINO ACIDS

Reticulocytes form ammonia in varying amounts, depending on their degree of maturity and the experimental conditions. The formation of ammonia is greatest in immature reticulocytes.[2] It is diminished in the presence of glucose and is increased greatly in anaerobiosis or in the presence of 2,4-DNP.

A large part of the ammonia may be accounted for by the deamination of nucleotides, particularly those of adenine. This pathway was elucidated some time ago. The lack of rephosphorylation of ADP to ATP leads to the formation of AMP by means of adenylate kinase. From ATP leads to the formation of AMP by means of adenylate kinase. From AMP two pathways are available: the first starts with deamination of AMP to IMP followed by dephosphorylation. The inosine is converted to hypoxanthine and ribose-1-phosphate by means of inosine phosphorylase. The other pathway begins with the dephosphorylation of AMP to adenosine which is deaminated by the powerful specific adenosine deaminase.[24-26]

There is, of course, ammonia formation from glutamine by means of glutaminases.[14,15] Glutamate dehydrogenase apparently plays only a minor role, if any, in ammonia formation, since there is only a small liberation of ammonia (3%) from ^{15}N-glutamate under conditions in which glutamate furnishes 30% of the respiratory substrate.

Since amino acids constitute the main metabolic substrate of the reticulocyte under endogenous conditions, one would expect a sizable formation of ammonia. Given a respiratory quotient of about 1, which indicates a preferential oxidation of aliphatic amino acids with less than six carbon atoms, and considering the minor contribution of non-nitrogenous substrates one would expect a ratio of 5:1 between oxygen consumption and NH_3 formation.

There is in fact time-dependent NH$_3$ formation which originates largely from the deamination of nucleotides, particularly of those of adenine. If one corrects for these processes, only a very small liberation of ammonia from amino acids, which falls far behind that expected from the share of amino acids in the respiratory substrate mixture, is left.

As described previously, it was demonstrated that the NH$_3$ potentially available from the oxidation of amino acids by the reticulocyte is utilized by the reticulocyte for the new synthesis of serine via transamination reactions with hydroxypyruvate and phosphohydroxypyruvate.

In other work it was shown that more than 90% of the lysine liberated from mitochondria by proteolysis were reutilized for globin synthesis.[27] Thus, direct proof was obtained for the highly efficient nitrogen economy of reticulocyte metabolism.

VII. *DE NOVO* SYNTHESIS OF PURINE NUCLEOTIDES

Mature erythrocytes are incapable of total synthesis of purine but do contain limited portions of its multistep pathway.[28] They can, however, perform the final step, i.e., the incorporation of the one-carbon unit, which implies that they are also able to synthesize formyl-tetrahydrofolate. Mature erythrocytes also contain adenylosuccinase activity.[29] Rabbit but not human erythrocytes can convert IMP into AMP by a pathway in which glutamine serves as the donor of the amino group.[30] Adenylosuccinate synthetase apparently is missing in the human erythrocyte.[29] The reticulocyte, on the other hand, is capable of *de novo* synthesis of both adenine and guanine nucleotides as indicated by the incorporation of (2-^{14}C)-glycine and ^{14}C-formate.[31,32] The extent of this synthesis is small and was estimated to amount to approximately 0.1 μmol $\times$ mℓ cells^{-1} $\times$ hr^{-1} of purine.[7] The decay of whichever step leads to interruption of *de novo* synthesis of purine nucleotides during maturation has thus far not been pinpointed.

REFERENCES

1. **Warburg, O., Kubowitz, F., and Christian, W.,** Über die Wirkung von Phenylhydrazin und Phenylhydroxylamin auf den Stoffwechsel der roten Blutzellen, *Biochem. Z.,* 242, 170, 1931.
2. **Schweiger, H. G., Rapoport, S., and Schölzel, E.,** Der N-Stoff-wechsel bei der Erythrocyten-Reifung; Reststickstoffbuidung und Hämoglobinsynthese, *Hoppe-Seyler's Z. Physiol. Chem.,* 306, 33, 1957.
3. **Schweiger, H. G. and Rapoport, S.,** Der N-Stoffwechsel bei der Erythrocytenreifung: Die N-Bilanz unter endogenen Bedingungen, *Hoppe-Seyler's Z. Physiol. Chem.,* 313, 97, 1958.
4. **Schultze, M., Rapoport, S., Scholz, I., and Lach, A.,** Die Veratmung von Aminosäuren im Retikulozyten, *Folia Haematol.,* 83, 361, 1965.
5. **Buchmann, R. and Rapoport, S.,** Freie Aminosäuren im Retikulozyten; das Verhalten von Glycin, Alanin und Glutamat bei kurzzeitigen Inkubationen, *Acta Biol. Med. Ger.,* 13, 819, 1964.
6. **Antonioli, J. A. and Christensen, H. N.,** Differences in schedules of regression of transport systems during reticulocyte maturation, *J. Biol. Chem.,* 244, 1505, 1969.
7. **Rapoport, S., Müller, M., Dumdey, R., and Rathmann, J.,** Nitrogen economy and the metabolism of serine and glycine in reticulocytes of rabbits, *Eur. J. Biochem.,* 108, 449, 1980.
8. **Rapoport, S.,** Molekularbiologische Probleme der Reifung von Erythrozyten, *Folia Haematol.,* 89, 105, 1968.
9. **Rapoport, S., Rost, J., and Schultze, M.,** Glutamine and glutamate as respiratory substrates of rabbit reticulocytes, *Eur. J. Biochem.,* 23, 166, 1971.
10. **Rost, J. and Rapoport, S. M.,** The pathway of glutamate oxidation in rabbit reticulocytes, *Eur. J. Biochem.,* 26, 106, 1972.
11. **Müller, M., Thamm, R., and Rapoport, S.,** On the formation of acetyl-CoA from glutamine and the mutual interconversion between oxaloacetate and pyruvate in rabbit reticulocytes, *FEBS Lett.,* 42, 279, 1974.

12. **Rost, J., Müller, M., Schultze, M., and Rapoport, S.**, The formation of free ammonia from glutamate oxidation in rabbit reticulocytes; evidence for the insignificance of oxidative deamination, *FEBS Lett.*, 24, 15, 1972.

13. **Heinz, E., Pichler, A. G., and Pfeiffer, B.**, Studies on the transport of glutamate in Ehrlich cells — inhibition by other amino acids and stimulation by H-ions, *Biochem. Z.*, 342, 542, 1965.

14. **Ababei, L.**, Über eine Erythrozyten-Glutaminase; ihr Verhalten bei der Reifung der roten Blutzelle, *Acta Biol. Med. Ger.*, 5, 630, 1960.

15. **Sandring, K. H., Rohde, A., and Rost, J.**, Glutaminase in roten Blutzellen, *Folia Haematol.*, 89, 208, 1968.

16. **Urbahn, H. and Rapoport, S.**, Über den Abbau von Glyzin in roten Blutzellen, *Acta Biol. Med. Ger.*, 6, 16, 1961.

17. **Bertino, J. R., Simmons, B., and Donohne, D. M.**, Purification and properties of the formate-activating enzyme from erythrocytes, *J. Biol. Chem.*, 237, 1314, 1962.

18. **Wilmanns, W. and Jaenicke, L.**, Die Bedeutung des Folsäurestoffwechsels für die normale und pathologische Reifung von Blutzellen, *Klin. Wochenschr.*, 41, 1075, 1963.

19. **Rapoport, S., Müller, M., and Knöfel, R.**, $^{14}CO_2$-Bildung aus [3-^{14}C] Serin in roten Blutzellen, *Eur. J. Biochem.*, 10, 207, 1969.

20. **Shemin, D.**, Biosynthesis of porphyrins, *Erg. Physiol.*, 49, 299, 1957.

21. **Rapoport, S., Müller, M., Hartwig, A., and Dumdey, R.**, Antimyzin A-resistenter Sauerstoffverbrauch in Kaninchenretikulozyten, *Acta Biol. Med. Ger.*, 34, 1301, 1975.

22. **Rapoport, S. and Müller, M.**, CO_2 formation from formate in red blood cells, *Acta Biochim. Pol.*, 14, 143, 1967.

23. **Rapoport, S., Hartwig, A., and Gross, J.**, Abhängigkeit der Katalaseaktivität von der Reifung roter Blutzellen der Ente, *Acta Biol. Med. Ger.*, 32, 601, 1974.

24. **Rapoport, S. M., Rapoport, I., Schauer, M., and Heinrich, R.**, The effect of pyruvate on glycolysis and the maintenance of adenine nucleotides in red cells, *Acta Biol. Med. Ger.*, 40, 669, 1981.

25. **Schauer, M., Heinrich, R., and Rapoport, S. M.**, Mathematische Modellierung der Glykolyse und des Adeninnukleotidstoffwechsels menschlicher Erythrozyten. I. Reaktionskinetische Ansätze, Analyse des in vivo Zustandes und Bestimmung der Anfangsbedingungen für in vitro Experimente, *Acta Biol. Med. Ger.*, 40, 1659, 1981.

26. **Schauer, M., Heinrich, R., and Rapoport, S. M.**, Mathematische Modellierung der Glykolyse und des Adeninnukleotidstoffwechsels menschlicher Erythrozyten. II. Simulation des Adeninnukleotidabbaus bei Glukoseverarmung, *Acta Biol. Med. Ger.*, 40, 1683, 1981.

27. **Siems, W., Dubiel, W., Dumdey, R., Müller, M., and Rapoport, S. M.**, Accounting for the ATP-consuming processes in rabbit reticulocytes, *Eur. J. Biochem.*, 139, 101, 1984.

28. **Lowy, B. A. and Williams, M. K.**, The presence of a limited portion of the pathway *de novo* of purine nucleotide biosynthesis in the rabbit erythrocyte in vitro, *J. Biol. Chem.*, 235, 2924, 1960.

29. **Lowy, B. A. and Dorfman, B.-Z.**, Adenylosuccinase activity in human and rabbit erythrocyte lysates, *J. Biol. Chem.*, 245, 3043, 1970.

30. **Lowy, B. A., Williams, M. K., and London, I. M.**, Enzymatic deficiencies of purine nucleotide synthesis in the human erythrocyte, *J. Biol. Chem.*, 237, 1622, 1962.

31. **Kruh, J. and Borsook, H.**, In vitro synthesis of ribonucleic acid in reticulocytes, *Nature (London)*, 175, 386, 1955.

32. **Lowy, B. A., Cook, J. L., and London, I. M.**, The biosynthesis of purine nucleotides *de novo* in the rabbit reticulocyte in vitro, *J. Biol. Chem.*, 236, 1442, 1961.

Chapter 9

HEME SYNTHESIS

I. INTRODUCTION

Since several extensive reviews on heme synthesis are available,[1,2] I shall limit myself to a brief review focusing on later publications and on the aspects that are germane to erythroid cells.

Heme synthesis occurs in a highly regulated manner by a pathway that involves eight enzymes. Four are located in the mitochondria including the initial step, the synthesis of 5-ALA, and the last three steps (Figure 25 and Table 14), whereas the intermediate reactions are cytosolic. For synthesis, several transports between mitochondria and cytosol are obligatory. They include (1) the uptake of glycine or serine, which is converted to glycine in the mitochondrial matrix. Both transports are concentrative and require the integrity of the mitochondria; (2) the export of 5-ALA to become a substrate of porphobilinogen synthase (5-ALA dehydratase); (3) the uptake of coproporphyrinogen III by the mitochondria; and (4) the transport of heme from the mitochondria to the cytosol in order to be incorporated into hemoglobin. There is very little information about the characteristics of these transport processes which may be sites of regulation. The movement of heme out of the mitochondria depends on the presence of cytosolic proteins.[1]

The differences in cellular localization of enzymes appear to be causally related to their contrasting fates. While enzymes that occur in mitochondria are reduced by two orders of magnitude and even to zero in mature erythrocytes, those in the cytosol persist, so that these cells can be good sources for their purification.

II. 5-ALA SYNTHASE

The initial step in the pathway leading to heme is the formation of 5-ALA. Because of its position in the reaction chain, its relatively low activity, and its susceptibility to repression and feedback inhibition, 5-ALA synthase is considered to be the main site of pathway control. This enzyme occurs in the mitochondrial matrix,[3] having been synthesized in the cytosol as a precursor with a larger molecular mass of 75 kD.[3-5] The import of the enzyme into mitochondria occurs in the same manner as that of other proteins in the mitochondrial matrix and involves the fission of an N-terminal signal peptide and also requires energy. Although no exact data are available, the enzyme appears to have a much longer half-life than that in liver mitochondria, in which a half-life of 1 to 3 hr has been determined.[1] It remains an open question what kind of mechanism is operative either to specifically break down this enzyme in liver mitochondria or to stabilize it in other types of cells. Aoki et al.[6] found a specific seryl-protease (molecular mass of 18 kD) which destroys the apo-form of 5-ALA synthase but does not act on other pyridoxalphosphate enzymes in their apo-form. This protease was obtained from human bone marrow. Its activity was high in erythroblasts and young granulocytes and apparently decreased during maturation of erythroid cells. Marked increases in enzyme activity were observed in the bone marrow of patients with rheumatoid arthritis and in those with chronic myelogenous leukemia. The significance of these observations for maturational events in red cells remains doubtful, since the protease was not isolated from pure erythroid cells. The conditions by which its activity was found to be increased are not characterized by great changes in erythropoiesis.

In the course of experimental anemia by bleeding in rabbits, the activity of 5-ALA synthase followed a course similar to that of cytochrome oxidase (see Chapter 3), declining by more than one order of magnitude during maturation of reticulocytes.[7]

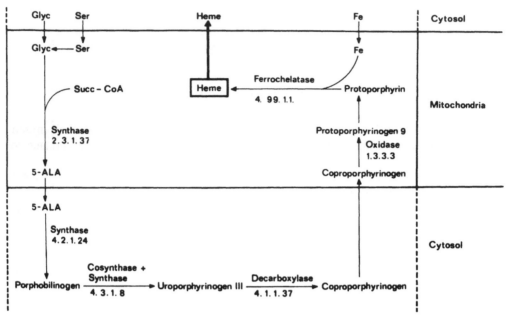

FIGURE 25.   Survey of heme synthesis enzymes and their intracellular localization.

Table 14
SOME PROPERTIES OF HEME SYNTHESIS ENZYMES[a]

Enzyme	Enzyme nomenclature (EC)	Location	Mol mass (kD)	Substrate K_m (μM)
5-ALA Synthase	2.3.1.37	Mitochondria	68	Gly: 4000; Succ-CoA: 5; Pyridoxal-P: 5
Porphobilinogen synthase (5-ALA dehydratase)	4.2.1.24	Cytosol	250 (octamer)	5-ALA: 270
Uroporphyrinogen synthase (porphobilinogen deaminase)	4.3.1.8	Cytosol	37	Porphobilinogen: 6
Uroporphyrinogen cosynthase		Cytosol	40	
Uroporphyrinogen decarboxylase	4.1.1.37	Cytosol	80 (dimer)	Uroporphyrinogen: 0.35; pentacarboxylic porphyrinogen: 0.05
Coproporphyrinogen oxidase (coproporphyrinogen decarboxylase)	1.3.3.3	Mitochondria	72	Coproporphyrinogen: 20
Protoporphyrinogen oxidase		Mitochondria	35	Protoporphyrinogen: 11
Ferrochelatase (heme synthase)	4.99.1.1	Mitochondria	40	Protoporphyrin: 30(1); Fe^{2+} 160

[a]   Listed in order of reaction steps.

5-ALA synthase has been purified to homogeneity from both rabbit reticulocytes[8] and from chick embryo liver mitochondria.[9,10] According to the most recent report, the enzyme has a molecular mass of 68 kD.[10] The lower molecular mass of 50 kD reported earlier appears to have been due to proteolytic degradation.[9]

Studies on the mechanism of action of 5-ALA synthase have elucidated a detailed reaction sequence. The first step is the formation of a Schiff base between glycine and the pyridoxal

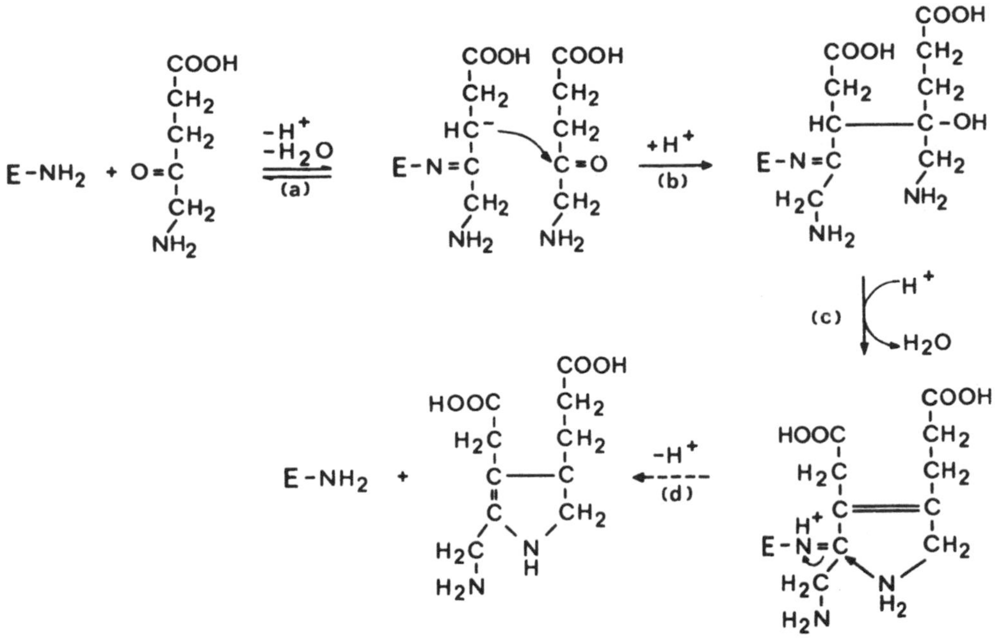

$$E-NH_2 + O=C \overset{-H^+}{\underset{(a)}{\rightleftharpoons}} \quad E-N=C \xrightarrow[(b)]{+H^+} \quad E-N=C$$

(with substituents: COOH, CH$_2$, CH$_2$, C=O, NH$_2$ etc. as shown)

(c) with $+H^+$ / $-H_2O$

$$E-NH_2 + \quad \xleftarrow[(d)]{-H^+}$$

FIGURE 26. Enzymatic mechanism of porphobilinogen synthase (5-ALA dehydratase).

enzyme, followed by the removal of a proton from the C-2 of the glycine. This is followed by condensation with succinyl-CoA and decarboxylation of the intermediate, and, finally, by addition of a proton and release of the free 5-ALA.[1] The enzyme has very low affinity for glycine with a K_m of about 5 mM, whereas its affinities for succinyl-CoA and pyridoxal phosphate, both with K_m-values of 5 μM, are very high. In isolated form the enzyme exhibits Michaelis-Menten kinetics.

It is interesting that 5-ALA synthase is activated by Pb^{2+} ions, which may explain the overproduction of 5-ALA and porphobilinogen in lead poisoning.[11]

The following four enzymes are cytosolic. Recently three of them have been isolated in homogeneous form from human erythrocytes.

III. FROM 5-ALA TO UROPORPHYRINOGEN III

Porphobilinogen synthase (5-ALA dehydratase) is the most abundant among the enzymes of heme synthesis. It was purified 38,000-fold from human erythrocytes. It has a molecular weight of 250 kD and is an octamer.[12] It appears to be a zinc enzyme and may contain a pyridoxal derivative.[13] In keeping with this postulate pyridoxal-5-phosphate was a strong competitive inhibitor. The K_m of the enzyme for its substrate was found to be 0.27 mM.

In the reticulocytes of anemic mutant strains of mice, porphobilinogen synthase activity has been estimated to be about 40 times higher than in mature erythrocytes.[14,15]

Porphobilinogen is formed by the condensation of two 5-ALA molecules with the loss of two molecules of water (Figure 26). According to detailed studies,[16] the first molecule of 5-ALA is bound in the form of a Schiff base between the amino group of a lysine residue and the keto group of 5-ALA with a loss of one water molecule. An aldol condensation with the subsequent elimination of the second molecule of water follows.

The formation of uroporphyrinogen III involves a position-specific condensation of four molecules of porphobilinogen. It is catalyzed by the cooperative action of two enzymes: uroporphyrinogen synthase and cosynthase.[17] Only the first has been isolated in a homogeneous form from human erythrocytes and was well characterized.[18] It has a molecular mass of 37 kD and appears to be a monomer. It has high affinity for its substrate with a K_m of 6 μM.

The formation of the tetrapyrrole involves a sequential addition of porphobilinogens; accordingly, four forms of enzyme-substrate intermediates are to be expected and have actually been found. The enzyme by itself catalyzes only the formation of uroporphyrinogen I. For the formation of the III-form, the activity of heat-labile uroporphyrinogen III cosynthase is necessary. The two enzymes must interact with each other during tetrapyrrole formation since the end product of the synthase, uroporphyrinogen I, is not a substrate for the cosynthase. Several schemes on the mechanism of the combined action of these two enzymes have been proposed.[1] According to the extensive work of Battersby[19] a head-to-tail condensation of all four porphobilinogens occurs, followed by intramolecular rearrangement in the resulting enzyme-bound straight-chain bilane, which involves only the D ring and the two adjunct C-atoms, catalyzed by the simultaneous action of both enzymes.

The properties of cosynthase have been delineated only to a small extent.[20] Only recently have there been major advances in our knowledge. The enzyme has been purified from rat liver to such an extent that its molecular mass could be determined. It was recognized to be a folate-binding protein which contains tetrahydropteroyl-heptaglutamic acid.[21] The most remarkable observation was that the coenzyme alone sufficed to direct the action of uroporphyrinogen I synthase to the formation of the III-isomer. It was postulated that the folate participates in the one-carbon transfer in the scheme of Battersby. However, a uroporphyrin III synthase (cosynthase) isolated from *Euglena gracilis* lacked any cofactors and did not require them for its activity.[21a]

IV. FROM UROPORPHYRINOGEN III TO PROTOPORPHYRIN*

The next step in the synthesis of porphyrine is the removal of the four carboxyl groups of the acetic acid side chains, one from each of the pyrroles. In this way coproporphyrinogen III is formed. The sequential removal of all four carboxyls is catalyzed by a single enzyme, uroporphyrinogen decarboxylase. This enzyme recently has been purified to homogeneity both from human and chicken erythrocytes.[22-24] Decarboxylase occurs apparently as a dimer with an 80 kD molecular mass. It does not seem to contain either a metal or a cofactor. It has high affinity for all its substrates with a K_m for uroporphyrinogen of 0.35 μM and 0.05 μM for the ultimate substrate, pentacarboxylic protoporphyrinogen.

None of the following three mitochondrial enzymes leading from coproporphyrinogen to protoporphyrin have been purified from reticulocytes so far. Homogeneous preparations have been obtained from mitochondria of liver. In view of the similarity of properties the data on liver enzymes should be applicable to erythroid cells.

The next step in the pathway is the oxidative decarboxylation of the propionic acid groups of pyrrole rings A and B with the formation of two vinyl groups, thus forming protoporphyrinogen 9. This reaction proceeds stepwise, catalyzed by a single enzyme, coproporphyrinogen oxidase (decarboxylase). The enzyme has been obtained in a homogeneous form from bovine liver mitochondria.[25] It is a monomer with a molecular mass of 72 kD which does not seem to contain either a metal or a prosthetic group.

The K_m for its substrate is about 20 μM. The enzyme requires molecular oxygen. The detailed mechanism of the reaction catalyzed by it is unknown; the intermediate formation of hydroxy-propionates has been shown. This enzyme may be located in the intermembrane space of mitochondria.[26]

The formation of protoporphyrin involves the oxidation of protoporphyrinogen 9 with the formation of methine bridges. In this manner porphyrin acquires its aromatic character. The reaction requires molecular oxygen. The enzyme from rat liver[27] has a molecular mass of 35 kD and a K_m for its substrate of 11 μM. It is likely that the reaction proceeds stepwise

---

* See Figure 27.

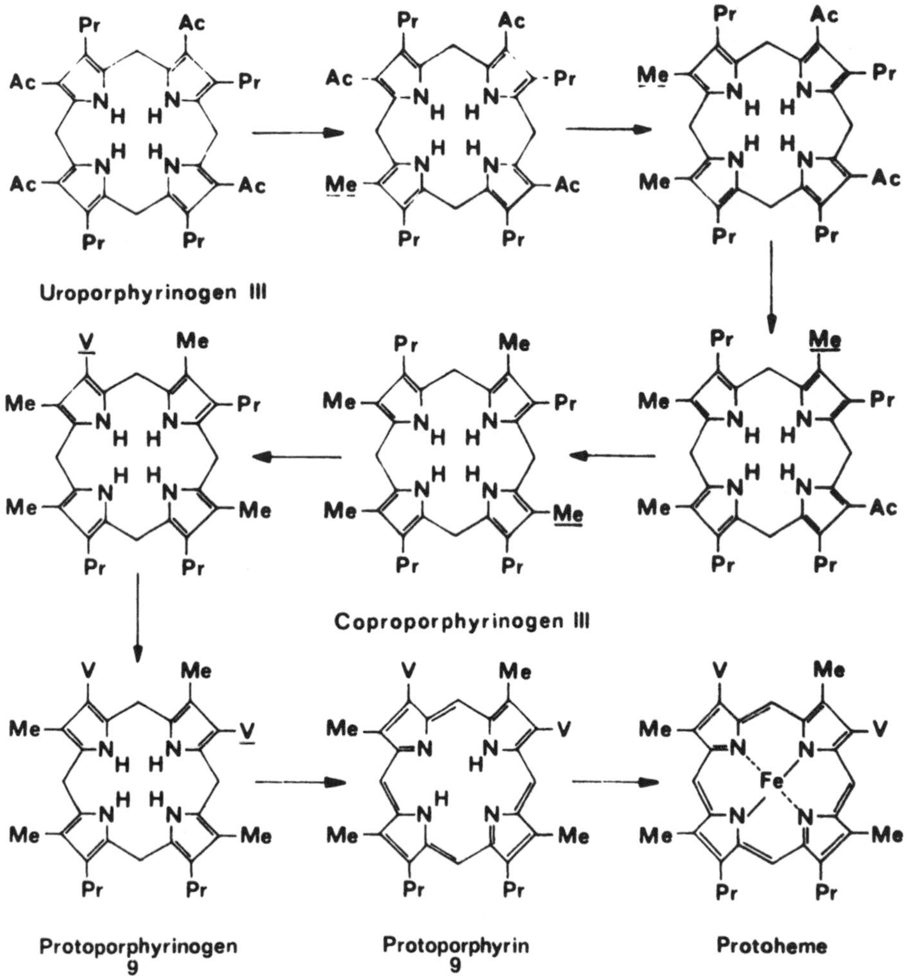

FIGURE 27. The pathway from uroporphyrinogen III to protoheme. Pr, Ac, Me, V = propionyl, acetyl, methyl, and vinyl side chains, respectively.

with the formation of partially oxidized intermediates. Again there does not appear to be evidence for the participation of a metal or cofactor. The enzyme is located in the lipid layer of the inner membrane.[41]

V. FERROCHELATASE

The final reaction of the pathway consists of the insertion of iron into protoporphyrin to form heme. The enzyme responsible is ferrochelatase, which is located in the inner membrane of mitochondria.[28] It has been obtained in homogeneous form from both beef and rat liver,[29-31] chicken erythrocyte mitochondria, as well as from *Rhodopseudomonas sphaeroides*.[29-33]

Ferrochelatase from all sources has a molecular mass of 40 kD. Again no metal or cofactor has been found. It is highly sensitive to Pb^{2+} and Hg^{2+}, which indicates the importance of SH-groups for its function. The affinity constants have been determined for the microbial enzyme and found to be 30 μM for both porphyrin and iron. Nearly identical values were found for ferrochelatase of human liver.[34] For the enzyme from chicken erythrocyte mitochondria, a similar K_m for protoporphyrin but a higher K_m of 160 μM for Fe^{2+} was reported.[32] The reaction may be written as follows:

$$Fe^{2+} + protoporphyrin = heme + H^+$$

Evidence has been obtained for a sequential order of iron binding preceding porphyrin and heme release occurring before proton release. Based on studies on the modification of sulfhydryl groups of ferrochelatase, it has been suggested that two vicinal sulfhydryls are important for Fe^{2+} binding and that they are the source of the liberated protons.[35]

The activity of ferrochelatase has been studied in human and rat erythroid cells.[36] Distinct differences were found which are in keeping with advanced maturity of human reticulocytes compared to those of experimental animals. Activity was highest in the bone marrow in the rat and within an order of magnitude as high in reticulocytes, but was less than $^1/_{100}$ in mature erythrocytes. Activity was $^1/_{300}$ in reticulocytes as compared with that in bone marrow. There was little if any difference between reticulocytes and erythrocytes. Other indicators like heme synthesis corresponded to great differences in the enzyme activity between human and rat reticulocytes. In this study a higher affinity for protoporphyrin with a K_m of about 1 μ*M* was found in a crude system.

It is uncertain whether the data indicate a difference in structure between the liver enzymes in mammals or in chickens.[32] There may also be a difference between erythroid ferrochelatase and that of liver, since protoporphyria (which is caused by a malfunction of this enzyme) is a disease of the erythroid system and does not affect the liver.

VI. CHANGES DURING DIFFERENTIATION AND MATURATION

During differentiation various enzymes undergo characteristic changes. In erythroleukemia cells from mice transformed by Friend virus as well as from normal human bone marrow cells in culture, sequential increases in enzyme activities in the heme biosynthesis pathway were observed. Accumulation of the mRNAs for porphyrobilinogen deaminase and uroporphyrinogen dicarboxylase were found along with those for α- and β-globulin.[37] Correspondingly, 5-ALA synthase, porphobilinogen synthase, and porphobilinogen deaminase increased early, 2 days prior to the appearance of hemoglobin, while incorporation of radioactive iron, which ferrochelatase activity, occurred late, indicating a critical role for that enzyme.[14,15,37,38] In a study on induced erythroleukemia cells,[36] particular attention was paid to 5-ALA synthase and ferrochelatase, which have the lowest activity.[39] In noninduced cells, neither enzyme was detectable. After induction with dimethyl sulfoxide all enzymes increased. 5-ALA synthase made its appearance within 24 hr after induction, long before heme synthesis was found, whereas ferrochelatase became manifest after 72 hr, synchronous with the onset of heme synthesis. Thus, it seems that ferrochelatase is the main enzyme controlling heme biosynthesis during differentiation. The time courses of heme and globin formation continue in a parallel manner during the further stages of differentiation and reach the highest level in the proerythroblast.

In studies on mice with hereditary hemolytic diseases, it was found that the homozygotes with a reticulocytosis of about 50% and an extremely short lifespan for red cells exhibited 10- to 20-fold increases in 5-ALA dehydratase, uroporphyrinogen I synthase, and protoporphyrin.[14,15] Even in heterozygote red cells with 4 to 8% reticulocytes and a compensated hemolysis, increases by about 50% of all characteristics studied were found. There was a good correlation between the degree of reticulocytosis and the extent of elevation of both enzyme activities and of protoporphyrin. It was estimated that the levels of 5-ALA dehydratase and uroporphyrinogen-I synthase, as well as of protoporphyrin, were by more than one order of magnitude higher in reticulocytes than in erythrocytes. A further result of this study was the observation of a strong developmental trend in these indexes. Both enzymes and protoporphyrin declined strongly during the first 4 to 6 weeks of life in normal mice. In rats with phenylhydrazine-induced anemia, a tenfold increase in 5-ALA dehydratase activity was found, which was strictly correlated with the percentage of reticulocytes.[40]

REFERENCES

1. **Granick, S. and Beale, S. I.,** Hemes, chlorophylls and related compounds: biosynthesis and metabolic regulation, *Adv. Enzymol.,* 46, 33, 1978.
2. **Granick, S. and Sassa, S.,** δ-Aminolevulinic acid synthetase and the control of heme and chlorophyll synthesis, in *Metabolic Pathways,* Vol. 5, Vogel, H. J., Ed., Academic Press, New York, 1971, 79.
3. **Patton, G. M. and Beattle, D. S.,** Studies on hepatic δ-aminolevulinic acid synthetase, *J. Biol. Chem.,* 248, 4467, 1973.
4. **Ades, I. Z. and Harpe, K. G.,** Biogenesis of mitochondrial proteins, *J. Biol. Chem.,* 256, 9329, 1981.
5. **Hay, R., Böhni, P., and Gasser, S.,** How mitochondria import proteins, *Biochim. Biophys. Acta,* 779, 65, 1984.
6. **Aoki, Y., Urata, G., Takaku, F., and Katunuma, N.,** A new protease inactivating δ-aminolevulinic acid synthetase in mitochondria of human bone marrow cells, *Biochem. Biophys. Res. Commun.,* 65, 567, 1975.
7. **Rosenthal, S., Gross, J., Grauel, E. L., Papies, B., Schulz, W., Belkner, J., Botscharowa, L., Coutelle, Ch., Hawemann, M., Nieradt-Hiebsch, Ch., Müller, M., Opitz, M., Prehn, S., Schultze, M., Staak, R., and Wiesner, R.,** Leitkriterien der Reticulocytenreifung, in *7th Internationales Symposium über Struktur und Funktion der Erythrocyten,* Rapoport, S. and Jung, F., Eds., Akademie-Verlag, Berlin, 1972, 513.
8. **Aoki, Y., Wada, O., Urata, G., Takaku, F., and Nakao, K.,** Purification and some properties of δ-aminolevulinate (ALA) synthetase in rabbit reticulocytes, *Biochim. Biophys. Res. Commun.,* 42, 568, 1971.
9. **Whiting, M. J. and Granick, S.,** δ-Aminolevulinic acid synthetase from chick embryo liver mitochondria, *J. Biol. Chem.,* 251, 1340, 1976.
10. **Borthwick, I. A., Srivastava, G., Brooker, D., May, B. K., and Elliott, W. H.,** Purification of 5-aminolevulinate synthase from liver mitochondria of chick embryo, *Eur. J. Biochem.,* 129, 615, 1983.
11. **Pirola, B. A., Borthwick, I. A., Srivastava, G., May, B. K., and Elliott, W. H.,** Effect of lead ions on chick-embryo liver mitochondria δ-aminolevulinate synthase, *Biochem. J.,* 222, 627, 1984.
12. **Anderson, P. M. and Desnick, R. J.,** Purification and properties of δ-aminolevulinate dehydrogenase from human erythrocytes, *J. Biol. Chem.,* 254, 6924, 1979.
13. **Bevan, D. R., Bodlaender, P., and Shemin, D.,** Mechanism of porphobilinogen synthase, *J. Biol. Chem.,* 255, 2030, 1980.
14. **Sassa, A. and Bernstein, S. E.,** Levels of δ-aminolevulinate dehydratase, uroporphyrinogen-I synthase, and protoporphyrin IX in erythrocytes from anemic mutant mice, *Proc. Natl. Acad. Sci. U.S.A.,* 74, 1181, 1977.
15. **Sassa, S., Kappas, A., Bernstein, S. E., and Alvares, A. P.,** Heme biosynthesis and drug metabolism in mice with hereditary hemolytic anemia, *J. Biol. Chem.,* 254, 729, 1979.
16. **Nandi, D. L. and Shemin, D.,** δ-Aminolevulinic acid dehydrogenase of *Rhodopseudomonas spheroides;* mechanism of porphobilinogen synthesis, *J. Biol. Chem.,* 243, 1236, 1968.
17. **Frydman, R. B. and Feinstein, G.,** Studies on porphobilinogen deaminase and uroporphyrinogen III cosynthase from human erythrocytes, *Biochim. Biophys. Acta,* 350, 358, 1974.
18. **Anderson, P. M. and Desnick, R. J.,** Purification and properties of uroporphyrinogen I synthase from human erythrocytes, *J. Biol. Chem.,* 255, 1993, 1980.
19. **Battersby, A. R.,** The discovery of nature's biosynthetic pathways, *Experientia,* 34, 1, 1978.
20. **Frydman, B., Frydman, R. B., Valasinas, A., Levy, E. S., and Feinstein, G.,** Biosynthesis of uroporphyrinogens from porphobilinogen: mechanism and nature of the process, *Phil. Trans. R. Soc. London Ser. B,* 273, 137, 1976.
21. **Kohashi, M., Clement, R. P., Tse, J., and Piper, W. N.,** Rat hepatic uroporphyrinogen III cosynthase, *Biochem. J.,* 220, 775, 1984.
21a. **Hart, G. J., and Battersby, A. R.,** Purification and properties of uroporphyrinogen III synthase (cosynthetase) from *Euglena gracilis, Biochem. J.,* 232, 151, 1985.
22. **Verneuil, H. D., Sassa, S., and Kaplan, A.,** Purification and properties of uroporphyrinogen decarboxylase from human erythrocytes, *J. Biol. Chem.,* 258, 2454, 1983.

23. **Elder, G. H., Tovey, J. A., and Sheppard, D. M.,** Purification of uroporphyrinogen decarboxylase from human erythrocytes, *Biochem. J.,* 215, 45, 1983.

24. **Kawanishi, S., Seki, Y., and Sano, S.,** Uroporphyrinogen decarboxylase, *J. Biol. Chem.,* 258, 4285, 1983.

25. **Yoshinaga, T. and Sano, S.,** Coproporphyrinogen oxidase. I. Purification, properties, and activation by phospholipids, *J. Biol. Chem.,* 255, 4722, 1980.

26. **Elder, G. H. and Evans, J. O.,** Evidence that the coproporphyrinogen oxidase activity of rat liver is situated in the intermembrane space of mitochondria, *Biochem. J.,* 172, 345, 1978.

27. **Poulson, R.,** The enzymatic conversion of protoporphyrinogen IX to protoporphyrin IX in mammalian mitochondria, *J. Biol. Chem.,* 251, 3730, 1976.

28. **Jones, M. S. and Jones, O. T. G.,** The structural organization of haem synthesis in rat liver mitochondria, *Biochem. J.,* 113, 507, 1969.

29. **Taketani, S. and Tokunaga, R.,** Rat liver ferrochelatase, *J. Biol. Chem.,* 256, 12748, 1981.

30. **Taketani, S. and Tokunaga, R.,** Purification and substrate specificity of bovine liver-ferrochelatase, *Eur. J. Biochem.,* 127, 443, 1982.

31. **Dailey, H. A. and Fleming, J. E.,** Bovine ferrochelatase, *J. Biol. Chem.,* 258, 11453, 1983.

32. **Hanson, J. W. and Dailey, H. A.,** Purification and characterization of chicken erythrocyte ferrochelatase, *Biochem. J.,* 222, 695, 1984.

33. **Dailey, H. A.,** Purification and characterization of membrane-bound ferrochelatase from Rhodopseudomonas sphaeroides, *J. Biol. Chem.,* 257, 14714, 1982.

34. **Camadro, J.-M., Ibraham, N. G., and Levere, R. D.,** Kinetic studies of human liver ferrochelatase, *J. Biol. Chem.,* 259, 5678, 1984.

35. **Dailey, H. A.,** Effect of sulfhydryl group modification on activity of bovine ferrochelatase, *J. Biol. Chem.,* 259, 2711, 1984.

36. **Verhoef, N. J., Noordeloos, P. J., and Leijnse, B.,** Heme synthetase activity in normal human and rat erythroid cells and in sideroblastic anemia, *Clin. Chim. Acta,* 82, 45, 1978.

37. **Grandchamp, B., Beaumont, C., de Verneuil, H., and Nordman, Y.,** Accumulation of porphobilinogen deaminase, uroporphyrinogen decarboxylase, and α- and β-globin mRNAs during differentiation of mouse erythroleukemic cells. Effects of succinylacetone, *J. Biol. Chem.,* 260, 9630, 1985.

38. **Sassa, S. and Urabe, A.,** Uroporphyrinogen-I synthase induction in normal human bone marrow cultures: an early and quantitative response of erythroid differentiation, *Proc. Natl. Acad. Sci. U.S.A.,* 76, 5321, 1979.

39. **Rutherford, T., Thompson, G. G., and Moore, M. R.,** Heme biosynthesis in Friend erythroleukemia cells: control by ferrochelatase, *Proc. Natl. Acad. Sci. U.S.A.,* 76, 833, 1980.

40. **Abdulla, M. and Svensson, S.,** Effect of phenylhydrazine on delta-aminolevulinic acid dehydratase in red blood cells, *Enzyme,* 23, 164, 1978.

41. **Deybach, J. C., de Silva, V., Grandchamp, B., and Nordman, Y.,** The mitochondrial location of protoporphyrinogen oxidase, *Eur. J. Biochem.,* 149, 431, 1985.

145

Chapter 10

UPTAKE AND FATE OF IRON: THE TRANSFERRIN RECEPTOR

I. INTRODUCTION

Iron is an essential nutrient for cellular growth and development, particularly in erythroid cells. The propensity of this element to assume the ferric form, the salts of which are insoluble or have a high tendency to hydrolyze, leads one to expect that the equilibrium concentration of free aquated Fe^{3+} cannot exceed 10^{-17} M at neutrality. Specific iron-sequestering plasma proteins, the transferrins, have evolved to transport iron to and from cells in vertebrates.

For the same physical-chemical reasons, the uptake, intracellular transport, and storage of iron also require special mechanisms and forms. In early work it was established that immature erythroid cells are able to take up transferrin-bound iron and utilize it for hemoglobin synthesis.[1] This process involves binding of the iron-transferrin complex, release of the iron, and its incorporation into heme in the mitochondria.[2,3] Metabolic energy is necessary for uptake and utilization.[1,4-7] Transferrin receptors were later identified.[8] As discussed in the chapter on heme synthesis, iron entering the mitochondria is incorporated into heme by action of the ferrochelatase followed by the release of heme to the cytosol where it combines with globin to form the final product, hemoglobin.

Over the years several controversial questions developed: (1) is transferrin uptake only receptor-mediated?; (2) does the pathway of iron utilization involve ferritin?; (3) what are the details in the release of iron from its transferrin complex?; (4) what are the intermediate steps between membrane and mitochondria?; (5) do the mitochondria play an active role in the intracellular movement of iron in addition to that of incorporating iron into heme?

In the following account the course of iron from the blood plasma to the site of heme synthesis will be followed.

II. TRANSFERRIN

There is no doubt that iron may be taken up by reticulocytes even in the absence of transferrin if offered in soluble form in vitro. On the other hand it is well known that the concentration of free iron in the plasma is negligible and that it is almost completely complexed to transferrin. Therefore physiologic iron uptake by the red cell can only be in the form of iron-transferrin.

The serum transferrin (for a review see Aisen and Listowsky[9]) is a single-chain glycoprotein with a molecular mass of about 80 kD. It contains two complex branched carbohydrate chains, N-asparagine-linked to mannose, acetylglucosamine, galactose, and terminal sialic acid. It appears doubtful that the carbohydrate chains are significant for recognition of the transferrin receptors in the cell membrane.[10-12] The microheterogeneity of transferrin of human blood plasma, which is due to differences in sialic acid content, has been utilized to investigate possible functional diversity of these transferrin forms with respect to their iron-donating properties, with the result that no significant differences in either membrane binding or in iron uptake by rat reticulocytes could be observed. Hence the conclusion was drawn that carbohydrate chains are not directly involved in the process of iron delivery to reticulocytes.[13] The transferrin molecule consists of two domains, both structurally and functionally, that are virtually independent but are not completely equivalent. At pH 6.7 the binding site located in the N-terminal half of the protein has less than $1/20$ the affinity for iron as compared with the other site. However, under physiological conditions more iron is bound to the site with lower affinity, probably for kinetic reasons. The binding of iron is obligatorily connected with bicarbonate, probably in form of carbonate iron. Under physiologic conditions in the blood plasma the effective affinity constants of the sites range from 1 to 6 × 10^{22} M^{-1}.

The existence of two forms of transferrin, monoferric and diferric, is the basis for a hypothesis according to which there should be a functional difference between the two binding sites of transferrin.[14,15] It involves the postulate that the iron-binding site in the C-terminal domain delivers its iron mainly to erythroid cells, whereas the other iron-binding site would mainly supply iron storage pools. Several studies were directed toward answering the question as to whether mono- and diferric transferrin differ in their delivery of iron to reticulocytes. It was found that the iron uptake from rat plasma transferrin by rat reticulocytes did not exhibit any differences between mono- and diferric transferrin.[16] However, a competitive advantage of diferric transferrin in delivering iron was found with human reticulocytes. From mixtures of the two transferrins, seven times more iron was delivered from the diferric form.[17] In a recent study the uptake of iron from pure human monoferric transferrins by human erythroid bone marrow cells and stimulated T-lymphocytes as well as from rat hepatocytes were compared.[18]

Ferrokinetic studies were also performed on healthy volunteers. The results were clearcut. There was no difference between the two monotransferrins, i.e., the one containing the iron in the N-terminal and the one in the C-terminal domain, with respect to the delivery of iron. Thus it appears that both of the iron-binding sites of transferrin are functionally equal.

III. THE TRANSFERRIN RECEPTOR

With the identification of transferrin receptors in reticulocytes[8,19] and the development of a general understanding of the function, properties, and recycling of receptors, a new phase in the studies on iron uptake was initiated.

A. Characteristics of the Transferrin Receptor

In kinetic and equilibrium studies, the first step, binding of diferric transferrin by human reticulocytes, was found to be energy-independent and reversible.[20] The same had been shown for the interaction of transferrin with solubilized receptors from reticulocytes and it was suggested that this reaction may be entropy-driven.[21]

Great progress has been made in the characterization of the transferrin receptor. It is a transmembrane protein, which under physiological conditions occurs as a disulfide-linked dimer with a molecular mass of 180 kD.[22-26] It contains N-asparagine-linked carbohydrates including mannose, galactose, N-acetylglucosamine, and sialic acid, which are arranged in two high-mannose chains and one complex chain. Furthermore, it contains phosphorylserine and covalently bound fatty acids. It appears likely that the highly charged phosphorylserine-bearing terminus of about 5 kD is located on the cytoplasmic side of the cell membrane. Each subunit binds one molecule of iron-transferrin with high affinity (K_d 10^{-9} M). The transferrin receptor can be reconstituted in lecithin vesicles.[18,27] The binding of transferrin to the receptor as well as its uptake and release appear to be modulated by the lipids in the plasma membrane.[27,28] After induction the transferrin receptors in erythroid progenitors reach a maximum of about 400,000 per cell at the orthochromatic normoblast or reticulocyte stage.[29-34]

B. Cycling of the Transferrin Receptor*

At any given moment about one half or more of the receptor protein is located intracellularly. The process of transferrin binding to the receptor, removal of its iron, and return to

---

* See Figure 28.

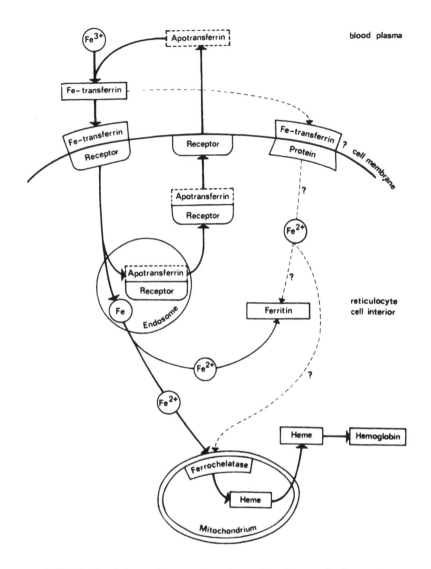

FIGURE 28.   Scheme of iron uptake and recycling of the transferrin receptor.

the incubation medium is fairly rapid;[32-35] 4 min after internalization undegraded apoferritin may be detected in the outside medium.[32,33,35,37] The release of the transferrin depends on the presentation of new ferric transferrin, a reflection of the ligand-dependence of the cycling process. The kinetics of internalization and recycling of transferrin and its receptors were studied thoroughly in a human hepatoma cell line.[38,39] Evidence was presented that the transferrin remains bound to the receptor during cycling, even though the iron is rapidly released in a special organelle, the acid endosome. Apotransferrin is released from the receptor upon reaching the cell membrane, since under prevailing neutral conditions it is only loosely bound and is displaced by ferric transferrin. From this work it would also appear that the acid endosome is essential only for the release of the iron but does not play a direct role in the cycling process.

In a study on human erythroleukemic cells induced by hemin, changes in receptor proteins were demonstrated.[34] A complex mixture was found with several phosphorylated forms, all of which contain N-acetylneuraminic acid. Internalized transferrin receptor appear to be more highly phosphorylated, a possible indication for the driving force of internalization. Hemin reduces the utilization of transferrin iron, probably by inhibition of the internalization of transferrin, and at the same time promotes the phosphorylation of the membrane transferrin receptors.[40,41] The internalization of the transferrin appears to occur by way of clathrin-coated pits preceded by clustering of the receptors.[34] With differentiation the number of receptors in coated pits increases.[42]

The results of a study of the intracellular movement of the transferrin receptor after internalization suggest that the receptor may also be transported from the cell surface to the Golgi complex.[43] This pathway appears to serve the sialylization of the transferrin receptor and is considerably slower than that involved in iron uptake.

The synthesis of the transferrin receptor is regulated by the iron supply at a transcriptional level, increasing with iron deficiency and being suppressed by an excess supply.[44]

How is the iron released from ferric transferrin in the cell? Earlier assumptions implicated lysosomes — likely candidates due to their low pH and their involvement in receptor-mediated uptake in fibroblasts, as judged by biochemical and morphological criteria.[45-47] Several studies, however, have clearly shown the absence of lysosomes in reticulocytes, which confirms earlier work.[48-51] Apparently a specific organelle is involved that is similar to lysosomes, with respect to its pH of about 5.0. This organelle has been called a "recep-tosome" or "acid endosome".[48,52-55] This type of endosome has been shown to be instru-mental in the uptake of α_2-macro globulin.[55] The inhibition of iron uptake in erythroid cells by ammonia and methylamine, well-known lysosomotropic agents,[47,32] may well be caused by an alkaline shift of the pH in the receptosome and therefore is not indicative of a lysosomal process.

Antibodies to receptors appear to exert contradictory effects. Inhibition of iron transferrin binding by specific antibodies was observed with rat reticulocytes[56] and with monoclonal antibodies to the human transferrin receptor with human reticulocytes.[57] On the other hand stimulation of transferrin and iron uptake in both rat and rabbit reticulocytes was found.[58,59] It was demonstrated that stimulation was due to an increase in the turnover rate of receptors. Electronmicroscopic examinations showed that the number of coated pits and membrane invaginations were increased in the antibody-treated cells paralleling the stimulation of iron uptake. The authors suggest that the antibody acted by promoting the formation of patches by the receptors and in this way facilitating the formation of coated pits containing the transferrin-receptor complex.

Receptor-enriched domains appear to form buds that are released into the environment without degradation of receptors; these are still able to bind transferrin or receptor antibodies.

C. Changes during Maturation

During maturation there is a loss in transferrin receptors which could be observed under in vitro conditions.[60] It occurs by way of selective externalization in the form of vesicles.[60] This process requires energy and is decreased by ATP production inhibitors or low tempera-tures.[61-63] The loss of receptors is retarded by antibodies against them but is increased in the presence of transferrin or calf serum. Methylamine and chloroquine, which may alkalinize the acid endosome and inhibitors of transglutaminase block the externalization, but inhibitors of proteinases fail to do so. The authors suggest that the selective externalization may be controlled by active changes in the membrane skeleton. However, it remains to be clarified which alteration in the receptor or membrane trigger the externalization.

The number of receptors may be regulated by the supply of iron. In a recent study[64] it was found that in the presence of iron or hemin the number of receptors as well as their rate of synthesis and turnover decreased, whereas the opposite, a fivefold increase in the number of receptors and in their synthesis was observed upon addition of protoporphyrin and picolinic acid, which is a good iron chelator. However, these results are at variance with those of other workers[34] in which the same human cell line (K 562) was used.

D. Nonreceptor Iron Uptake

Is endocytotic internalization the only process of iron uptake performed by the reticulocyte? Originally it was proposed that transferrin may deliver its iron by binding to its receptor on the cell membrane without physical internalization[2] (see also Aisen and Listowsky[9]). This assumption appeared to be supported by several lines of evidence: for one, there is more iron than transferrin taken up by rat reticulocytes.[65,66] It was further reported that iron can be taken up from sepharose-bound transferrin.[65] In addition, monoferric transferrins are found in the medium within 15 sec after presentation of diferric transferrin to the cells.[67-69]

It was argued that both the rapidity of release and the unlikelihood that iron transferrin could withstand the acid pH of the endosome would make it improbable that transferrin should have passed through the endocytotic cycle. Further support of this hypothesis was provided in the studies on the inhibition of iron uptake by α, α'-dipyridyl.[70] The authors reason that both the quickness of the onset of inhibition and that the inhibitor accumulated almost entirely in the cell membrane militate against the endocytosis theory. They postulate instead the function of an unspecified plasma membrane protein as an intermediate in iron uptake. While none of this evidence is incontrovertible, it is possible that two pathways exist, one involving endocytosis and release of the iron by acid endosomes, the other one by way of delivery of the iron from cell surface-bound transferrin. The second type of pathway may be functional in the uptake of inorganic iron salts.

IV. WHAT IS THE ROLE OF FERRITIN IN RED CELLS?

The possible role of ferritin as an intermediate in the intracellular traffic of iron remains controversial. Particles containing inorganic iron in erythrocytes have been observed for a long time. Such cells have been designated siderocytes.[71-74] Iron is present in the form of ferritin. Various morphological and clinical studies have indicated that the ferritin content in reticulocytes and erythroblasts is considerably higher than in mature erythrocytes[75] and that ferritin is abundant in primitive old cells of chicks, mice, and frogs and in erythroid precursors.[9] Content may be influenced by experimental procedures or by clinical conditions. Under normal conditions in man, all normoblasts and most reticulocytes, but only a few erythrocytes, contain ferritin. After blood loss the percentage of siderocytes increases to nearly 100%.[76] In view of its widespread occurrence, ferritin was supposed to be an intermediate in the intracellular delivery of iron to ferrochelatase.[77-80] Various types of evidence contradict this suggestion and support the view that the availability of ferritin-iron is low and may relfect a disturbance of hemoglobin synthesis or a disproportion between uptake and utilization of iron for heme synthesis.[81-85] In early work[76] it was shown that in red cells of pigs with pyridoxine deficiency and decreased synthesis of heme, the number of siderocytes increased considerably. Borova et al.[83] demonstrated that in the presence of heme synthesis inhibitors that act on pyridoxalphosphate enzymes, ferritin accumulated and was not available for utilization of iron. In recent work a selective inhibitor of 5-ALA dehydratase, succinylacetone, was employed.[86] This compound inhibited iron incorporation into heme and even stimulated its uptake from the medium by 25 to 30%. In cells treated with succinylacetone, iron accumulated both in a fraction containing plasma membranes and mitochondria as well as in cytosolic ferritin. It was demonstrated that cytosolic ferritin was mostly unavailable for delivery of iron to heme.

In a study on bullfrogs, insights were obtained in the relationship between provision of iron from ferric transferrin and the accumulation of ferritin in red cells.[87] At low levels of transferrin saturation reticulocytes of tadpoles utilized about one half of the iron taken up for heme synthesis; this process, however, had a low capacity. With increasing provision of iron, at higher degrees of transferrin saturation the iron share entering into ferritin rose to more than 90%. Thus it would appear that the fate of cellular iron is determined by the affinity and capacity of the two systems — heme synthesis and ferritin formation — which compete with each other. The low ferritin content of reticulocytes in adult bullfrogs appears explicable by low saturation of transferrin in their plasma.[88]

Similar results were obtained by pulse chase analysis of iron uptake in newt erythroblasts.[89] In these experiments no transfer of ^{59}Fe from ferritin to hemoglobin could be detected. In puromycin-inhibited cells with reduced iron uptake but continuing heme synthesis, radio-activity appeared in the heme without any prior labeling of ferritin.

From these studies it would appear that ferritin is not an obligatory intermediate in the main pathway of iron utilization for heme synthesis or cytosolic transport. On the other hand it is likely that during differentiation ferritin may be utilized over longer periods of time. The best evidence for this contention are the early morphological studies that used light or electron microscopy in which it was demonstrated that all normoblasts and a great majority of reticulocytes contain ferritin, whereas mature erythrocytes are devoid of it. Ferritin appears to accumulate in overflow as a mechanism for long-term storage, or pathologically, whenever there is a dissociation between iron uptake and utilization by a red cell. This may occur if the supply of iron is excessive, the membrane transport of iron is reduced, or if, conversely, heme synthesis is diminished or disturbed.

V. THE PATHWAY OF IRON FROM THE ACID ENDOSOME TO THE MITOCHONDRIA

The fate of iron after release from endosomes is obscure as yet. Manyfold attempts to separate and characterize putative intermediates have not met with success so far.[79,82-85,90-94] The most active fraction in the delivery of iron which is also the first to be labeled is a mixture of membranes and mitochondria that could contain unutilized transferrin or mito-chondrial iron, either or both of which may be released under the experimental conditions employed and would not represent the cytosolic intermediate sought.

The mitochondria appear to play an active role in iron uptake. An energy dependent accumulation of iron by isolated mitochondria from rabbit reticulocytes has been demon-strated.[95] The suggestion by Egyed[96] that the membrane potential is the driving force for iron uptake is reasonable, even though the agents used to demonstrate this mechanism were not adequate. It may be significant that inorganic iron can accumulate in mitochondria, as was found in early morphological work indicating that the influx of iron into mitochondria may exceed the capacity of ferrochelatase.[74] This iron may be released if an uncoupler of oxidative phosphorylation is added and the membrane potential of the mitochondria is dissipated.

From the account so far given it follows that there are at least two energy-requiring steps in the utilization of iron by the cell: one is the internalization of the transferrin-receptor complex, which requires energy in a manner thus far unspecified; the other is the uptake by mitochondria, which appears to depend on their energized state.

The question of the state of iron oxidation in the cell deserves further consideration. In transferrin as well as in ferritin, iron is in a tervalent state and both proteins have no affinity for the divalent iron; on the other hand ferrochelatase requires Fe^{2+}. Therefore it is of great significance — according to a recent study by Egyed and Saltman[97] — that iron in erythroid cell is maintained in the divalent form even under aerobic conditions. This condition may speed up the release of the iron from either the acid endosome or from ferritin. One would surmise that the intracellular transport form of iron involves Fe^{2+}.

The system for iron uptake and utilization appears to have several self-regulatory features. These include cycling of transferrin, regulation of the number of transferrin receptors, and feedback inhibition or iron uptake by intracellular heme.[98-100]

On the basis of model experiments in the iron uptake of rat liver mitochondria from transferrin in the presence of inorganic pyrophosphate, it was hypothesized that inorganic pyrophosphate may be the putative intermediate ligand for delivery of iron to mitochondria *in situ.*[101-104] Iron uptake in these experiments involved transfer of ferric iron from transferrin to pyrophosphate with subsequent cleavage of the ferric-iron-pyrophosphate complex by reduction and uptake of ferrous iron by the mitochondria.[102,104] Iron uptake and heme synthesis are coupled, thereby protecting mitochondria from an overload of iron. It has also been suggested that uptake of iron and protoporphyrin by mitochondria are coupled with heme efflux.[102,105]

REFERENCES

1. **Jandl, J. H., Inman, J. K., Simmons, R. L., and Allen, D. W.,** Transfer of iron from serum iron-binding protein to human reticulocytes, *J. Clin. Invest.,* 38, 161, 1959.
2. **Jandl, J. H. and Katz, J. H.,** The plasma-cell cycle of transferrin, *J. Clin. Invest.,* 42, 314, 1963.
3. **Jones, M. S. and Jones, O. T. G.,** The structural organization of haem synthesis in rat liver mitochondria, *Biochem. J.,* 113, 507, 1969.
4. **Morgan, E. H. and Baker, E.,** The effect of metabolic inhibitors on transferrin and iron uptake and transferrin release from reticulocytes, *Biochim. Biophys. Acta,* 184, 442, 1969.
5. **Egyed, A.,** On the mechanism of uptake of iron by reticulocytes, *Acta Biochim. Biophys. Acad. Sci. Hung.,* 9, 43, 1974.
6. **Kailis, S. G. and Morgan, E. H.,** Iron uptake by immature erythroid cells. Mechanism of dependence of metabolic energy, *Biochim. Biophys. Acta,* 464, 389, 1977.
7. **Martinez-Medellin, J. and Benavides, L.,** The rate limiting step in the reticulocyte uptake of transferrin and transferrin iron, *Biochim. Biophys. Acta,* 584, 84, 1979.
8. **van Bockxmeer, F. M. and Morgan, E. H.,** Identification of transferrin receptors in reticulocytes, *Biochim. Biophys. Acta,* 468, 437, 1977.
9. **Aisen, P. and Listowsky, I.,** Iron transport and storage proteins, *Annu. Rev. Biochem.,* 49, 357, 1980.
10. **Kornfeld, S.,** The effects of structural modifications on the biologic activity of human transferrin, *Biochemistry,* 7, 945, 1968.
11. **Kornfeld, R. and Kornfeld, S.,** The structure of a phytohemagglutinin receptor site from human erythrocytes, *J. Biol. Chem.,* 245, 2536, 1970.
12. **Hemmaplardh, D. and Morgan, E. H.,** Transferrin uptake and release by reticulocytes treated with proteolytic enzymes and neuraminidase, *Biochim. Biophys. Acta,* 426, 385, 1976.
13. **Dekker, C. J., Kroos, M. J., van der Heul, C., and van Eijk, H. G.,** Iron uptake from sialo transferrins by rat reticulocytes, *Int. J. Biochem.,* 16, 787, 1984.
14. **Fletcher, J. and Huehns, E. R.,** Significance of the binding of iron by transferrin, *Nature (London),* 215, 584, 1967.
15. **Fletcher, J. and Huehns, E. R.,** Function of transferrin, *Nature (London),* 218, 1211, 1968.
16. **Huebers, H., Huebers, E., Csiba, E., and Finch, C. A.,** Iron uptake from rat plasma transferrin by rat reticulocytes, *J. Clin. Invest.,* 62, 944, 1978.
17. **Huebers, H. A., Csiba, E., Huebers, E., and Finch, C. A.,** Competitive advantage of diferric transferrin in delivering iron to reticulocytes, *Proc. Natl. Acad. Sci. U.S.A.,* 80, 300, 1983.
18. **van der Heul, C., Kroos, M. J., van Noort, W. L., and van Eijk, H. G.,** In vitro and in vivo studies of iron delivery by human monoferric transferrins, *Br. J. Haematol.,* 56, 571, 1984.
19. **Newman, R., Schneider, C., Sutherland, R., Vodinelich, L., and Greaves, M.,** The transferrin receptor, *Trends Biochem. Sci.,* 7, 397, 1982.
20. **Woodworth, R. C., Brown-Mason, A., Christensen, T. G., Witt, D. P., and Comeau, R. D.,** An alternative model for the binding and release of differric transferrin by reticulocytes, *Biochemistry,* 21, 4220, 1982.
21. **van Bockxmeer, F. M., Yates, G. K., and Morgan, E. H.,** Interaction of transferrin with solubilized receptors from reticulocytes, *Eur. J. Biochem.,* 92, 147, 1978.
22. **Seligman, P. A., Schleicher, R. B., and Allen, R. H.,** Isolation and characterization of the transferrin receptor from human placenta, *J. Biol. Chem.,* 254, 9943, 1979.
23. **Schneider, C., Sutherland, R., Newman, R., and Greaves, M.,** Structural features of the cell surface receptor for transferrin that is recognized by the monoclonal antibody OKT9, *J. Biol. Chem.,* 257, 8516, 1982.

24. **Wada, H. G., Hass, P. E., and Sussman, H. H.,** Transferrin receptor in human placental brush border membranes, *J. Biol. Chem.,* 254, 12629, 1979.

25. **Omary, M. B. and Trowbridge, I. S.,** Biosynthesis of the human transferrin receptor in cultured cells, *J. Biol. Chem.,* 256, 12888, 1981.

26. **Enns, C. A. and Sussman, H. H.,** Physical characterization of the transferrin receptor in human placentae, *J. Biol. Chem.,* 256, 9820, 1981.

27. **Nunez, M. T. and Glass, J.,** Reconstitution of the transferrin receptor in lipid vesicles: effect of cholesterol on the binding of transferrin, *Biochemistry,* 21, 4139, 1982.

28. **Hemmaplardh, D., Morgan, R. G. H., and Morgan, E. H.,** Role of plasma membrane phospholipids in the uptake and release of transferrin and its iron by reticulocytes, *J. Membrane Biol.,* 33, 195, 1977.

29. **van Bockxmeer, F. M. and Morgan, E. H.,** Transferrin receptors during rabbit reticulocyte maturation, *Biochim. Biophys. Acta,* 584, 76, 1979.

30. **Frazier, J. L., Caskey, D. U., Yaffe, M., and Seligman, P. A.,** The transferrin receptor on human reticulocytes and nucleated human cells in culture, *J. Clin. Invest.,* 69, 853, 1982.

31. **Jacopetta, B. J., Morgan, E. H., and Yeoh, G. C. T.,** Transferrin receptors and iron uptake during erythroid cell development, *Biochim. Biophys. Acta,* 687, 204, 1982.

32. **Klausner, R. D., van Renswoude, J., Ashwell, G., Kempf, C., Schechter, A. N., Dean, A., and Bridges, K. R.,** Receptor-mediated endocytosis of transferrin in K562 cells, *J. Biol. Chem.,* 258, 4715, 1983.

33. **Karin, M. and Mintz, B.,** Receptor-mediated endocytosis of transferrin in developmentally totipotent mouse teratocarcinoma stem cells, *J. Biol. Chem.,* 256, 3245, 1981.

34. **Hunt, R. C., Ruffin, R., and Yang, Y. S.,** Alterations in the transferrin receptor of human erythroleukemic cells after induction of hemoglobin synthesis, *J. Biol. Chem.,* 259, 9944, 1984.

35. **Jacopetta, B. S., Morgan, E. H., and Yeoh, G. C. T.,** Receptor-mediated endocytosis of transferrin by developing erythroid cells from the fetal rat liver, *J. Histochem. Cytochem.,* 31, 336, 1983.

36. **Nunez, M. T. and Glass, J.,** The transferrin cycle and iron uptake in rabbit reticulocytes, *J. Biol. Chem.,* 258, 9676, 1983.

37. **Jacopetta, B. J. and Morgan, E. H.,** The kinetics of transferrin endocytosis and iron uptake from transferrin in rabbit reticulocytes, *J. Biol. Chem.,* 258, 9108, 1983.

38. **Ciechanover, A., Schwartz, A. L., Dautry-Varsat, A., and Lodish, H. F.,** Kinetics of internalization and recycling of transferrin and the transferrin receptor in a human hepatoma cell line, *J. Biol. Chem.,* 258, 9681, 1983.

39. **Dautry-Varsat, A., Ciechanover, A., and Lodish, H. F.,** pH and the recycling of transferrin during receptor-mediated endocytosis, *Proc. Natl. Acad. Sci. U.S.A.,* 80, 2958, 1983.

40. **Ponka, P. and Schulman, H. M.,** Regulation of heme synthesis in erythroid cells; hemin inhibits transferrin iron utilization but not protoporphyrin synthesis, *Blood,* 65, 850, 1985.

41. **Lox, T., Donnell, M. W., Aisen, P., and London, M.,** Hemin inhibits internalization of transferrin by reticulocytes and promotes phosphorylation of the membrane transferrin receptor, *Proc. Natl. Acad. Sci. U.S.A.,* 82, 5170, 1985.

42. **Lesley, J., Hyman, R., Schulte, R., and Trotter, J.,** Expression of transferrin receptor on neurine hematopoietic progenitors, *Cell. Immunol.,* 83, 14, 1984.

43. **Snider, M. D. and Rogers, O. C.,** Intracellular movement of cell surface receptors after endocytosis: resialylation of asialotransferrin receptor in human erythroleukemia cells, *J. Cell. Biol.,* 100, 826, 1985.

44. **Louache, F., Pelosi, E., Titeyse, M., Peschle, C., and Testa, U.,** Molecular mechanisms regulating the synthesis of transferrin receptors and ferritin in human erythroleukemic cell lines, *FEBS Lett.,* 183, 223, 1985.

45. **Schneider, Y. J., Tulkens, P., DeDuve, C., and Trouet, A.,** Fate of plasma membrane during endocytosis. II. Evidence for recycling (shuttle) of plasma membrane constituents, *J. Cell Biol.,* 82, 466, 1979.

46. **Octave, J.-N., Schneider, Y.-J., Chrichton, R. R., and Trouet, A.,** Transferrin protein and iron uptake by isolated rat erythroblasts, *FEBS Lett.,* 137, 119, 1982.

47. **Morgan, E. H.,** Inhibition of reticulocyte iron uptake by NH_4Cl and CH_3NH_2, *Biochim. Biophys. Acta,* 642, 119, 1981.

48. **Van Renswoude, J., Bridges, K. R., Harford, J. B., and Klausner, R. D.,** Receptor-mediated endocytosis of transferrin and the uptake of Fe in K562 cells: identification of a nonlysosomal acidic compartment, *Proc. Natl. Acad. Sci. U.S.A.,* 79, 6186, 1982.

49. **Veldman, A., Kroos, M. J., van der Heul, C., and van Eijk, H. G.,** Are lysosomes directly involved in the iron uptake by reticulocytes?, *Int. J. Biochem.,* 16, 39, 1984.

50. **Rapoport, S. M., Rosenthal, S., Schewe, T., Schultze, M., and Müller, M.,** The metabolism of the reticulocyte, in *Cellular and Molecular Biology of Erythrocytes,* Yoshikawa, H. and Rapoport, S. M., Eds., University of Tokyo Press, Tokyo, 1974, 93.

51. **Poenaru, L. and Dreyfus, J. C.,** α-Mannosidase in human red cells, *Biochim. Biophys. Acta,* 566, 67, 1979.

52. **Light, A. and Morgan, E. U.**, Transferrin endocytosis in reticulocytes: an electron microscope study using colloidal gold, *Scand. J. Haematol.*, 28, 205, 1982.

53. **Armstrong, N. J. and Morgan, E. H.**, The effect of lysosomotrophic bases and inhibitors of transglutaminase on iron uptake by immature erythroid cells, *Biochim. Biophys. Acta*, 762, 175, 1983.

54. **Paterson, S., Armstrong, N. J., Jacopetta, B. J., McArdle, H. J., and Morgan, E. H.**, Intravesicular pH and iron uptake by immature erythroid cells, *J. Cell. Physiol.*, 120, 225, 1984.

55. **Tycko, B. and Maxfield, F. R.**, Rapid acidification of endocytic vesicles containing α_2-macroglobulin, *Cell*, 28, 643, 1982.

56. **van der Heul, C., Kroos, M. J., and van Eijk, H. G.**, Binding sites of iron transferrin on rat reticulocytes inhibition by specific antibodies, *Biochim. Biophys. Acta*, 511, 430, 1978.

57. **Trowbridge, I. S. and Lopez, F.**, Monoclonal antibody to transferrin receptor blocks transferrin binding and inhibits human tumor cell growth in vitro, *Proc. Natl. Acad. Sci. U.S.A.*, 79, 1175, 1982.

58. **McArdle, H. J. and Morgan, E. H.**, The effect of monoclonal antibodies to the human transferrin receptor on transferrin and iron uptake by rat and rabbit reticulocytes, *J. Biol. Chem.*, 259, 1629, 1984.

59. **Verhoef, N. J. and Noordeloos, P. J.**, Binding of transferrin and uptake of iron by rat erythroid cells in vitro, *Clin. Sci. Mol. Med.*, 52, 87, 1977.

60. **Pan, B.-T., Blostein, R., and Johnstone, R. M.**, Loss of transferrin receptor during the maturation of sheep reticulocytes in vitro, *Biochem. J.*, 210, 37, 1983.

61. **Pan, B.-T. and Johnstone, R. M.**, Fate of the transferrin receptor during maturation of sheep reticulocytes in vitro; selective externalization of the receptor, *Cell*, 33, 967, 1983.

62. **Pan, B.-T. and Johnstone, R.**, Selective externalization of the transferrin receptor by sheep reticulocytes in vitro. Response to ligands and inhibitors of endocytosis, *J. Biol. Chem.*, 259, 9776, 1984.

63. **Weigensberg, A. M. and Blostein, R.**, Energy depletion retards the loss of membrane transport during reticulocyte maturation, *Proc. Natl. Acad. Sci. U.S.A.*, 80, 4978, 1983.

64. **Louache, F., Testa, U., Pelicci, P., Thomopoulos, P., Titeux, M., and Rochant, H.**, Regulation of transferrin receptors in human hematopoietic cell lines, *J. Biol. Chem.*, 259, 11576, 1984.

65. **Loh, T. T., Young, Y. G., and Young, D.**, Transferrin and iron uptake by rabbit reticulocytes, *Biochim. Biophys. Acta*, 471, 118, 1977.

66. **Zaman, Z., Heynen, M.-J., and Verwilghen, R. L.**, Studies on the mechanism of transferrin iron uptake by rat reticulocytes, *Biochim. Biophys. Acta*, 632, 553, 1980.

67. **van der Heul, C., Veldman, A., Kroos, J. M., and van Eijk, H. G.**, Two mechanisms are involved in the process of iron-uptake by rat reticulocytes, *Int. J. Biochem.*, 16, 383, 1984.

68. **Young, S. P.**, Evidence for the functional equivalence of the iron-binding sites of rat transferrin, *Biochim. Biophys. Acta*, 718, 35, 1982.

69. **Loh, T. T.**, Studies on the forms of iron-transferrin released from rabbit reticulocytes, *Life Sci.*, 32, 915, 1983.

70. **Nunez, M.-T., Cole, E. S., and Glass, J.**, The reticulocyte plasma membrane pathway of iron uptake as determined by the mechanism of α,α'-dipyridyl inhibition, *J. Biol. Chem.*, 258, 1146, 1983.

71. **Bessis, M.**, *Living Blood Cells and Their Ultrastructure*, Translated by Weed, R. I., Springer-Verlag, Berlin, 1973.

72. **Bessis, M. and Breton-Gorius, J.**, Iron particles in normal erythroblasts and normal and pathological erythrocytes, *J. Biophys. Biochem. Cytol.*, 3, 503, 1957.

73. **Bessis, M. and Breton-Gorius, J.**, Différences entre sidéroblastes normaux et pathologiques. Étude au microscope électronique, *Nouv. Rev. Fr. Hematol.*, 2, 629, 1962.

74. **Bessis, M. C. and Jensen, W. N.**, Sideroblastic anemia, mitochondria and erythroblastic iron, *Br. J. Haematol.*, 11, 49, 1965.

75. **Ali, F., Leison, M. K., Jones, B. M., and Jacobs, A.**, Enrichment of erythroblasts from human bone marrow using complement-mediated lysis: measurement of ferritin, *Br. J. Haematol.*, 53, 227, 1983.

76. **Deiss, A., Kurth, D., Cartwright, G. E., and Wintrobe, M. M.**, Experimental production of siderocytes, *J. Clin. Invest.*, 45, 353, 1966.

77. **Fielding, J. and Speyer, B. E.**, Iron transport intermediates in human reticulocytes and the membrane binding site of iron-transferrin, *Biochim. Biophys. Acta*, 363, 387, 1974.

78. **Speyer, B. E. and Fielding, J.**, Ferritin as a cytosol iron transport intermediate in human reticulocytes, *Br. J. Haematol.*, 42, 255, 1979.

79. **Mazur, A. and Carleton, A.**, Relation of ferritin iron to heme synthesis in marrow and reticulocytes, *J. Biol. Chem.*, 238, 1817, 1963.

80. **Nunez, M. T., Glass, J., and Robinson, S. H.**, Mobilization of iron from the plasma membrane of the murine reticulocyte. The role of ferritin, *Biochim. Biophys. Acta*, 509, 170, 1978.

81. **Zail, S. S., Charlton, R. W., Torrance, J. D., and Bothwell, T. H.**, Studies on the formation of ferritin in red cell precursors, *J. Clin. Invest.*, 43, 670, 1964.

82. **Primosigh, J. V. and Thomas, E. D.**, Studies on the partition of iron in bone marrow cells, *J. Clin. Invest.*, 47, 1473, 1968.

83. **Borova, J., Ponka, P., and Neuwirt, J.**, Study of intracellular iron distribution in rabbit reticulocytes with normal and inhibited heme synthesis, *Biochim. Biophys. Acta*, 320, 143, 1973.

84. **Storring, P. L. and Fatih, S.**, Erythropoietic effects on iron metabolism in rat bone marrow cells, *Biochim. Biophys. Acta*, 392, 26, 1975.

85. **Pollack, S. and Campana, T.**, Early events in guinea pig reticulocyte iron uptake, *Biochim. Biophys. Acta*, 673, 366, 1981.

86. **Ponka, P., Wilczyska, A., and Schulman, H. M.**, Iron utilization in rabbit reticulocytes. A study using succinylacetone as an inhibitor of heme synthesis, *Biochim. Biophys. Acta*, 720, 96, 1982.

87. **Valaitis, A. P. and Theil, E. C.**, Developmental changes in plasma transferrin concentrations related to red cell ferritin, *J. Biol. Chem.*, 259, 779, 1984.

88. **Schaefer, F. V. and Theil, E. C.**, The effect of iron on the synthesis and amount of ferritin in red blood cells during ontogeny, *J. Biol. Chem.*, 256, 1711, 1981.

89. **Grasso, J. A., Hillis, T. J., and Mooney-Frank, J. A.**, Ferritin is not a required intermediate for iron utilization in heme synthesis, *Biochim. Biophys. Acta*, 797, 247, 1984.

90. **Egyed, A.**, Studies on the partition of transferrin-donated iron in rabbit reticulocytes. II. Distribution and kinetics of non-haem iron in cytosol, *Br. J. Haematol.*, 53, 217, 1983.

91. **Nunez, M. T., Cole, E. S., and Glass, J.**, Cytosol intermediates in the transport of iron, *Blood*, 55, 527, 1980.

92. **Ponka, P., Borova, J., and Neuwirt, J.**, Iron and transferrin distribution in reticulocytes incubated with heme synthesis inhibitors, *Biochim. Biophys. Acta*, 632, 527, 1980.

93. **Workman, E. F., Jr. and Bates, G. W.**, Mobilization of iron from reticulocyte ghosts by cytoplasmic agents, *Biochem. Biophys. Res. Commun.*, 58, 787, 1974.

94. **Blackburn, G. W. and Morgan, E. H.**, Factors affecting iron and transferrin release from rabbit reticulocyte ghosts to cytosol, *Biochim. Biophys. Acta*, 497, 728, 1977.

95. **Romslo, L.**, Energy dependent accumulation of iron by isolated rabbit reticulocyte mitochondria, *Biochim. Biophys. Acta*, 357, 34, 1974.

96. **Egyed, A.**, Studies on the partition of transferrin-donated iron in rabbit reticulocytes. I. The kinetics of iron distribution between stroma and cytosol, *Br. J. Haematol.*, 52, 475, 1982.

97. **Egyed, A. and Saltman, P.**, Iron is maintained as Fe (II) under aerobic conditions in erythroid cells, *Biol. Trace Element Res.*, 6, 357, 1984.

98. **Ponka, P. and Neuwirt, J.**, Regulation of iron entry into reticulocytes. I. Feedback inhibitory effect of heme on iron entry into reticulocytes and on heme synthesis, *Blood*, 33, 690, 1969.

99. **Ponka, P., Neuwirt, J., and Borova, J.**, The role of heme in the release of iron from transferrin in reticulocytes, *Enzyme*, 17, 91, 1974.

100. **Schulman, H. M., Martinez-Medellin, J., and Sidloi, R.**, The reticulocyte-mediated release of iron and bicarbonate from transferrin: effect of metabolic inhibitors, *Biochim. Biophys. Acta*, 343, 529, 1974.

101. **Konopka, K. and Romslo, I.**, Uptake of iron from transferrin by isolated rat-liver mitochondria mediated by phosphate compounds, *Eur. J. Biochem.*, 107, 433, 1980.

102. **Konopka, K. and Romslo, I.**, Studies on the mechanism of pyrophosphate-mediated uptake of iron from transferrin by isolated rat-liver mitochondria, *Eur. J. Biochem.*, 117, 239, 1981.

103. **Nilsen, T. and Romslo, I.**, Pyrophosphate as a ligand for delivery of iron to isolated rat-liver mitochondria, *Biochim. Biophys. Acta*, 766, 233, 1984.

104. **Nilsen, T. and Romslo, I.**, Transferrin as a donor of iron to mitochondria. Effect of pyrophosphate and relationship to mitochondrial metabolism and heme synthesis, *Biochim. Biophys. Acta*, 802, 448, 1984.

105. **Husby, P. and Romslo, I.**, Studies on the efflux of metalloporphyrin from rat liver mitochondria. Effect of K^+ and other cations, *Biochem. J.*, 196, 451, 1981.

Chapter 11

PROTEIN SYNTHESIS

I. INTRODUCTION

It cannot be the purpose of this chapter to discuss all of the problems involved in protein synthesis, even though most work has been carried out on reticulocytes or their lysates (for reviews see Jagus et al.,[1] Voorma et al.,[2] and Marks et al.[3]). Instead, this chapter will deal with selected topics which are specific for the reticulocyte: (1) the type of proteins synthetized in reticulocytes; (2) the balance of the synthesis of α- and β-globin chains; (3) the questionable role of membrane-bound ribosomes; (4) the role of heme and the effects of its absence; (5) modifications of translation products — phosphorylations and dephosphorylations of proteins; and (6) synthesis and modification of RNA. The maturational changes of the translation system including ribosomes will be discussed in Chapter 15.

II. PROTEINS SYNTHESIZED IN THE RETICULOCYTE

There is a discrepancy between the fairly large number of mRNA species detectable in reticulocytes (see Chapters 2 and 3) and that of identified translation products. About 90% represent globin; after unmasking of its mRNA, the synthesis of lipoxygenase takes second place with up to 4%.[4] Considerable new formation of catalase, carboanhydrase, 2,3 PG synthase, and poly-A-binding protein occurs, with shares ranging from 0.05 to 0.8%.[5-8] The synthesis of several membrane-attached proteins can be detected, particularly the 4.1 band;[9,10] (see Chapter 4). This list is by no means exhaustive and further additions can be expected with the refinement of protein separation and detection methodology.

III. THE BALANCE OF SYNTHESIS OF α- AND β-GLOBIN CHAINS

In the normal reticulocytes of adults, the synthesis of α- and β-globin chains is well balanced which means that equal numbers of both types of chains are produced to form hemoglobin A. In striking contrast, there are duplicate α-globin genes, but only single β-globin genes in many mammals. Thus one would expect a twofold production of α-globin mRNA. The question therefore arises as to how this discrepancy can be explained. This can be answered by the following circumstances: first, in the greater lability of the α-globin mRNA compared to β-globin mRNA, and, second, in its lower initiation efficiency.

The greater lability of α-globin mRNA in the stages preceding the reticulocyte has been discussed in Chapter 2.[11] Even so, the reticulocytes of mice, rabbits, and sheep still contain approximately one and one half times as much α-globin mRNA as β-globin mRNA.[12-15] This difference is compensated for by the higher rate of initiation of β-globin chain translation as proved by Lodish and Jacobson.[16] This explains the observation that on the average four ribosomes are attached to α-globin mRNA and about six to β-globin mRNA.[17] The reason for the lower initiation efficiency of α-globin mRNA was found to be its low binding to the initiation factor eIF-4B, which is only $1/_{50}$ that of β-globin mRNA.[18] The differences in binding efficiency are expressed under competitive circumstances, i.e., at higher concentrations of the mRNAs, in which they exceed the initiation factor.

Even with this basic clarification of the mechanism by which the balanced synthesis of α- and β-globin chains is achieved, the close adjustment in the formation of final products of gene expression still remains to be clarified. One is forced to assume that some kind of feedback signal originating from the free α- or β-globin chains modulates the initiation efficiency of the mRNAs in such a way that a 1:1 ratio of globin chains is released. Nevertheless, it would appear that the twofold provision of α- vs. β-globin genes is a dramatic example of evolutionary adaptation.

IV. THE ROLE OF MEMBRANE-BOUND RIBOSOMES

Schreml and Burka originally reported that about 30% of the cellular RNA of reticulocytes is found in their stroma in the form of membrane-bound ribosomes, which differ from those in the cytosol both in physicochemical and functional characteristics.[19] In later communications, variable amounts of stroma-bound RNA were reported.[20] It was claimed that membrane-bound ribosomes were involved in the synthesis of special nonglobin proteins and that free ribosomes synthesized mainly globin chains. These results were not confirmed by Woodward et al.[21] or by Lodish and Desalu.[22] It was shown that membrane-free reticulocyte lysates produced the same nonglobin proteins as whole cells and in approximately the same proportions. However, in these studies the products of membrane-bound ribosomes were not directly determined. Both Bulova and Burka[20] as well as Woodward et al.[21] noted that ribosomes, when liberated from membranes by deoxycholate, were much less active in translation in a cell-free system than free cytosolic ribosomes. The reason for this low activity was found to be the degradation of messenger and ribosomal RNAs; this breakdown could be greatly reduced by raising the ionic strength of the ribosomes during liberation.[23] A comparison of translation products of free and membrane-bound ribosomes indicated that globin was the predominant protein produced by both types of ribosomes, but that differences were noticeable in kind and amount of nonglobin proteins formed.

Nevertheless, one must conclude that the overwhelming majority of ribosomes bound to stroma of reticulocytes are functionally no different from those found free in the cytosol. Moreover, it was demonstrated by Lodish and Small[9] that two proteins found attached to the cytosolic side of the cell membrane were synthesized on free ribosomes. The possibility still remains that a small proportion of stroma-bound ribosomes are attached to vestigial remnants of the endoplasmic reticulum and might produce proteins in an abortive manner, which in earlier stages of differentiation were transported to the cell membranes.[24] Overall, one finds a variety of relics from the system by which proteins destined for the membrane were translocated and modified. Such a remnant is the signal recognition particle which has been found in the cytosol of reticulocytes by Meyer et al.[25] This particle represents a protein complex which establishes contact between the ribosomes and the endoplasmic membrane on emergence of a signal sequence from the large ribosomal subunit. Furthermore, one may cite the reports on the glycosylation of endogenous proteins by dolichol derivatives in plasma membranes of reticulocytes, the existence of mannosyl-phosphoryl polyisoprenol synthesis in reticulocytes, and incorporation of labeled glucosamine into the membrane proteins;[26-29] glycoproteins are found exclusively in the plasma membranes of red cells (see Chapter 4).

V. THE ROLE OF HEME AND THE EFFECTS OF HEME DEPRIVATION

Protein synthesis in reticulocytes and their lysates ceases within a short time in the absence of added heme.[30-32,34] This effect is caused by the action of an inhibitor, later shown to be a protein kinase which phosphorylates the α-subunit of the initiation factor eIF-2 (for reviews see Ochoa;[35] and London et al.[36]). The mechanism of protein synthesis inhibition turned out to be a highly refractory problem which required the competing and concerted efforts of many research workers. This work was intimately connected with the unraveling of the complexities of protein synthesis initiation. The sequence of events which forms the initiation complex requires the participation of at least nine protein factors, denominated eIF-1, eIF-2, eIF-3, eIF-4A, eIF-4B, eIF-4C, eIF-4D, eIF-4E, eIF-4F, and eIF-5. The formation of

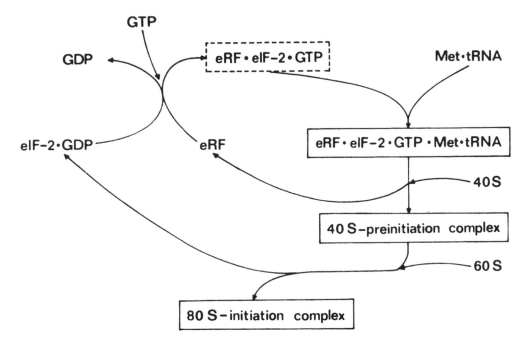

FIGURE 29.   Recycling scheme of eIF-2 and eRF during initiation of translation.

the initiation complex is preceded by several steps including a ternary complex between eIF-2, Met-tRNA, and GTP, and its subsequent combination with another preformed complex consisting of the 40S ribosomal subunit, eIF-3 and eIF-4C. The binding of mRNA follows which involves a variety of factors (eIF-1, eIF-4A, eIF-4B, eIF-4E, and eIF-4F); finally the 40S complex combines with the 60S ribosomal subunit, which is mediated by eIF-5. This sequence of events requires recycling of the initiation factors and renewal of GTP, which is converted to GDP during the formation of the initiation complex.

The molecular events which lead to the inhibition of protein synthesis in heme-deficient lysates have only recently been clarified. Apparently the phosphorylation of eIF-2 results in an interruption of its recycling, which appears to be a rate-controlling step in the initiation of protein synthesis[37-43] (Figure 29).

In order for eIF-2 to recycle, GDP must be released from it in exchange for a new molecule of GTP. However, the eIF-2 affinity for GDP is much higher than for GTP. A recently discovered protein factor now called eRF, the eIF-2-recycling factor, plays an important role (for a review see Voorma et al.[2]). It forms a complex with eIF-2, whereby the affinity for GDP is decreased, thus facilitating the GDP/GTP exchange.[44-57] The eRF consists of three subunits with molecular masses of 41, 60, and 81 kD, respectively. Its effects formerly escaped detection since the assay systems used did not involve recycling. In such systems no difference between phosphorylated and unphosphorylated eIF-2 could be detected,[58] even though the addition of purified eIF-2 released the inhibition in a heme-deprived lysate.[46-48] Complete inhibition of protein synthesis occurs even though only a small portion of eIF-2 is phosphorylated.[42,46,59] This circumstance is explained by the fact that the ratio of eRF to eIF-2 is about 1:6 to 1:10 and that eRF forms an inactive complex with phosphorylated eIF-2 + GDP.[54,55,57,60-62] Whether or not the GDP in the complex is dissociable is unclear.[63] Recent work, in which the 80 kD component of eRF was isolated, indicates that the mechanism of the eIF-2 recycling may be more complicated than assumed by previous schemes.[33] This is also suggested by the observation that the primary effect of phosphorylation of eIF-2 is the inhibition of the release of its complex with GDP from the 60S ribosomal subunit.[64]

How does heme deficiency activate the eIF-2α kinase? There is some evidence that the kinase is converted from an inactive to an active form by phosphorylation, which occurs in the absence of heme or after treatment with sulfhydryl reagents.[66,67] It is suggested that it might be an autophosphorylation and that heme combines with the kinase, thereby suppressing its activity.[36,65]

In addition to the mechanism of activation of eIF-2 kinase produced by heme deficiency, there may be another type of activation caused by oxidation processes, which may lead to the conversion of certain SH groups of the enzyme to disulfides. The inhibitory effects of oxidized glutathione or NADPH-deficiency are explicable on this basis.[66,67,79,80] The inhibition of translation caused by the addition of Ca^{2+} and phospholipids may also be explained in this manner.[67,68] Further investigation of the effect revealed that free polyunsaturated fatty acids, such as arachidonic acid, were as effective in promoting the inhibition of translation and the phosphorylation of eIF-2α. It appears that the active agent in either case may be the lipoperoxides of the polyunsaturated fatty acids, presumably produced by lipoxygenase, which would exert an oxidative action. It is possible that the oxidative inhibition of translation mediated by lipoxygenase may be part of the maturational process of reticulocytes.

Phosphorylation of eIF-2α may occur even in the presence of hemin. After addition of Ca^{2+} or phospholipid, inhibition of translation was observed, which may be due to the activation of Ca^{2+}- and phospholipid-dependent protein kinase.[68-69] It was suggested that this system acts like heme deficiency and could be significant for translation control.

Inhibition of protein synthesis in reticulocyte lysates can also be brought about by the activation of another eIF-2 kinase by double-stranded RNA.[63,70-73] This eIF-2α kinase differs in various respects in its properties from heme-controlled enzyme. It is (1) associated with ribosomes in contrast to the cytosolic location of the latter; (2) less selective and phosphorylates histones; (3) inhibited rather than activated by SH-reagents; and (4) immunologically distinct and its molecular mass is somewhat smaller.[72,74] In contrast to an earlier report, both heme-regulated and double-stranded RNA-dependent eIF-2α kinases occur in human reticulocytes.[75,76] It is remarkable that in many other types of cells the formation of the ribosome-associated eIF-2α-prokinase is induced by interferon, whereas in reticulocytes it only requires the addition of the double-stranded RNA.[77]

The mechanism of action of both heme-controlled and double-stranded RNA-dependent eIF-2α kinases appears to be identical.[63,73,78]

Further proof for the key role of eRF was furnished by the fact that dephosphorylation of inactive eRF-eIF-2 by a phosphatase results in liberation of functional eRF. It is suggested that the recommencement of protein synthesis upon addition of hemin to a heme-deprived lysate may be due to the action of physiological phosphatase.

In summary it appears that absence of heme or presence of double-stranded RNA cause, in an as yet undefined manner, the activation of two specific protein kinases which lead to dead end inactive complexes of eRF with eIF-2 phosphorylated in the α-subunit.

One may question the biologic significance that heme regulation of translation might have. It appears plausible that it is a mechanism which synchronizes the synthesis of heme and globin in such a way that a lack of heme would reduce the initiation rate of globin translation and an excess of heme would have the opposite effect. Yet even if the phenomenon of protein synthesis inhibition by heme deprivation is an artifact, the unravelment of its mechanism signifies a great advance in the general understanding of protein synthesis initiation.

It should be mentioned that oxidized glutathione or other means of oxidation of an eIF-2 cystein group also decreased the efficient use of this factor in protein synthesis.[79-81]

The study of other possible effects on protein synthesis by phosphorylation of components of the translational system has been complicated by both the variety of protein kinases and the large number of constituents, which incorporate phosphate. The major site of phosphorylation of ribosomes is protein S6 in the 40S subunit.[82] A comparison of the effects of three protein kinases on ribosomal function indicated significant differences. With a cAMP-dependent protein kinase, phosphorylation led to inhibition of AUG and poly(A, U, G) binding and of translation, whereas one cAMP-independent kinase (protease-activated kinase II) had the opposite effect.[83]

Of possible biological interest is the observation that polyamines in concentrations such as occur in reticulocytes produce a 2,5-fold activation of the kinase and reverse inhibition by 2,3-bisphosphoglycerate.[81]

VI. MODIFICATIONS OF PROTEINS

Posttranslational processing may occur in reticulocytes. For one, N-terminal methionine can be split off by cathepsin-B-like peptidases (see Chapter 13). Amino-terminal acetylation and processing of actin has been reported to occur in reticulocyte lysates.[84] A ribosome-associated transacetylase has also been demonstrated.[85] The synthesis of dolichol derivatives and the glycosylation of endogenous proteins by them should also be mentioned;[26-29] these modifications presumably occur on proteins destined for the cell membrane.

Phosphorylation by protein kinases and dephosphorylation by protein phosphatases are much more active in reticulocytes than in erythrocytes. Phosphorylation may involve cell membrane proteins, ribosomes, and initiation factors, particularly eIF-2. Some of the protein kinases are cAMP-dependent and may be assumed to serve in the effectuation chain of hormone actions. A variety of cAMP-independent kinases have been found, the physiological function of which is uncertain.[86-91] Some are activated by proteases. They differ also in their pH optima and their site specificity.

The counterpart of protein kinases are protein phosphatases which also form a large family (see Chapter 3). As mentioned previously, one of their identifiable functions may be the dephosphorylation of eIF-2. Both protein kinases and protein phosphatases are cytosolic enzymes.

VII. THE MACHINERY OF PROTEIN SYNTHESIS IN THE RETICULOCYTE

A. Initiation Factors

It is of course of interest to compare the reticulocyte with other cells with respect to its equipment for protein synthesis and with regard to special properties of its mRNAs. This comparison faces the difficulty that very little work in this direction has been performed on other cells. From the universal use of the reticulocyte lysate system one may conclude that there are few, if any, qualitatively discriminating features. In one study, five initiation factors, namely, eIF-2, eIF-3, eIF-4A, eIF-4B, and eIF-5, were purified from human HeLa cells and were compared to those of rabbit reticulocytes.[92] The protein structures, activities, and immunochemical properties of the factors were found to show great similarities.

B. Translation Factors

Elongation factor 1 (eEF-1) has been purified from rabbit reticulocytes and its properties compared with a variety of analogous factors of eukaryotes and *Escherichia coli*. Depending on the purification method used, several forms can be isolated. The aggregated form with a molecular mass of more than 500 kD, called eEF-1H, contains four distinct polypeptides, α-, β-, γ- and δ-; it is possible that the δ-subunit is a breakdown product of γ-polypeptide. Thus, eEF-1H of reticulocytes may contain only three types of subunits like the corresponding factors from pig liver.[93-95]

The eEF-1α unit with a molecular mass of about 50 kD is very similar in its size and amino acid composition to the preparations from other eukaryotic sources; on the other hand, the subunits eEF-1β, eEF-1γ, and eEF-1δ exhibited significant differences.[93]

The second elongation factor eEF-2 has been isolated in pure form from reticulocytes.[96] Like the corresponding factors from other eukaryotic sources, it is a single peptide with a mass of about 100 kD but exhibits some differences, which may be species-related (for a review see Bermek[97]).

C. The tRNAs

Another question which has generated some interest is whether the distribution of transfer ribonucleic acids in reticulocytes is adapted to the synthesis of globin as a predominant product.[98-100] It was found that, in a general sense, the distribution of tRNAs in rabbit reticulocytes corresponds to that of hemoglobin with some deviations. Particularly interesting was the observation that the specialization of tRNA distribution in erythroid cells coincides with their hemoglobinization. The conclusion was drawn that tRNA utilization controls the profile of tRNAs in a cell.

A study of the amino acid acceptance of tRNAs in which human and rabbit reticulocytes were compared confirmed these conclusions; four tRNAs were found to be adapted to the specific requirements for the synthesis of the globin chains.[101]

It is well known that tRNAs contain many modified nucleosides at specific sites. During sequence analyses a previously unknown nucleoside, later named "queuosine", was found in the first position of the anticodon (for a review see Nishimura[102]). Its purine skeleton lacks N-7, which is replaced by C. Queuosine occurs in reticulocytes, in which a tRNA-guanine transglycosylase was found that catalyzes the exchange of queuine for guanine. Queuine is not synthesized in mammals; its source is nutritional. Possible effects of its presence are quite unclear.

D. The mRNAs

Translation of mRNA requires the interactions of its specific sites with other macromolecules. mRNAs appear to differ with respect to their efficiency of translation.[16] One possible cause might be differences in their configuration. In a study on the secondary structure of purified α- and β-globin mRNAs of mouse and rabbit, the exposure to nucleases was tested and it was suggested that the accessibility of the initial AUG codon could be responsible for the differences in their initiation efficiency.[103] The question of the role of proteins in modifying configuration or accessibility was not discussed in this work. An investigation of mRNPs isolated from dissociated polyribosomes of chicken reticulocytes, with respect to their susceptibility to micrococcal nuclease, indicated that most of the coding region of β-globin mRNA — except for a short region near the initiation site — did not contain bound proteins.[104]

Masked and exposed sites of rabbit β-globin mRNA mapped with RNase T₁ gave somewhat different results.[105] It was found that the first 150 nucleotides of the coding region were remarkably resistant to the action of the enzyme, which was taken an an indication of a high degree of folding. Most of the 3'-region, on the other hand, was well accessible. A protecting function of the poly-A sequence was suggested since mRNAs with short poly-A regions were cleaved preferentially. Specific proteins binding to poly-A of mRNA have been demonstrated and purified.[8,106,107] They are part of one type of mRNP.

Reconstitution of functional mRNA-protein complexes from protein-free mRNA and cytosolic reticulocyte proteins has also been achieved.[108a,108b]

VIII. RNA SYNTHESIS

Some activities related to the maturation and modification of RNA are present in reticulocytes. First, it has been shown that the pathway leading to the complex structure known as "cap" at the N-terminus of mRNA is partially present in the reticulocytes of mice. This structure is required for the formation of the mRNA-ribosome complex and consequently for the initiation of protein synthesis in eukaryotes. The enzymatic reactions include transfer of GMP to the 5'-terminus of the mRNA followed by methylations of the terminal guanosine — the penultimate adenosine — and of cytidine. Sequential methylation of all four methyl sites at the 5'-capped structure of globin mRNA was found in mouse reticulocytes. This methylation was only 5 to 8% as efficient as in the erythroid precursor cell. Pulse chase experiments demonstrated that the methyl groups were added to the N-terminus of the mRNA in an orderly sequence with guanosine being first and cytidine last to be methylated. The N^7-methyl group on the 5'-terminal guanosine exhibited a certain degree of turnover, which was not the case with respect to the other methyl groups.[108]

It is of great interest that (2'-5') oligoadenylate synthetase, which is dependent on double-stranded RNA and had originally been found in cells treated with interferon, also occurs in rabbit reticulocytes and in avian erythrocytes but not in mature red cells of mammals.[109-111] This enzyme may disrupt protein synthesis by degradation of mRNA. The activity of the enzyme reached a high level shortly after the peak of reticulocytosis in anemic rabbits to decline with a half-life of 3 days. It was suggested that the erythrocyte originating from stress reticulocytes may still contain significant enzyme activity.[109]

REFERENCES

1. **Jagus, R., Anderson, W. F., and Safer, B.,** The regulation of initiation of mammalian protein synthesis, *Prog. Nucleic Acid Res. Mol. Biol.,* 25, 128, 1981.
2. **Voorma, H. O., Goumans, H., Amesz, H., and Benne, R.,** The control of the rate of initiation of eukaryotic protein synthesis, *Curr. Top. Cell. Regul.,* 22, 51, 1983.
3. **Marks, P. A., Rifkind, R. A., and Danon, D.,** Polyribosomes and protein synthesis during reticulocyte maturation in vitro, *Proc. Natl. Acad. Sci. U.S.A.,* 50, 336, 1963.
4. **Thiele, B. J., Belkner, J., Andree, H., Rapoport, T. A., and Rapoport, S. M.,** Synthesis of non-globin proteins in rabbit erythroid cells, *Eur. J. Biochem.,* 96, 563, 1979.
5. **Cramer, F., Gould, H., Barlow, S., and Carter, N.,** Synthesis of carbonic anhydrase in rabbit and chicken reticulocyte lysates, *Eur. J. Biochem.,* 95, 99, 1979.
6. **Boyer, S. H., Smith, K. D., Noyes, A. N., and Young, K. E.,** Adjuvants to immunological methods for mRNA purification, *J. Biol. Chem.,* 258, 2068, 1983.
7. **Narita, H., Yanagawa, S., Sasaki, R., and Chiba, H.,** Synthesis of 2,3-bisphosphoglycerate synthase in erythroid cells, *J. Biol. Chem.,* 256, 7059, 1981.
8. **Vincent, A., Akhauat, O., Goldenberg, S., and Scherrer, K.,** Differential repression of specific mRNA in erythroblast cytoplasm: a possible role for free mRNP proteins, *EMBO J.,* 2, 1869, 1983.
9. **Lodish, M. F. and Small, B.,** Membrane protein synthesized by rabbit reticulocytes, *J. Cell Biol.,* 65, 51, 1975.
10. **Chang, H., Langer, P. J., and Lodish, H. F.,** Asynchronous synthesis of erythrocyte membrane proteins, *Proc. Natl. Acad. Sci. U.S.A.,* 73, 3206, 1976.
11. **Orkin, S. H., Swan, D., and Leder, P.,** Differential expression of α- and β-globin genes during differentiation of cultured erythroleukemic cells, *J. Biol. Chem.,* 250, 8753, 1975.
12. **Lodish, H. F.,** Model for the regulation of mRNA translation applied to haemoblobin synthesis, *Nature (London),* 251, 385, 1974.
13. **Lodish, H. F.,** Alpha and beta globin messenger ribonucleic acid. Different amounts and rates of initiation of translation, *J. Biol. Chem.,* 246, 7131, 1971.
14. **Phillips, J. A., Snyder, P. G., and Kazazian, H. H.,** Ratios of α- to β-globin mRNA and regulation of globin synthesis in reticulocytes, *Nature (London),* 269, 442, 1977.
15. **Mezl, V. A., Kawasaki, E. S., and Hunt, J. A.,** Analysis of the ratio of to β-globin and globin messenger ribonucleic acid content of fractionated rabbit erythroid bone-marrow cells, *Biochem. J.,* 179, 525, 1979.
16. **Lodish, H. F. and Jacobson, M.,** Regulation of hemoglobin synthesis. Equal rates of translation and termination of α- and β-globin chains, *J. Biol. Chem.,* 247, 3622, 1972.
17. **Lodish, H. F.,** Translational control of protein synthesis, *Annu. Rev. Biochem.,* 45, 39, 1976.
18. **Kabat, D. and Chappell, M. R.,** Competition between globin messenger ribonucleic acids for a discriminating initiation factor, *J. Biol. Chem.,* 252, 2684, 1977.

19. **Schreml, W. and Burka, E. R.**, Properties of membrane-bound ribosomes in reticulocytes, *J. Biol. Chem.*, 243, 3573, 1968.

20. **Bulova, S. I. and Burka, E. R.**, Biosynthesis of nonglobin protein by membrane-bound ribosomes in reticulocytes, *J. Biol. Chem.*, 245, 4907, 1970.

21. **Woodward, W. R., Adamson, D. S., McQueen, H. M., Larson, J. W., Estvanik, S. M., Wilairat, P., and Herbert, E.**, Globin synthesis on reticulocyte membrane-bound ribosomes, *J. Biol. Chem.*, 248, 1556, 1973.

22. **Lodish, H. F. and Desalu, O.**, Regulation of synthesis of non-globin proteins in cell-free extracts of rabbit reticulocytes, *J. Biol. Chem.*, 248, 3520, 1973.

23. **Lemieux, R. and Godin, C.**, Proteins synthetized by rabbit reticulocytes membrane-bound ribosomes, *Can. J. Biochem.*, 60, 580, 1982.

24. **Sullivan, A. L., Grasso, J. A., and Weintraub, L. R.**, Micropinozytoses of transferrin by developing cells: an electronmicroscopic study utilizing ferritin-conjugated transferrin and ferritin-conjugated antibodies to transferrin, *Blood*, 47, 133, 1976.

25. **Meyer, D. I., Krause, E., and Dobberstein, B.**, Secretory protein translocation across membranes — the role of the "docking protein", *Nature (London)*, 297, 647, 1982.

26. **Parodi, A. J. and Martin-Barrientos, J.**, Glycosylation of endogenous proteins through dolichol derivatives in reticulocyte plasma membranes, *Biochim. Biophys. Acta*, 500, 80, 1977.

27. **Martin-Barrientos, J. and Parodi, A. J.**, Synthesis of dolichol derivatives in human erythrocyte membranes, *Mol. Cell. Biochem.*, 16, 111, 1977.

28. **Lucas, J. J. and Nevar, C.**, Loss of mannosyl phosphoryl polyisoprenol synthesis upon conversion of reticulocytes to erythrocytes, *Biochim. Biophys. Acta*, 528, 475, 1978.

29. **Harris, E. D. and Johnson, C. A.**, Incorporation of glucosamine-^{14}C into membrane proteins of reticulocytes, *Biochemistry*, 8, 512, 1969.

30. **Kruh, J. and Borsook, H.**, Hemoglobin synthesis in rabbit reticulocytes in vitro, *J. Biol. Chem.*, 220, 905, 1956.

31. **Bruns, G. D. and London, I. M.**, The effect of hemin on the synthesis of globin, *Biochem. Biophys. Res. Commun.*, 18, 236, 1965.

32. **Zucker, W. V. and Schulman, H. M.**, Stimulation of globin-chain initiation by hemin in the reticulocyte cell-free system, *Proc. Natl. Acad. Sci. U.S.A.*, 59, 582, 1968.

33. **Chakravaty, I., Bagchi, M. R., Roy, R., and Gupta, N. K.**, Protein synthesis in rabbit reticulocytes. Purification and properties of an M. 80,000 polypeptide (Co-eIF-2A^{80}) with Co-eIF-2A activity, *J. Biol. Chem.*, 260, 6945, 1985.

34. **Legon, S., Jackson, R., and Hunt, T.**, Control of protein synthesis in reticulocyte lysates by haemin, *Nature (London)*, 241, 150, 1973.

35. **Ochoa, S.**, Regulation of protein synthesis initiation in eucaryotes, *Arch. Biochem. Biophys.*, 223, 325, 1983.

36. **London, I. M., Fagard, R., Leroux, A., Levin, D. H., Matts, R., and Petryshyn, R.**, The regulation of hemoglobin synthesis by heme and protein kinases, in *Regulation of Hemoglobin Biosynthesis*, Goldwasser, E., Ed., Elsevier Biomedical, New York, 1983, 165.

37. **Adamson, S. D., Godchaux, W., III, and Herbert, E.**, Factors affecting the rate of protein synthesis in lysate systems from reticulocytes, *Arch. Biochem. Biophys.*, 125, 671, 1968.

38. **Maxwell, C. R. and Rabinowitz, M.**, Evidence for an inhibitor in the control of globin synthesis by hemin in reticulocyte lysates, *Biochem. Biophys. Res. Commun.*, 35, 79, 1969.

39. **Howard, G. A., Adamson, S. D., and Herbert, E.**, Studies on cessation of protein synthesis in a reticulocyte lysate cell-free system, *Biochim. Biophys. Acta*, 213, 237, 1970.

40. **Kramer, G., Cimadevilla, M., and Hardesty, B.**, Specificity of the protein kinase activity associated with the hemin-controlled repressor of rabbit reticulocytes, *Proc. Natl. Acad. Sci. U.S.A.*, 73, 3078, 1976.

41. **Levin, D. H., Ranu, R. S., Ernst, V., and London, I. M.**, Regulation of protein synthesis in reticulocyte lysates: phosphorylation of methionyl-tRNA$_f$ binding factor by protein kinase activity of the translational inhibitor isolated from heme-deficient lysates, *Proc. Natl. Acad. Sci. U.S.A.*, 73, 3112, 1976.

42. **Ranu, R. S. and London, I. M.**, Regulation of protein synthesis in rabbit reticulocyte lysates: purification and initial characterization of the cyclic 3′:5′-AMP-independent protein kinase of the heme-regulated translational inhibitor, *Proc. Natl. Acad. Sci. U.S.A.*, 73, 4349, 1976.

43. **Gross, M. and Mendelewski, J.**, Additional evidence that the hemin-controlled translational repressor from rabbit reticulocytes is a protein kinase, *Biochem. Biophys. Res. Commun.*, 74, 559, 1977.

44. **Gross, M.**, Reversal of the inhibitor action of the hemin-controlled translation repressor by a post-ribosomal supernatant factor from rabbit reticulocyte lysates, *Biochem. Biophys. Res. Commun.*, 67, 1507, 1975.

45. **Gross, M.**, Control of protein synthesis by hemin: isolation and characterization of a supernatant factor from rabbit reticulocyte lysates, *Biochim. Biophys. Acta*, 447, 445, 1976.

46. **Clemens, M. J.,** Functional relationships between a reticulocyte polypeptide-chain-initiation factor (IF-MP) and the translational inhibitor involved in regulation of protein synthesis by haemin, *Eur. J. Biochem.,* 66, 413, 1976.

47. **Ranu, R. S., Levin, D. H., Delaunay, J., Ernst, V., and London, I. M.,** Regulation of protein synthesis in reticulocyte lysates: characteristics of inhibition of protein synthesis by a translational inhibitor from heme-deficient lysates and its relationship to the initiation factor which binds Met-t RNA$_f$, *Proc. Natl. Acad. Sci. U.S.A.,* 73, 2720, 1976.

48. **Amesz, H., Goumans, H., Haubrich-Morree, T., Voorma, H. O., and Benne, R.,** Purification and characterization of a protein factor that reverses the inhibition of protein synthesis by the heme-regulated translational inhibitor in rabbit reticulocyte lysates, *Eur. J. Biochem.,* 98, 513, 1979.

49. **Ralston, R. O., Das, A., Dasgupta, A., Roy, R., Palmier, S., and Gupta, N. K.,** Protein synthesis in rabbit reticulocytes: characteristics of a ribosomal factor that reverses inhibition of protein synthesis in heme-deficient lysates, *Proc. Natl. Acad. Sci. U.S.A.,* 75, 4858, 1978.

50. **Ralston, R., Das, A., Grace, M., Das, H. K., and Gupta, N. K.,** Protein synthesis in rabbit reticulocytes: characteristics of a postribosomal supernatant factor that reverses inhibition of protein synthesis in heme-deficient lysates and inhibition of ternary complex (Met-tRNA$_f$^{MET} · e IF-2 · GTP) formation by heme-regulated inhibitor, *Proc. Natl. Acad. Sci. U.S.A.,* 76, 5490, 1979.

51. **Grace, M., Bagchi, M., Ahmad, F., Yeager, T., Olson, C., Chakravarty, I., Nasrin, N., Banerjee, A., and Gupta, N. K.,** Protein synthesis in rabbit reticulocytes: a study of the mechanism of action of the protein factor RF that reverses protein synthesis inhibition in heme-deficient reticulocyte lysates, *Proc. Natl. Acad. Sci. U.S.A.,* 81, 5379, 1984.

52. **Siekierka, J., Mitsui, K.-I., and Ochoa, S.,** Mode of action of the heme-controlled translational inhibitor: relationship of eukaryotic initiation factor 2-stimulating protein to translation restoring factor, *Proc. Natl. Acad. Sci. U.S.A.,* 78, 220, 1981.

53. **Siekierka, J., Mauser, L., and Ochoa, S.,** Mechanism of polypeptide chain initiation in eukaryotes and its control by phosphorylation of the α subunit of initiation factor 2, *Proc. Natl. Acad. Sci. U.S.A.,* 79, 2537, 1982.

54. **Konieczny, A. and Safer, B.,** Purification of the eukaryotic initiation factor 2-eukaryotic initiation factor 2B complex and characterization of its guanine nucleotide exchange activity during protein synthesis initiation, *J. Biol. Chem.,* 258, 3402, 1983.

55. **Matts, R. L., Levin, D. H., and London, I. M.,** Effect of phosphorylation of the α-subunit of eukaryotic initiation factor 2 on the function of reversing factor in the initiation of protein synthesis, *Proc. Natl. Acad. Sci. U.S.A.,* 80, 2559, 1983.

56. **Pain, V. M. and Clemens, M. J.,** Assembly and breakdown of mammalian protein synthesis initiation complexes: regulation by guanine nucleotides and by phosphorylation of initiation factor eIF-2, *Biochemistry,* 22, 726, 1983.

57. **Panniers, R. and Henshaw, E. C.,** A GDP/GTP exchange factor essential for eukaryotic initiation factor 2 cycling in Ehrlich ascites tumor cells and its regulation by eukaryotic initiation factor 2 phosphorylation, *J. Biol. Chem.,* 258, 7928, 1983.

58. **Trachsel, H., Ranu, R. S., and London, I. M.,** Regulation of protein synthesis in rabbit reticulocyte lysates: purification and characterization of heme-reversible translational inhibitor, *Proc. Natl. Acad. Sci. U.S.A.,* 75, 3654, 1978.

59. **Farrell, P. J., Balkow, J., Hunt, T., Jackson, R. J., and Trachsel, H.,** Phosphorylation of initiation factor eIF-2 and the control of reticulocyte protein synthesis, *Cell,* 11, 187, 1977.

60. **Matts, R. L. and London, I. M.,** The regulation of initiation of protein synthesis by phosphorylation of eIF-2 (α) and the role of reversing factor in the recycling of eIF-2, *J. Biol. Chem.,* 259, 6708, 1984.

61. **Siekierka, J., Manne, V., and Ochoa, S.,** Mechanism of translational control by partial phosphorylation of the α-subunit of eukaryotic initiation factor 2, *Proc. Natl. Acad. Sci. U.S.A.,* 81, 352, 1984.

62. **Salimans, M., Goumans, H., Amesz, H., Benne, R., and Voorma, H. O.,** Regulation of protein synthesis in eukaryotes. Mode of action of eRF an EIF-2-recycling factor from rabbit reticulocytes involved in GDP/GTP exchange, *Eur. J. Biochem.,* 145, 91, 1984.

63. **Thomas, N. S., Matts, R. L., Petryshyn, R., and London, I. M.,** Distribution of reversing factor in reticulocyte lysates during active protein synthesis and on inhibition by heme deprivation or double-stranded RNA, *Proc. Natl. Acad. Sci. U.S.A.,* 81, 6998, 1984.

64. **Gross, M., Redman, R., and Kaplansky, D. A.,** Evidence that the primary effect of phosphorylation of eukaryotik initiation factor 2 (α) in rabbit reticulocyte lysate is inhibition of the release of eukaryotik initiation factor-2 · GDP from 60 S ribosomal subunits, *J. Biol. Chem.,* 260, 9491, 1985.

65. **Fagard, R. and London, I. M.,** Relationship between the phosphorylation and activity of the heme-regulated eIF 2 α-kinases, *Proc. Natl. Acad. Sci. U.S.A.,* 78, 866, 1981.

66. **Trachsel, H., Ranu, R. S., and London, I. M.,** Regulation of protein synthesis in rabbit reticulocyte lysates: purification and characterization of heme-reversible translational inhibitor, *Proc. Natl. Acad. Sci. U.S.A.,* 75, 3654, 1978.

164 *The Reticulocyte*

67. **Ernst, V., Levin, D. H., and London, I. M.,** In situ phosphorylation of the α-subunit of eukaryotic initiation factor 2 in reticulocyte lysates inhibited by heme-deficiency, double-stranded RNA, oxidized glutathione, or heme-regulated protein kinase, *Proc. Natl. Acad. Sci. U.S.A.,* 76, 2118, 1979.
68. **de Haro, C., de Herreros, A. G., and Ochoa, S.,** Activation of a heme-stabilized translational inhibitor of reticulocyte lysates by calcium ions and phospholipids, *Proc. Natl. Acad. Sci. U.S.A.,* 80, 6843, 1983.
69. **de Herreros, A. G., de Haro, C., and Ochoa, S.,** Mechanism of activation of the heme-stabilized translational inhibitor of reticulocyte lysates by calcium ions and phospholipid, *Proc. Natl. Acad. Sci. U.S.A.,* 82, 3119, 1985.
70. **Hathaway, G. M. and Trangh, J. A.,** Kinetics of activation of casein kinase II by polyamines and reversal of 2,3-bisphosphoglycerate inhibition, *J. Biol. Chem.,* 259, 7011, 1984.
71. **Ehrenfeld, E. and Hunt, T.,** Double-stranded polio virus RNA inhibits initiation of protein synthesis by reticulocyte lysates, *Proc. Natl. Acad. Sci. U.S.A.,* 68, 1075, 1971.
72. **Grosfeld, H. and Ochoa, S.,** Purification and properties of the double-stranded RNA-activated eukaryotic initiation factor kinase from rabbit reticulocytes, *Proc. Natl. Acad. Sci. U.S.A.,* 77, 6526, 1980.
73. **Ernst, V., Levin, D. H., Leroux, A., and London, I. M.,** Site-specific phosphorylation of the α-subunit of eukaryotic initiation factor eIF-2 α by the heme-regulated and double-stranded RNA-activated eIF-2α kinase from rabbit reticulocyte lysates, *Proc. Natl. Acad. Sci. U.S.A.,* 77, 1286, 1980.
74. **Petryshyn, R., Trachsel, H., and London, I. M.,** Regulation of protein synthesis in reticulocyte lysates: immune serum inhibits heme-regulated protein kinase from double-stranded RNA-induced protein kinase, *Proc. Natl. Acad. Sci. U.S.A.,* 76, 1575, 1979.
75. **Franco, R. S., Roberts, S. K., and Marlebo, O. J.,** Lack of inhibition of protein synthesis by double-stranded RNA in a cell-free system prepared from human reticulocytes, *Arch. Biochem. Biophys.,* 214, 186, 1982.
76. **Petryshyn, R., Rosa, F., Fagard, R., Levin, D., and London, I. M.,** Control of protein synthesis in human reticulocytes by heme-regulated and double-stranded RNA dependent eIF-2α kinases, *Biochem. Biophys. Res. Commun.,* 119, 891, 1984.
77. **Revel, M. and Groner, Y.,** Post-transcriptional and translational controls of gene expression in eukaryotes, *Annu. Rev. Biochem.,* 47, 1079, 1978.
78. **Ranu, R. S. and London, I. M.,** Regulation of protein synthesis in rabbit reticulocyte lysates: additional initiation factor required for formation of ternary complex (eIF-2-GTP-Met-tRNA$_f$) and demonstration of inhibitory effect of heme-regulated protein kinase, *Proc. Natl. Acad. Sci. U.S.A.,* 76, 1079, 1979.
79. **Kosower, N. S., Vanderhoff, G. A., and Kosower, E. M.,** The effect of glutathione disulfide on initiation of protein synthesis, *Biochim. Biophys. Acta,* 272, 623, 1972.
80. **Jagus, R. and Safer, B.,** Activity of eukaryotic initiation factor 2 is modified by processes distinct from phosphorylation. II. Activity of eukoryotic initiation factor 2 in lysate is modified by oxidation-reduction state of its sulfhydryl groups, *J. Biol. Chem.,* 256, 1324, 1981.
81. **Safer, B., Jagus, R., Konieczny, A., and Crouch, D.,** Catalytic utilization of eIF-2: regulation by covalent modification and recycling factors, in *Regulation of Hemoglobin Biosynthesis,* Goldwasser, E., Ed., Elsevier Biomedical, New York, 1983, 185.
82. **Burkhard, S. J. and Traugh, J. A.,** Changes in ribosome function by cAMP-dependent and cAMP-independent phosphorylation of ribosomal protein S6, *J. Biol. Chem.,* 258, 14003, 1983.
83. **Perisic, O. and Traugh, J. A.,** Protease-activated kinase II mediates multiple phosphorylation of ribosomal protein S6 in reticulocytes, *J. Biol. Chem.,* 258, 13998, 1983.
84. **Redman, K. L. and Rubenstein, P. A.,** Actin amino-terminal acetylation and processing in a rabbit reticulocyte lysate, *Methods Enzymol.,* 106 (Part A), 179, 1984.
85. **Traugh, J. A. and Sharp, S. B.,** Protein modification enzymes associated with the protein synthesizing complex from rabbit reticulocytes. Protein kinase, phosphoprotein phosphatase, and acetyltransferase, *J. Biol. Chem.,* 252, 3738, 1977.
86. **Hathaway, G. M. and Traugh, J. A.,** Cyclic-nucleotide-independent protein kinases from rabbit reticulocytes, *J. Biol. Chem.,* 254, 762, 1979.
87. **Tuazon, P. T., Bingham, E. W., and Traugh, J. A.,** Cyclic-nucleotide-independent protein kinases from rabbit reticulocytes. Site-specific phosphorylation of casein variants, *Eur. J. Biochem.,* 94, 497, 1979.
88. **Takara, S. M. and Traugh, J. A.,** Cyclic-nucleotide-independent protein kinases from rabbit reticulocytes. Identification and characterization of a protein kinase activated by proteolysis, *J. Biol. Chem.,* 256, 11558, 1981.
89. **Takara, S. M. and Traugh, J. A.,** Differential activation of 2 protease-activated protein kinases from reticulocytes by a calcium-stimulated protease and identification of phosphorylated translational components, *Eur. J. Biochem.,* 126, 395, 1982.
90. **Lubben, T. H. and Traugh, J. A.,** Cyclic nucleotide-independent protein-kinases from rabbit reticulocytes, *J. Biol. Chem.,* 258, 13992, 1983.

165

91. **Fairbanks, G., Palek, J., Dino, J. E., and Liu, P. A.,** Protein kinases and membrane protein phosphorylation in normal and abnormal human erythrocytes. Variation related to mean cell age, *Blood,* 61, 850, 1983.
92. **Brown-Luedi, M. L., Meyer, L. J., Milburn, S. C., Mo-Ping Yau, P., Corbett, S., and Hershey, J. W. B.,** Protein synthesis initiation factors from human HeLa cells and rabbit reticulocytes are similar: comparison of protein structure, activities, and immunochemical properties, *Biochemistry,* 21, 4202, 1982.
93. **Carralho, J. F., Carralho, M., and Merrick, W. C.,** Purification of various forms of elongation factor 1 from rabbit reticulocytes, *Arch. Biochem. Biophys.,* 234, 591, 1984.
94. **Caralho, M., Carralho, J. F., and Merrick, W. C.,** Biological characterization of various forms of elongation factor 1 from rabbit reticulocytes, *Arch. Bioch. Biophys.,* 234, 603, 1984.
95. **Hattori, S. and Iwasaki, K.,** Studies on the high molecular weight form of polypeptide chain elongation factor-1 from pig liver. I. Purification and subunit structure, *J. Biochem.,* 88, 725, 1980.
96. **Merrick, W. C., Kemper, W. M., Kantor, J. A., and Anderson, W. F.,** Purification and properties of rabbit reticulocyte protein synthesis elongation factor 2, *J. Biol. Chem.,* 250, 2620, 1975.
97. **Bermek, E.,** Mechanisms in polypeptide chain elongation in ribosomes, *Prog. Nucleic Acid Res. Mol. Biol.,* 21, 64, 1978.
98. **Smith, D. W. E. and McNamara, A. L.,** The distribution of transfer ribonucleic acid in rabbit reticulocytes, *J. Biol. Chem.,* 249, 1330, 1974.
99. **Smith, D. W. E.,** Reticulocyte transfer RNA and hemoglobin synthesis, *Science,* 190, 529, 1975.
100. **Smith, D. W. E., Randazzo, R. F., and McNamara, A. L.,** The tRNA content of non-hemoglobinized red cell precursors: evidence that tRNA content is controlled by tRNA utilization, *Biochem. Biophys. Res. Commun.,* 95, 468, 1980.
101. **Hatfield, D., Varricchio, F., Rice, M., and Forget, B. G.,** The aminoacyl transfer RNA population of human reticulocytes, *J. Biol. Chem.,* 257, 3183, 1982.
102. **Nishimura, S.,** Structure, biosynthesis, and function of queuosine in transfer RNA, *Prog. Nucleic Acid Res. Mol. Biol.,* 28, 50, 1983.
103. **Parlakis, G. N., Lockard, R. F., Vamrakopoulus, N., Rieser, L., RajBhandary, U. L., and Vournakis, J. N.,** Secondary structure of mouse and rabbit α- and β-globin mRNAs: Differential accessibility of α and β-initiator AUG codons towards nucleases, *Cell,* 19, 91, 1980.
104. **Chae, C.-B. and Patton, J. R.,** Chicken reticulocyte polysomal messenger RNA-protein complex: absence of bound proteins in most of the coding region of β-globin mRNA, *Nucleic Acid Res.,* 12, 5693, 1984.
105. **Albrecht, G., Krowczynska, A., and Brawerman, G.,** Configuration of β-globin messenger RNA in rabbit reticulocytes. Identification of sites exposed to endogenous and exogenous nucleases, *J. Mol. Biol.,* 178, 881, 1984.
106. **Blobel, G.,** Protein tightly bound to globin mRNA, *Biochem. Biophys. Res. Commun.,* 47, 88, 1972.
107. **Blobel, G.,** A protein of molecular weight 78,000 bound to the polyadenylate region of eukaryotic messenger RNAs, *Proc. Natl. Acad. Sci. U.S.A.,* 70, 924, 1973.
108. **Cheng, T. C. and Kazazian, H. H., Jr.,** Sequential methylation of globin mRNA in nucleated erythroid cells and reticulocytes of mice, *J. Biol. Chem.,* 253, 246, 1978.
108a. **Gaedigk, R., Oehler, S., Köhler, K., and Setyono, B.,** In vitro reconstitution of messenger ribonucleoprotein particles from globin messenger RNA and cytosol proteins, *FEBS Lett.,* 179, 201, 1985.
108b. **Jay, R. et al.,** Reconstitution of functional mRNA-protein complexes in a rabbit reticulocyte cell-free translation system, *Mol. Cell Biol.,* 5, 342, 1985.
109. **Hovanessian, A. G. and Kerr, T. M.,** Synthesis of an oligonucleotide inhibitor of protein synthesis in rabbit reticulocyte lysates analogous to that formed in extracts from interferon-treated cells, *Eur. J. Biochem.,* 84, 149, 1978.
110. **Ferbus, D., Justesen, J., Bertrand, H., and Thang, M. N.,** (2'-5') Oligoadenylate synthetase in the maturation of rabbit reticulocytes, *Mol. Cell. Biochem.,* 62, 51, 1984.
111. **Sokawa, J., Shimizu, N., and Sokawa, Y.,** Presence of (2'-5') oligoadenylate synthetase in avian erythrocytes, *J. Biochem.,* 96, 215, 1984.

178

Chapter 12

LIPOXYGENASE AND THE MATURATIONAL PROGRAM OF THE RETICULOCYTE

I. INTRODUCTION

This chapter will discuss in some detail one single enzyme since it plays a crucial role in the maturation of the reticulocyte. Its discovery was based upon the consideration that rapid and selective disappearance of cellular respiration may well be the common denominator of a variety of maturational changes. It was assumed that the reduction of ATP-formation by two orders of magnitude would in itself — regardless of other specific changes — lead to a loss in protein synthesis and other ATP-requiring processes. The disappearance of cellular respiration goes hand in hand with the destruction of mitochondria. Based on this conception, a successful search was undertaken for a factor that inactivates the respiratory chain.[1,2]

A protein was found in the cytosol of rabbit reticulocytes which inhibits the respiratory chain of mitochondria in a quasi-stoichiometric reaction. The extent of inhibition of the succinate- or NADH-oxidase activity of submitochondrial particles was found to be suitable for quantification of the biological activity of the inhibitory factor. At first with this test system, only a partial purification and preliminary characterization of the inhibitory protein was achieved. Its nature as a non-heme iron protein and the irreversible inactivation by phospholipids were recognized and the time course of the appearance and disappearance of the inhibitor during the course of anemia could be determined.[3-5] In the course of these studies, other maturation-dependent activities were found, such as the inhibition of cytochrome oxidase and the lysis of mitochondria. All these activities turned out to be properties of a single protein, which was identified as a lipoxygenase specific for erythroid cells.[6,7] The following aspects will be discussed in this chapter: the properties of lipoxygenase, its actions, the evidence for its functioning in the intact cell, and its biological dynamics and molecular biology.

II. PROPERTIES OF THE LIPOXYGENASE OF RETICULOCYTES*

A. General

Lipoxygenases (EC 1.13.11.12) generally catalyze in the presence of O_2 the dioxygenation of polyenoic fatty acids or their derivatives, with at least one 1,4-*cis,cis*-pentadiene system. The resulting product is a conjugated hydroperoxy-2,4 *trans,cis*-pentadiene system (Figure 30). Various lipoxygenases differ with respect to three characteristics: (1) the site of primary hydrogen abstraction; (2) the direction of the double bond shift in the primary radical; and (3) the stereospecificity of both hydrogen abstraction and dioxygen insertion. According to the location of the dioxygen insertion, the reticulocyte enzyme may be classified as an n-6 lipoxygenase, with numbering beginning at the methyl group.

The first reaction step is a hydrogen abstraction from the C-atom between two double bonds by which the tervalent enzyme iron is reduced to the divalent form. It is followed by a rearrangement of the radical with a shift of the free electron and the formation of the *cis-, trans*-conjugated double bond; at this point the insertion of dioxygen occurs, followed by a rapid inner electron transfer. In this manner the iron acquires again the tervalent state and the hydroperoxy anion is liberated. With arachidonic acid as substrate, it yields primarily

* For a review, see Schewe et al.[8]

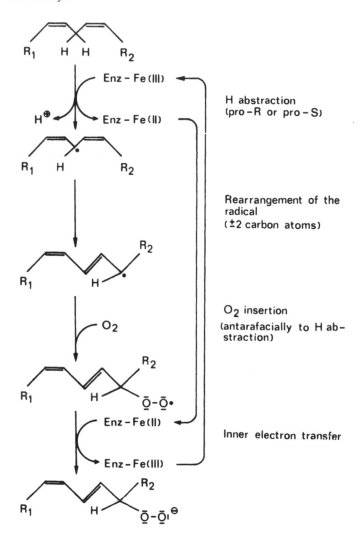

FIGURE 30.   Lipoxygenase reaction scheme.

15 L$_S$-HPETE (15 L$_S$-hydroperoxy-5,8,11,13 *cis,cis*-transeicosatetraenoic acid), and to a minor extent, 12 L$_S$ HPETE.

This dual positional specificity, which was first demonstrated with the reticulocyte enzyme, is also shown by other lipoxygenases.[9,10] Under anaerobic conditions the lipoxygenases catalyze the so-called lipohydroperoxidase reaction (Figure 31). In this reaction the conversion of one polyenic acid molecule is stoichiometrically coupled to the utilization of one molecule of hydroperoxy fatty acid.[11,12] The reaction cycle involves a valency change in the enzyme iron. The primary product of the hydrogen abstraction is a fatty acid radical, while an alkoxy radical originates from a homolytic cleavage of the O—O bond of the hydroperoxy group. A variety of products result from secondary reactions of the primary radicals which include, among others, oxodienoic acids, fatty acid dimers, and pentane.[13] Lipoxygenases have long been recognized as constituents of plants with an unclear biological function but have gained, however, new biological and medical interest in the last decade in connection with the stormy progress of prostaglandin and leukotriene research. The reticulocyte lipoxygenase which was obtained in a homogeneous form from rabbit reticulocytes is so far the only lipoxygenase of animal origin to be purified and characterized.[7,14,15] Its properties are summarized in Table 15.

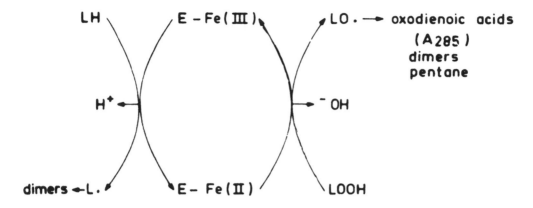

FIGURE 31. Lipohydroperoxidase reaction scheme.

Table 15
CHARACTERISTICS OF RETICULOCYTE LIPOXYGENASE

Molecular mass	78 kD
N-terminus	Gly
C-terminus	His, Ile, Asp
Structure	One chain
Nonpolypeptide constituents	Fe (one atom per mol); carbohydrates
Iron-chelating inhibitors	Nordihydroguaiaretic acid, 4-nitrocatechole, 3-t-butyl-4-hydroxy-anisole, salicylhydroxamic acid
Substrates	Esterified polyenic acids
Positional specificity	n-6 ≫ n-9
Inactivators	Polyacetylenic, hydroperoxy fatty acids
pH Optimum	≈8
$K_{m \text{ (linoleate)app.}}$	12 μM
$K_{m \text{ (O}_2\text{)app.}}$	7 μM

a   Both free and in membrane lipids.

Some remarkable features of this enzyme deserve discussion. It consists of a single polypeptide chain and contains one Fe per mole. The C-terminal heterogeneity may be due to endogenous posttranslational action of exopeptidases.[16] The presence of a carbohydrate moiety indicates glycosylation, in the absence of an endoplasmic reticulum and the Golgi apparatus, which is somewhat puzzling. Among the animal lipoxygenases hitherto known, the reticulocyte enzyme has the exclusive capability of attacking intact phospholipids and even biological membranes, leading to severe disturbance of the lipid bilayer.[6] The enzyme has an apparent high affinity for both O_2 and its organic substrates and is inhibited by some iron-chelating compounds. A specific property to be discussed later is self-inactivation by its products.

B. The Mechanism in the Lipoxygenase Reaction

The detailed mechanism in the lipoxygenase reaction has as yet not been fully clarified (see Schewe et al. for a review[8]). Most of the work has been done on lipoxygenase I from soybeans, which differs from the reticulocyte enzyme with respect to several properties including its pH optimum and its nonsusceptibility to self-inactivation.

A characteristic feature of lipoxygenases is the obligatory activation by the product hydroperoxy fatty acid. This activation may be due in part to the conversion of the ferrous form of the enzyme to a higher oxidized state. In the absence of hydroperoxypolyenic acids a lag phase is observed. Lipoxygenases are inhibited by substrate fatty acids depending on

Table 16
SECONDARY ACTIONS OF RETICULOCYTE LIPOXYGENASE

- Lysis of mitochondrial membranes
- Inhibition of cytochrome oxidase
- Inactivation of the respiratory chain at the level of Complexes I and II
- Uncoupling of oxidative phosphorylation
- Cooxidative destruction of Fe-S centers of the mitochondrial outer membrane
- Cooxidative inactivation of SH-enzymes
- Self-inactivation by lipohydroperoxides
- Triggering of ATP-dependent proteolysis

the concentration of oxygen, the second substrate. On the basis of kinetic studies, models with two binding sites have been proposed for which the substrate fatty acid and its product compete. One of the sites is assumed to be catalytic, the other the activator site. An analysis of the kinetics of progress curves in the reaction of reticulocyte lipoxygenase with linoleic acid was conducted by Ludwig et al.[17] Several models were fitted to the experimental data. According to the simplest model one may assume a valency change cycle of the iron between its divalent and an activated ferric form. Hydrogen abstraction is the rate-limiting step, as is to be expected. The true affinity for oxygen appears to be low, but since product formation is much faster than hydrogen abstraction, a low apparent K_m value results. The oxygen is bound by ferrous lipoxygenase. It is uncertain whether two fatty acid binding sites exist and several arguments may be adduced for a single site.[8] After the initial hydrogen abstraction by the activated ferric enzyme the fatty acid radical that is formed remains enzyme-bound. The resulting ferrous enzyme binds oxygen. Its reaction with the fatty acid radical yields a hydroperoxy radical, which in picking up an electron from the ferrous iron is converted to the hydroperoxy fatty acid anion.

III. THE ACTIONS OF LIPOXYGENASE

The various actions of reticulocyte lipoxygenase that are relevant to maturational processes are listed in Table 16.

A. Lysis of Mitochondria

The enzyme produces drastic lysis of both outer and inner membranes, as demonstrated on rat liver mitochondria (Figure 32).[18-21] These changes are accompanied by both the release of malate dehydrogenase, an enzyme of the mitochondrial matrix, as well as the formation of malonyl dialdehyde, which is a secondary product in the peroxidation of membrane phospholipids.[6,19] The attack of lipoxygenase depends on the functional state of the mitochondria. Mitochondria of immature reticulocytes appear resistant to its attack.[22] In model experiments a strong protective effect of ATP or ADP succinate was demonstrable.[20] One may assume that the protection was caused by energization of the mitochondrial inner membrane, which may influence the state of protein-lipid interactions in the membrane.

The lysis is not accompanied by release of free fatty acids or free amino groups. Hence there is no major involvement of phospholipases or proteinases in the primary attack.

It should be mentioned that the cytoplasmic membrane of red cells is much more resistant to the attack of lipoxygenase than that of the mitochondria.[23] While lipoxygenase from rabbit reticulocytes caused a large formation of malonyl dialdehyde in rat liver mitochondria, erythrocyte ghosts were only slightly attacked, independent of their type of preparation. The formation of malonyl dialdehyde was not enhanced by release of spectrin or actin from the ghosts. Lipoxygenase did not give rise to hemolysis of intact erythrocytes. There appear to

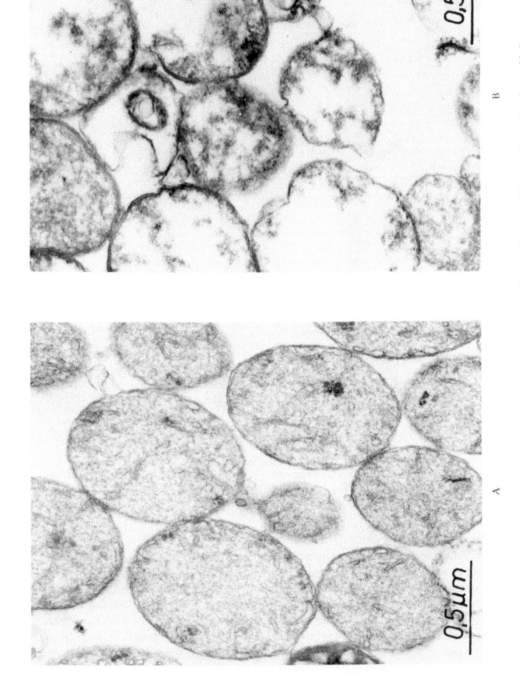

FIGURE 32. Lysis of rat liver mitochondria by lipoxygenase. (A) Control, (B) 5 min; (C) 15 min-incubation times with lipoxygenase.

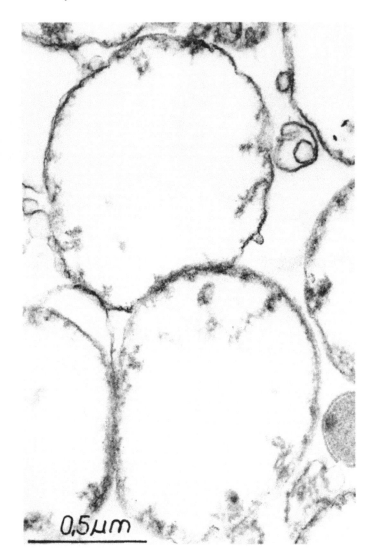

FIGURE 32C

be two reasons for plasma membrane resistance: (1) the protective effect of cholesterol, and (2) a different type of protein-lipid interaction which shields the phospholipids. This interaction is highly heat labile.

B. Inhibition of the Respiratory Chain

Mitochondria of widely varying origin including those from sea urchin eggs and cauliflower are sensitive to the attack by lipoxygenase of reticulocytes.[24,25] Thus, one may conclude that there are common features in the structure of mitochondria, or more precisely, in their protein-lipid interactions which makes them susceptible to this attack. In keeping with the general similarity between mitochondria and chloroplasts there are definite effects of lipoxygenase on the structure and the photosynthetic apparatus of plant organelles. Lipoxygenase exerts two types of inhibitory actions on the respiratory chain (Figure 33):

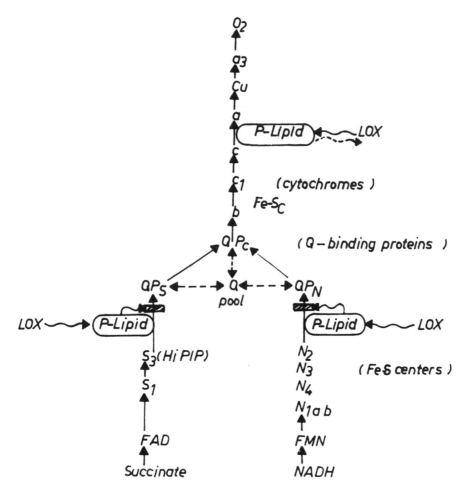

FIGURE 33. Action sites of reticulocyte lipoxygenase on the respiratory chain. S_1, S_3, N_1, N_2, N_3, N_4, Fe-S_C = Fe-S centers; Q = ubiquinone; QP$_S$, QP$_N$ = ubiquinone-binding proteins; LOX = lipoxygenase.

1. One is exerted on cyt c oxidase and is limited to phospholipids.[26] It can therefore be reversed by subsequent substitution of native phospholipids for the peroxidized ones. This attack on the phospholipid has two consequences: (1) a lowering of hydrophobicity owing to the introduction of hydroperoxy and secondary hydroxy groups, and (2) a decrease in membrane fluidity.
2. The other type of inhibition occurs at the sites between the proximate Fe-S centers and ubiquinone.[27] Its irreversibility indicates damage to proteins in addition to changes in phospholipids. A study of various subsystems of the respiratory chain indicated that ubiquinone-binding proteins may be the site of the irreversible attack of lipoxygenase.[27,28]

There is an uncoupling effect of lipoxygenase with a decline in active and an increase in the basal type of ATP formation.[29,30]

The actions of lipoxygenase are not limited to the inner membrane of mitochondria. Drastic changes in their outer membrane were observed, which was no longer visible in electron micrographs of rat liver mitochondria exposed to lipoxygenase of reticulocytes.[18] The Fe-S clusters were completely destroyed with loss in acid-labile sulfur in contrast to the Fe-S centers of the inner membrane which remained intact.[27] In model experiments the monoaminooxidase of brain mitochondria, which is a marker enzyme for the outer membrane, was converted from the A type to the B type. Such a change has been reported after nonenzymatic lipid peroxidation. On the other hand, rotenone-insensitive NADH-cyt c reductase, which is also an enzyme of the outer mitochondrial membrane, was not affected (Schewe, T., unpublished observation).

All these effects are potentiated by hemoglobin. It appears that this effect is due to the ability of hemoglobin, by virtue of its pseudolipohydroperoxidase activity, to decompose catalytically the hydroperoxy lipids generated by lipoxygenase.[31,32] It was suggested that the damaging effects of both lipoxygenase itself and in combination with hemoglobin are brought about by alkoxy radicals, the formation of which is generally assumed to be an initial step of the lipohydroperoxidase reactions.

The action of lipoxygenase on mitochondrial membranes is characterized by a strong burst of oxygen consumption which is prevented by salicylhydroxamic acid (SHAM), a selective inhibitor of lipoxygenase.[33]

What are the specific events connected to the action of reticulocyte lipoxygenase on mitochondria? One would expect changes in the physicochemical properties of the membranes such as membrane fluidity, surface charge, and ion permeability, which have been demonstrated for lipid peroxidation induced by ferrous ascorbate or ultraviolet irradiation (for a review see Vladimirov et al.[34]). In a study of the passive electrical properties of bovine heart submitochondrial particles it was found that there was a fourfold increase in the electrical capacity of the membrane. This effect was probably due to an increase in the dielectric constant of the membrane, caused by the formation of hydrophilic clusters consisting of lipid hydroperoxides or their decomposition products.[35] The formation of such polar regions within the hydrophobic bilayer may also be the reason for the twofold increase in membrane conductivity. Similar but even more pronounced changes were obtained with frozen and thawed beef heart mitochondria which exhibited a fivefold increase in membrane capacity and a sevenfold increase in membrane conductivity.

Considering the circumstances that the system became quickly anaerobic, that only about 0.01 mol of hydroperoxyphospholipids were found per mole phospholipid, and changes in the electrical properties of the membranes as well as the inhibition of respiratory enzymes progress under conditions of anaerobiosis, it follows that the oxygenation of membrane-bound lipids cannot be the main mechanism responsible for the changes. It was apparently the anaerobic hydroperoxidase activity of the reticulocyte lipoxygenase with hydroperoxyphospholipids as substrates which caused the effects described. This type of reaction has been found to be enhanced by hemoglobin, which by itself is also a pseudohydroperoxidase.[36,37]

The changes in the passive electrical properties demonstrated in the model system may be related to the observation of a maturational breakdown of mitochondria in reticulocytes in two respects: (1) the increase in the conductivity of mitochondrial membranes caused by the action of lipoxygenase may give rise to an enhanced proton permeability; indeed there is an uncoupling of oxidative phosphorylation in intact reticulocytes during maturation in vitro;[29] and (2) the changes in passive electric properties may reflect alterations in lipid-protein interactions which lead to an exposure of integral membrane proteins to extramitochondrial proteolytic systems. In this way one may rationalize that the action of the cytosolic ubiquitin-ATP-dependent proteolytic system is preceded and triggered by the lipoxygenase attack.

Lipoxygenase appears to be instrumental in the decline of enzymes with essential SH groups of the cytosol, which occurs during the maturation of reticulocytes.[3,4] Model experiments with muscle phosphoglyceraldehyde dehydrogenase would indicate that such effects are caused by cooxidation during the enzymatic reaction.[7]

C. The Self-Inactivation and Maturational Disappearance of Lipoxygenase

A remarkable feature of reticulocyte lipoxygenase is the suicidal character of the oxygenase reaction at temperatures above 20°C.[38] It was demonstrated that the reaction products, hydroperoxy fatty acids, cause the inactivation of the enzyme with the conversion of one single methionine — which is probably located in the active center of the enzyme — out of a total of 14 methionines, to methionine sulfoxide.[39] The inactivation of soybean lipoxygenase, by 5,8,11,14-eicosatetraynoic acid (ETYA), a commonly used lipoxygenase inhibitor, is also caused by the formation of 1 methionine sulfoxide per mole of enzyme, out of 17 methionines.[40] Thus the assumption appears justified that all lipoxygenases contain one reactive methionine in the active center.

The original assumption that the inactivation of lipoxygenase may be related to the lipohydroperoxidase reaction was refuted by the observation that the anaerobic inactivation by hydroperoxylinoleic acid does not require the presence of linoleic acid, which is an obligatory partner of the hydroperoxidase reaction. Furthermore 13- and 9-hydroperoxylinoleic acids exerted similar inactivating effects but differed greatly in their activity in the hydroperoxidase reaction.[41]

It is suggested that inactivation of the enzyme proceeds at the level of the ferric enzyme-hydroperoxy fatty acid complex. Such complexes have been postulated for soybean lipoxygenase on the basis of EPR spectral data.[42]

Inactivation per se is a localized event with little, if any, change of the three-dimensional structure of the enzyme, as judged from measurements of its circular dichroism and from the lack of change in its antigenicity.[43] It initiates, however, slow progressive changes leading to the denaturation of the enzyme. The self-inactivation accounts only partly for the disappearance of lipoxygenase during maturation. Proteolysis, presumably by plasma-membrane-bound proteases, appears to play a major role; at any rate the decline of lipoxygenase during in vitro maturation is strongly inhibited by leupeptin, an inhibitor of thiol proteinases. Some lipoxygenase escapes inactivation and degradation and may be found in mature erythrocytes of rabbits for prolonged periods after a forced bleeding anemia.

IV. LIPOXYGENASE ACTIVITY IN THE INTACT RETICULOCYTE

It appeared important to establish lipoxygenase functioning in the intact reticulocyte and to determine the share of the lipoxygenase pathway in the total oxygen consumption of the cell. Since lipoxygenase-mediated oxygen consumption has a nonrespiratory character, antimycin A was used to inhibit the respiratory chain and thereby to increase the sensitivity of the assay system. Antimycin-resistant oxygen uptake, normally about 25% of the total, was reduced about one fourth by a variety of inhibitors or inactivators of lipoxygenase, indicating a share of 5 to 7% for lipoxygenase of the total oxygen consumption of reticulocytes.[44] This value is an overestimate, since antimycin A, by inhibiting the utilization of fatty acids via respiration-dependent pathways, favors the lipoxygenase-catalyzed pathway.

Although it has been fully established that lipoxygenase attacks the phospholipids of mitochondria in a susceptible state, the question still remained whether free polyenoic fatty acids might not be better substrates. Various types of evidence indicate that this question can be answered in the affirmative. For one, it was found that free arachidonic acid is attacked at a rate about five times higher than the corresponding lecithin.[45] Second, under the influence of a calcium ionophore, which permits the influx of Ca^{2+} from the ambient medium, there is an increase in both the share of lipoxygenase-mediated oxygen consumption and in pentane formation;[13] this effect is explicable by the calcium stimulation of phospholipase A_2 with a consequently greater availability of free polyenic fatty acids arising from hydrolysis of phospholipids. Third, the same effects are achieved by the addition of arachidonic acid. The reason for this difference between free and phospholipid-bound substrates may be the poorer accessibility of the phospholipids in membrane-linked structures, but other explanations are conceivable.

With this in mind it may be mentioned that membranes exert a curious stimulatory effect on the activity of lipoxygenase from reticulocytes on free fatty acid substrates.[46,47] This effect is exerted by electron transport particles and microsomes as well as by various fractions of lipoproteins in the blood plasma.[48,49] Such effects are absent with soybean lipoxygenase. It would appear as if the monomeric form of the free fatty acid, rather than dimers or higher aggregates, may be favored in the membrane phase. It is likely that the monomer is the actual substrate of the enzyme.

In another series of experiments the pentane formation by reticulocytes was studied. During the hydroperoxidase reaction pentane is formed in a radical chain reaction. Intact reticulocytes produce pentane only if the lipoxygenase pathway is stimulated either by a calcium ionophore which increases the supply of fatty acids by activating phospholipase A_2, or by addition or arachidonic acid.[13] Under these conditions the cellular level of hydroperoxy acids rises sufficiently so that they partly escape reduction by glutathione peroxidase. It may be assumed that the hemoglobin-catalyzed peroxidase reaction is involved in pentane production. A further indication of the role of lipoxygenase is the inhibition of the decline of cytochrome oxidase during maturation in vitro by salicylhydroxamic acid. These studies demonstrate beyond a doubt the functioning of lipoxygenase in the intact reticulocyte.

V. LIPOXYGENASE mRNA

The absence of lipoxygenase activity in the bone marrow as well as in the first days of an anemia in rabbits made it imperative to study both the synthesis of the enzyme and its molecular biology. Cellular synthesis of erythroid lipoxygenase was studied in reticulocytes from bone marrow and in density-separated fractions from peripheral blood of anemic rabbits.[50] Synthesis of lipoxygenase was found to be absent in erythroblasts, in very young reticulocytes obtained from bone marrow, or in the lightest fractions of reticulocytes from the peripheral blood. More mature reticulocytes of the peripheral blood show considerable synthesis of the enzyme.

It was demonstrated that lipoxygenase mRNA is present in reticulocytes as a translationally inactive free cytoplasmic messenger ribonucleoprotein (mRNP) particle.[51] After deproteinization or treatment with proteases, isolated mRNA obtained from masked mRNA codes for authentic lipoxygenase in a cell-free protein synthesizing system of reticulocytes (Figure 34). It may be seen that the lightest fractions 1 and 2 of density-separated reticulocytes failed to exhibit cellular synthesis of lipoxygenase; in contrast the mRNA prepared from these fractions was fully active in directing the synthesis of lipoxygenase in a cell-free system. Lipoxygenase is one of the most abundant nonhemoglobin proteins which are synthesized in reticulocytes, accounting for about 3 to 4% of newly synthesized cytosolic proteins, including hemoglobin and about 30% of newly made nonhemoglobin proteins. The nature of the translationally inactive RNP complexes and the process, which causes a demasking of the mRNP, are being studied.

There are three questions to be asked: (1) is the lipoxygenase found in rabbit reticulocytes specific for erythroid cells?; (2) does the maturation in the normal steady state proceed according to the same mechanism?, and (3) can the results be generalized and do they apply to other species?

The answers to all three questions appear to be affirmative. The collaborative work of Drs. P. Harrison from the Beatson Institute in Glasgow and B. Thiele from our institute led to the isolation of a clone of cDNA derived from the lipoxygenase mRNA of rabbits.[52] By hybridization experiments with the cDNA it was found that lipoxygenase mRNA occurs only in the bone marrow and peripheral blood, but not in other tissues of rabbits, thus confirming earlier immunological data.[7] The mRNA is found in the bone marrow and to a small extent in the peripheral blood of non-anemic animals. The occurrence of lipoxygenase or its mRNA has been demonstrated in reticulocytes of rats and mice and in earlier work even in chickens.[53]

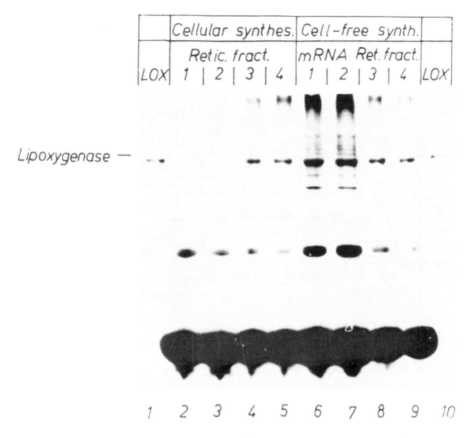

	Cellular synthes.				Cell-free synth.				
LOX	Retic. fract.				mRNA Ret. fract.				LOX
	1	2	3	4	1	2	3	4	

Lipoxygenase —

1 2 3 4 5 6 7 8 9 10

FIGURE 34. Cellular and cell-free synthesis of lipoxygenase in reticulocytes of different maturity. Peripheral reticulocytes were separated by density. One part of each fraction was incubated with ^{35}S-methionine for cellular synthesis; from the other poly-A mRNA was isolated and used for cell-free protein synthesis in the reticulocyte system. Lanes 1 to 10, ^{3}H-labeled marker lipoxygenase; Lanes 2 to 5, cytoplasm of cellularly labeled reticulocytes of different density; Lanes 6 to 9, cell-free protein synthesis with mRNA isolated from corresponding cell fractions. (From Thiele, B. J., et al., *Eur. J. Biochem.*, 129, 133, 1982. With permission.)

Hybridization experiments suggest a close phylogenetic relationship between erythroid cell-specific lipoxygenases of various species, including not only rodents, but man as well (Thiele, B. J., unpublished observation). Thus erythroid-specific lipoxygenases appear to be a highly conserved family of proteins. On the other hand this family differs widely from other animal lipoxygenases including those of white blood cells and thrombocytes, which originate from the same stem cell.

VI. BIOLOGICAL DYNAMICS OF LIPOXYGENASE

It should be emphasized that the biological dynamics of lipoxygenase, i.e., its appearance and disappearance in reticulocytes, exhibit great variations, both under different conditions in one type of organism and between animal species. Early studies demonstrated that the respiration-inhibitory activity of the lipoxygenase is absent in either the bone marrow or in the peripheral red cells of non-anemic rabbits.[1,7] The activity makes its appearance on the 3rd or 4th day and reaches peak levels on the 6th day or after of forced bleeding (Figure 35). In density-fractionated red cells a sharp maximum is found in reticulocytes of intermediate maturity with a steep decline on further maturation. Since changes in enzyme activity

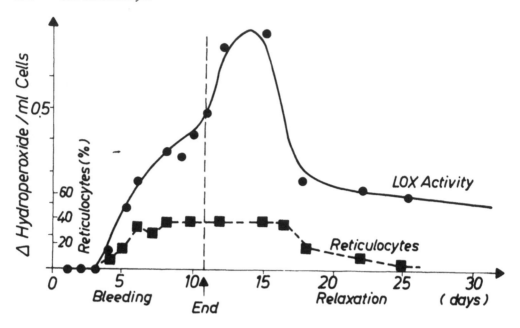

FIGURE 35.   Time course of lipoxygenase activity during bleeding anemia of a rabbit. LOX = lipoxygenase.

in vivo might be complicated by cellular dynamics, i.e., by influx and removal of reticulocytes from the circulation, the biological dynamics of lipoxygenase activity were studied during maturation in vitro.[54]

Parallel immunochemical determinations of lipoxygenase protein concentration and of its activity closely agreed, indicating that the observed changes result from synthesis and degradation of the enzyme. Maximum lipoxygenase activity coincides with the period of most rapid decrease in cyt c oxidase (Figure 36).[55] The role of lipoxygenase in the decay of the respiratory function of mitochondria was demonstrated by experiments on maturation of reticulocytes in vitro. As discussed in Chapter 6 there is a decline in respiratory enzymes under such conditions. This is prevented by lipoxygenase inhibitors.[56] In contrast to the rabbit, lipoxygenase is nearly undetectable in the mouse even if severe anemia is induced. Likewise, in man the appearance of lipoxygenase in the peripheral blood is observed only exceptionally. In chickens some lipoxygenase seems to occur during anemia as judged from the inhibitory activity of red cell cytosol on the succinate oxidase of a heart muscle preparation. Nevertheless it should be stressed that not only in the rabbit but in the mouse and in man as well, lipoxygenase mRNA is present in the bone marrow and in peripheral anemic blood. Taking together all of this evidence, it seems a reasonable conclusion that lipoxygenase plays an important role in the maturation of reticulocytes in all species, regardless of the detectability of free enzyme in the cytosol.

What could be the explanation for the differences? Three possibilities come to mind, none of which are as yet solidly founded. The most obvious explanation would be superinduction with increased transcription and consequent formation of supranormal amounts of lipoxygenase mRNA in the anemic rabbit; one would have to invoke selective effects on promoters. A superinduction, if occurring, would indicate both early onset of the effects of anemia and also that highly immature precursors are affected by it. A close relation between the appearance of lipoxygenase and the occurrence of megalo-reticulocytes appears to exist, which is in line with this hypothesis.

The second and third possibilities involve a change in the balance between synthesis and degradation of lipoxygenase. One possible explanation might be that the synthesis of

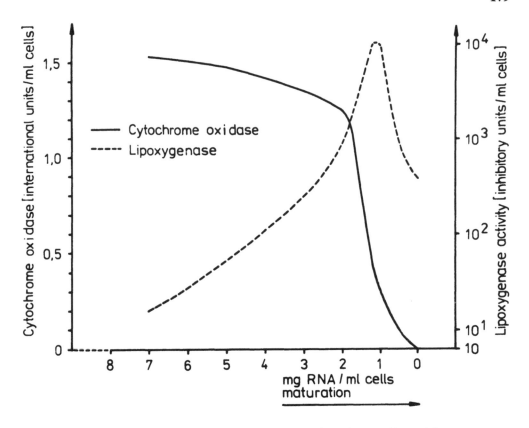

FIGURE 36.   The relation between lipoxygenase and cytochrome oxidase activity.

lipoxygenase is governed by the rate of unmasking of lipoxygenase mRNA. Under normal conditions this rate may be so low that at any given moment only little lipoxygenase is synthesized, which is then subject both to inactivation by interaction with susceptible mitochondria and by proteases in the cytosol. In the anemic rabbit, in which the entire erythropoiesis is accelerated, the unmasking step, presumably a localized proteolytic attack on the RNPs containing lipoxygenase mRNA, may also be speeded up. In this manner synthesis of lipoxygenase would occur at a much greater rate, transiently exceeding that of degradation.

A further possibility might lie in differences in the rate of degradation of either mRNA or lipoxygenase protein between normal steady-state conditions and those of anemic stress. One would have to assume a greater rate of degradation in normal reticulocytes as compared with anemic ones. This explanation is rendered less likely in view of the declining activities of proteinases and other degrading enzymes during maturation.

The first possibility, i.e., superinduction, would lead one to expect much larger amounts of mRNA in the bone marrow of anemic rabbits as compared with non-anemic animals. Such differences have not been found (Thiele, B. J., unpublished observation). This circumstance would favor the second proposed possibility, namely, that differences in unmasking are the cause for the change in lipoxygenase expression found in anemic rabbits as compared with normal ones. Similar arguments can be adduced to explain the differences between various species.

VII. THE SUSCEPTIBILITY OF MITOCHONDRIA TO LIPOXYGENASE DURING MATURATION

One further problem might be discussed at this juncture. What are the temporal and

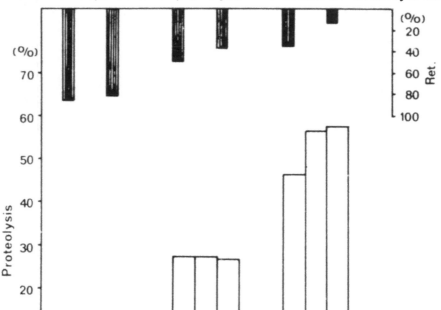

FIGURE 37.   Proteolysis during in vitro incubation of reticulocytes of different maturity. Fr. I = immature reticulocytes; Fr. II = intermediate maturity reticulocytes; Fr. III = mature reticulocytes.

functional relations between the synthesis of lipoxygenase on the one hand and the fate of mitochondria on the other?

It is reasonable to assume — and this has actually been found — that lipoxygenase may be present concomitantly with nonsusceptible mitochondria in immature reticulocytes. In fact, functional mitochondria are a prerequisite for the synthesis of proteins, including lipoxygenase, on account of the ATP requirement, which can only be supplied sufficiently by oxidative phosphorylation. Thus one may conclude that in any given cell the synthesis of lipoxygenase precedes the degradation of mitochondria.

It was demonstrated that immature reticulocytes incubated for as long as 24 hr in a culture medium without iron failed to exhibit proteolysis, which means that their mitochondria were not attacked by the lipoxygenase that was present (Figure 37).[57-59] It was an obvious question, whether a factor might be missing in the culture medium, in the absence of which the mitochondria would not acquire susceptibility to lipoxygenase. It turned out that the addition of iron, preferably in the form of its transferrin complex, rendered the mitochondria of immature reticulocytes susceptible to the attack of lipoxygenase. The hypothesis that the presence of iron had led to the synthesis of a "mitochondria susceptibility protein" was supported by further experiments.[60] Inhibition of protein synthesis by cycloheximide completely suppressed the effect of iron, i.e., the mitochondria remained resistant to degradation during in vitro maturation. Furthermore a factor was actually found in the cytosol of the immature reticulocyte fraction incubated for 10 hr in the presence of iron, which rendered the mitochondria-rich stroma of unincubated immature reticulocytes susceptible to degra-

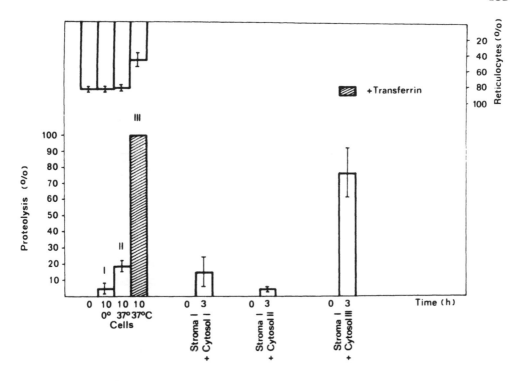

FIGURE 38.   The effect of iron on proteolysis during in vitro incubation of immature reticulocytes. Column I: 10 hr incubation at 0°C; Column II: 10 hr incubation at 37°C without iron; Column III: 10 hr incubation at 37°C with iron. Tests for mitochondria susceptibility protein were performed as follows: cytosols of incubations I to III were incubated with portions of stroma kept at 0°C for 3 hr in the presence of ATP.

dation. This factor was absent both in unincubated immature reticulocytes or in those incubated for 10 hr without iron (Figure 38).

On the basis of (1) the appearance of lipoxygenase, (2) the onset of susceptibility of mitochondria to the attack of lipoxygenase, and (3) the occurrence of ATP-dependent proteolysis, four classes of reticulocyte maturity may be defined, as described in Chapter 3.

How much does the interaction between the mitochondria and lipoxygenase contribute to the biological dynamics of the enzyme? The deleterious effects of the enzyme on mitochondrial function — indicated by the degree of uncoupling and inhibition of the respiratory chain — presumably play a major role in shutting off protein synthesis in addition to changes in the machinery of translation itself. With the termination of protein synthesis, therefore, the time course of lipoxygenase concentration during maturation is mainly determined by the inactivation and degradation of the enzyme. Experiments in which activity changes in lipoxygenase during in vitro maturation were compared with and without addition of iron permit an approximate answer to the question posed above. Lipoxygenase activity declined to one tenth its initial value in the iron-containing medium in which all mitochondria became susceptible and were completely degraded, whereas it decreased only to one third during maturation in a medium without iron (unpublished data). These observations indicate that the interaction between lipoxygenase and mitochondria contribute significantly, but not exclusively, to the decline of lipoxygenase.

VIII. SYNOPSIS OF THE DEGRADATION OF MITOCHONDRIA

At this point a synopsis of the processes instrumental in the degradation of mitochondria during maturation may be given. This serves to integrate existing knowledge but also points

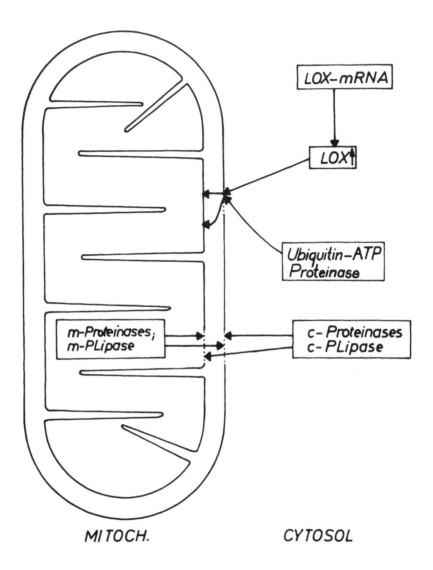

FIGURE 39.   Synopsis of the breakdown of reticulocyte mitochondria. LOX = lipoxygenase.

out the lacunae which remain to be explored. The discussion will be based on the accompanying scheme (Figure 39). The first step is the unmasking of mRNA for lipoxygenase by a hitherto unidentified process. It is most likely a limited hydrolysis of one or more proteins which are postulated to be constituents of the mRNPs specific for lipoxygenase mRNA and are supposed to be responsible for its translational inactivity. Unmasking is succeeded by synthesis of lipoxygenase. The third step, not represented in the scheme but previously discussed, is the transition of mitochondria from a resistant to a susceptible state of lipoxygenase attack. The mechanism of action of the "mitochondria susceptibility protein" has not been explored as yet. It may be presumed that in some manner the protein lipid interactions of the mitochondrial membrane are altered. Now the stage is set for the various deleterious effects of lipoxygenase. These include breaks in the physical integrity of outer and inner membranes of the mitochondria, changes in physicochemical properties, inhibition of respiratory enzymes, and loss of matrix enzymes. At this point ubiquitin-ATP-dependent proteolysis sets in, the details of which are discussed in the next chapter.

REFERENCES

1. **Rapoport, S. M. and Gerischer-Mothes, W.**, Biochemische Vorgänge bei der Erythrozytenreifung: über einen Hemmstoff des Succinatoxydase-Systems in Reticulocyten, *Hoppe-Seyler's Z. Physiol. Chem.*, 302, 167, 1955.
2. **Rapoport, S. M. and Nieradt-Hiebsch, C.**, Biochemische Vorgänge bei der Reticulocytenreifung: über den Angriffspunkt des Reticulocytenhemmstoffes in der Atmungskette, *Hoppe-Seyler's Z. Physiol. Chem.*, 302, 179, 1955.
3. **Rapoport, S.**, Reifung und Alterungsvorgänge in Erythrocyten, *Folia Haematol.*, 78, 96, 1962.
4. **Rapoport, S.**, Molekularbiologische Probleme der Reifung von Erythrozyten, *Folia Haematol.*, 89, 105, 1968.
5. **Rosenthal, S., Gross, J., Grauel, E. L., Papies, B., Schulz, W., Belkner, J., Botscharowa, L., Coutelle, C., Hawemann, M., Nieradt-Hiebsch, C., Müller, M., Opitz, M., Prehn, S., Schultze, M., Staak, R., and Wiesner, R.**, Leitkriterien der Reticulocytenreifung, in *6th Internationales Symposium über Struktur und Funktion der Erythrocyten*, Rapoport, S. and Jung, F., Eds., Akademie-Verlag, Berlin, 1972, 513.
6. **Schewe, T., Halangk, W., Hiebsch, Ch., and Rapoport, S. M.**, A lipoxygenase in rabbit reticulocytes which attacks phospholipids and intact mitochondria, *FEBS Lett.*, 60, 149, 1975.
7. **Rapoport, S. M., Schewe, T., Wiesner, R., Halangk, W., Ludwig, P., Janicke-Höhne, M., Tannert, Ch., Hiebsch, Ch., and Klatt, D.**, The lipoxygenase of reticulocytes. Purification, characterization and biological dynamics of the lipoxygenase, its identity with the respiratory inhibitors of the reticulocyte, *Eur. J. Biochem.*, 96, 545, 1979.
8. **Schewe, T., Rapoport, S. M., and Kühn, H.**, Enzymology and physiology of reticulocyte lipoxygenase: comparison with other lipoxygenases, *Adv. Enzymol.*, in press.
9. **Bryant, R. W., Bailey, J. M., Schewe, T., and Rapoport, S. M.**, Positional specificity of a reticulocyte lipoxygenase, *J. Biol. Chem.*, 257, 6050, 1982.
10. **Kühn, H., Wiesner, R., Schewe, T., and Rapoport, S. M.**, Reticulocyte lipoxygenase exhibits both n-6 and n-9 activities, *FEBS Lett.*, 153, 353, 1983.
11. **Garssen, G. J., Vliegenthart, J. F. G., and Boldingh, J.**, The origin and structures of dimeric fatty acids from the anaerobic reaction between soya-bean lipoxygenase, linoleic acid and its hydroperoxide, *Biochem. J.*, 130, 435, 1972.
12. **Verhagen, J., Veldink, G. A., Egmond, M. R., Vliegenthart, J. F. G., Boldingh, J., and van der Star, J.**, Steady-state kinetics of the anaerobic reaction of soybean lipoxygenase-1 with linoleic acid and 13-L-hydroxyperoxylinoleic acid, *Biochim. Biophys. Acta*, 259, 369, 1978.
13. **Salzmann, U., Kühn, H., Schewe, T., and Rapoport, S. M.**, Pentane formation during the anaerobic reactions of reticulocyte lipoxygenase. Comparison with lipoxygenases from soybeans and green pea seeds, *Biochem. Biophys. Acta*, 795, 535, 1984.
14. **Schewe, T., Wiesner, R., and Rapoport, S. M.**, Lipoxygenase from rabbit reticulocyte, *Methods Enzymol.*, 71, 430, 1981.
15. **Wiesner, R., Tannert, C., Hausdorf, G., Schewe, T., and Rapoport, S. M.**, Reinigung und Charakterisierung des Atmungshemmstoffes RF aus Kaninchenretikulozyten, *Acta Biol. Med. Ger.*, 36, 393, 1977.
16. **Wiesner, R., Hausdorf, G., Anton, M., and Rapoport, S. M.**, Lipoxygenase from rabbit reticulocytes: iron content, amino acid composition and C-terminal heterogeneity, *Biomed. Biochim. Acta*, 42, 431, 1983.
17. **Ludwig, P., Holzhütter, H.-G., and Colosimo, A.**, in preparation.
18. **Krause, W., Schewe, T., and Behrisch, D.**, Über das Vorkommen eines Lyse-Faktors der Mitochondrien in Kaninchenretikulozyten, *Acta Biol. Med. Ger.*, 34, 1609, 1975.
19. **Halangk, W., Schewe, T., Hiebsch, Ch., and Rapoport, S. M.**, Some properties of the lipoxygenase from rabbit reticulocytes, *Acta Biol. Med. Germ.*, 36, 405, 1976.
20. **Krause, W. and Halangk, W.**, Beziehungen zwischen der Konformation isolierter Rattenlebermitochondrien und ihrer Angreifbarkeit durch Lipoxygenase aus Kaninchenretikulozyten, *Acta Biol. Med. Ger.*, 36, 381, 1977.
21. **Schewe, T., Halangk, W., Hiebsch, C., and Rapoport, S. M.**, Degradation of mitochondria by cytosolic factors in reticulocytes, *Acta Biol. Med. Germ.*, 36, 363, 1977.
22. **Rapoport, S. M., Dubiel, W., and Müller, M.**, Proteolysis of mitochondria in reticulocytes during maturation is ubiquitin-dependent and is accompanied by a high rate of ATP hydrolysis, *FEBS Lett.*, 180, 249, 1985.
23. **Fritsch, B., Maretzki, D., Hiebsch, Ch., Schewe, T., and Rapoport, S.**, Zur Selektivität der Wirkung der Lipoxygenase aus Kaninchenretikulozyten auf Mitochondrien- und Erythrozytenmembranen, *Acta Biol. Med. Ger.*, 38, 1315, 1979.
24. **Rapoport, S. M., Hofmann, E. C. G., and Ghiretti-Magaldi, A.**, Über die Atmungsenzyme des Seeigeleies, *Experientia*, 14, 169, 1958.

25. **Schewe, T., Hiebsch, C., Garcia Parra, M., and Rapoport, S. M.,** Zur Wirkung von Atmungshemms-toffen aud die Atmungsenzyme der Mitochondrien aus Blumenkohl (Brassica oleracea L.), *Acta Biol. Med. Ger.*, 32, 419, 1974.
26. **Wiesner, R., Ludwig, P., Schewe, T., and Rapoport, S. M.,** Reversibility of the inhibition of cytochrome c oxidase by reticulocyte lipoxygenase, *FEBS Lett.*, 123, 123, 1981.
27. **Schewe, T., Albracht, S. P. J., and Ludwig, P.,** On the site of action of the inhibition of the mitochondrial respiratory chain by lipoxygenase, *Biochim. Biophys. Acta*, 636, 210, 1981.
28. **Schewe, T., Hiebsch, Ch., and Rapoport, S. M.,** Biochemische Vorgänge bei der Erythrozytenreifung, *Acta Biol. Med. Ger.*, 29, 189, 1972.
29. **Thilo, Ch., Schewe, T., Belkner, J., and Rapoport, S. M.,** In vitro-Reifung von Kaninchenretikulozyten: Verhalten des Sauerstoffverbrauchs, *Acta Biol. Med. Ger.*, 38, 1431, 1979.
30. **Schewe, T. and Rapoport, S. M.,** Role of a cell-specific lipoxygenase in the maturation of reticulocytes, *Acta Biol. Med. Ger.*, 40, 591, 1981.
31. **Schewe, T., Hiebsch, C., Ludwig, P., and Rapoport, S. M.,** Haemoglobin potentiates the respiration-inhibitory action of lipoxygenases via its pseudolipohydroperoxidase activity, *Biomed. Biochim. Acta*, 42, 789, 1983.
32. **Kühn, H., Schewe, T., and Rapoport, S. M.,** Catalytic activities of haemoglobin related to lipid per-oxidation, *Biomed. Biochim. Acta*, 42, S139, 1983.
33. **Rapoport, S. M., Schewe, T., Thiele, B., and Dubiel, W.,** The role of lipoxygenase and ATP-dependent proteolysis in the maturation of the reticulocyte, in *Cell Function and Differentiation*, Part A, Akoyunoglou, G., Evangelopoulos, A. E., Georgatsos, J., et al., Eds., Alan R. Liss, New York, 1982, 47.
34. **Vladimirov, Yu. A., Olenev, V. I., Suslova, T. B., and Cheremisina, Z. P.,** Lipid peroxidation in mitochondrial membrane, *Adv. Lipid Res.*, 17, 173, 1980.
35. **Kühn, H., Pliquett, F., Wunderlich, S., Schewe, T., and Krause, W.,** Reticulocyte lipoxygenase changes the passive electrical properties of bovine heart submitochondrial particles, *Biochim. Biophys. Acta*, 735, 283, 1983.
36. **Kühn, H., Götze, R., Schewe, T., and Rapoport, S. M.,** Quasi-lipoxygenase activity of haemoglobin. A Model for lipoxygenases, *Eur. J. Biochem.*, 120, 161, 1981.
37. **Kühn, H., Götze, R., Schewe, T., and Rapoport, S. M.,** The quasi-lipoxygenase activity of haemoglobin. Discrimination from the haemin-catalyzed lipid peroxidation, *Biomed. Biochim. Acta*, 43, S35, 1984.
38. **Härtel, B., Ludwig, P., Schewe, T., and Rapoport, S. M.,** Self-inactivation by 13-hydroperoxylinoleic acid and lipohydroperoxidase activity of the reticulocyte lipoxygenase, *Eur. J. Biochem.*, 126, 353, 1982.
39. **Rapoport, S., Härtel, B., and Hausdorf, G.,** Methionine sulfoxide formation: the cause of self-inactivation of reticulocyte lipoxygenase, *Eur. J. Biochem.*, 139, 573, 1984.
40. **Kühn, H., Holzhütter, H.-G., Schewe, T., Hiebsch, C., and Rapoport, S. M.,** The mechanism of inactivation of lipoxygenase by acetylenic fatty acids, *Eur. J. Biochem.*, 139, 577, 1984.
41. **Härtel, B., Kühn, H., and Rapoport, S. M.,** Methionine sulfoxide formation — the cause of self-inactivation of lipoxygenases, in *Proceedings of the 16th FEBS Meeting, Moscow, 1984*, Part A, VNU Science, Utrecht, 1985, 299.
42. **Slappendel, S., Veldink, G. A., Vliegenthart, J. F. G., Assa, R., and Malmstrom, B. G.,** A quantitative optical and EPR study on the interaction between soybean lipoxygenase-1 and 13-L-hydroxyperoxylinoleic acid, *Biochim. Biophys. Acta*, 747, 32, 1983.
43. **Ludwig, P., Tordi, M. G., and Colosimo, A.,** Circular dichroism observations on the lipoxygenase from reticulocytes, *Biochim. Biophys. Acta*, 830, 136, 1985.
44. **Salzmann, U., Ludwig, P., Schewe, T., and Rapoport, S. M.,** The share of lipoxygenase in the antimycin resistant oxygen uptake of intact rabbit reticulocytes, *Biomed. Biochim. Acta*, 44, 211, 1985.
45. **Lankin, V. Z., Tikhaze, A. K., Osis, Yu. G., Vikhert, A. M., Schewe, T., and Rapoport, S. M.,** Enzymatic regulation of lipid peroxidation in biomembranes: role of phospholipase A_2 and glutathione-S-transferase, *Dokl. Akad. Nauk SSSR*, 281, 204, 1985.
46. **Lankin, V. Z., Kühn, H., Hiebsch, C., Schewe, T., Rapoport, S. M., Tikhaze, A. K., and Gordeeva, N. T.,** On the nature of the stimulation of the lipoxygenase from rabbit reticulocytes by biological mem-branes, *Biomed. Biochim. Acta*, 44, 657, 1985.
47. **Lankin, V. Z., Tikhaze, A. K., Gordeeva, N. T., Schewe, T., and Rapoport, S. M.,** Animal lipox-ygenases. Activation of reticulocyte lipoxygenase by interaction with natural and artificial membranes, *Biokhimiya*, 48, 2009, 1983.
48. **Gordeeva, N. T., Lankin, V. Z., Osis, Yu. G., Budinzkaja, E. V., Vikhert, A. M., Schewe, T., and Rapoport, S.,** Activation of reticulocyte lipoxygenase by lipoproteins of blood plasma of the rabbit, *Dokl. Akad. Nauk SSSR*, 259, 736, 1981.
49. **Lankin, V. Z., Gordeeva, N. T., Osis, Y. G., Vikhert, A. M., Schewe, T., and Rapoport, S. M.,** Changes in the activity of lipoxygenase from rabbit reticulocytes during its interaction with blood plasma lipoproteins, *Biokhimiya*, 48, 914, 1983.

50. **Thiele, B. J., Belkner, J., Andree, H., Rapoport, T. A., and Rapoport, S. M.,** Synthesis of nonglobin proteins in rabbit erythroid cells. Synthesis of a lipoxygenase in reticulocytes, *Eur. J. Biochem.,* 96, 563, 1979.

51. **Thiele, B. J., Andree, H., Höhne, M., and Rapoport, S. M.,** Lipoxygenase mRNA in rabbit reticulocytes. Its isolation, characterization and translational repression, *Eur. J. Biochem.,* 129, 133, 1982.

52. **Thiele, B., Black, E., and Harrison, P. R.,** Cloning of a RBC-specific lipoxygenase, *Nucleic Acid. Res.,* in press.

53. **Augustin, H. W. and Rapoport, S.,** Über Atmung und Succinatoxydasesystem bei reifen und jugendlichen Hühnererythrozyten, *Acta Biol. Med. Germ.,* 3, 433, 1959.

54. **Höhne, M., Bayer, D., Prehn. S., Schewe, T., and Rapoport, S. M.,** In vitro maturation of rabbit reticulocytes. III. Response of lipoxygenase, *Biomed. Biochim. Acta,* 42, 1129, 1983.

55. **Wiesner, R., Rosenthal, S., and Hiebsch, C.,** Leitkriterien der Retikulozytenreifung. II. Das Verhalten von Zytochromoxydase und Hemmstoff F der Atmungskette bei der Retikulozytenreifung, *Acta Biol. Med. Ger.,* 30, 631, 1973.

56. **Schmidt, J., Prehn, S., and Rapoport, S. M.,** Proteolysis during in vitro maturation of rabbit reticulocyte, *Biomed. Biochim. Acta,* 44, 1429, 1985.

57. **Müller, M., Dubiel, W., Rathmann, J., and Rapoport, S. M.,** Determination and characteristics of energy-dependent proteolysis in rabbit reticulocytes, *Eur. J. Biochem.,* 109, 405, 1980.

58. **Rapoport, S. M., Dubiel, W., and Müller, M.,** The mechanism of maturation-dependent breakdown of mitochondria in reticulocytes, *Acta Biol. Med. Ger.,* 40, 1277, 1981.

59. **Rapoport, S. M., Schmidt, J., and Prehn, S.,** Maturation of rabbit reticulocytes: susceptibility of mitochondria to ATP dependent proteolysis is determined by the maturational state of reticulocytes, *FEBS Lett.,* 183, 1985.

60. **Rapoport, S., Schmidt, J., and Prehn, S.,** Fe-dependent formation of a protein that makes mitochondria lipoxygenase-susceptible during maturation of reticulocytes, in press.

Mechanisms of Maturation of the Reticulocyte

Chapter 13

PROTEOLYTIC PROCESSES

I. INTRODUCTION

Cellular proteolytic processes are essential for the dynamics of any living system. Their regulated activity is a necessary component of the steady state, which obviously represents a balance between synthesis and breakdown of the cell constituents. This balance encompasses a large range of turnover rates of individual proteins and their assemblies which largely take the form of organelles. The wide spectrum of lifetimes leads one to expect that a variety of proteolytic mechanisms are adapted to specific properties and functions of their substrates. In many cells there appears to be a distinction between proteins with long and short lifespans and with respect to the mechanism of their degradation (for reviews see Goldberg and St. John[1] and Hershko and Ciechanover[2]). Obviously this is a rather crude classification which cannot adequately cover the whole range of protein lifetimes.

Proteolysis is also important during the cell cycle. As will be discussed later, the cause of temperature-sensitivity in a cell cycle mutant is due to the thermolability of one component of a specific proteolytic system.[3] The participation of other types of proteolysis is likely.

A third, general role is played by proteolysis at various stages of differentiation. Space does not permit the discussion of a multitude of examples, some of which involve massive involution, e.g., tadpole transition in amphibia, developmental events during embryogenesis, etc.

It is obvious that erythropoiesis as one specific type of differentiation also involves a variety of proteolytic processes. These include losses in nuclear proteins related to the cessation of cell division and pycnosis of the nucleus,[4] elimination of the endoplasmic reticulum, and disappearance of lysosomes and cytoskeleton. Some proteolytic processes go hand in hand with the synthesis and assembly of cellular structures. One example is the formation of the membrane skeleton in the erythroid cell, which contains ankyrin and the tetramer α_2,β_2 spectrin in stoichiometric proportions. It has been demonstrated that the stability of this assembly depends on the presence of all its components.[5] β-Spectrin is usually produced in excess of the requirement. β-Spectrin not utilized, for the formation of the membrane skeleton is rapidly broken down. The importance of the susceptibility to proteolysis — which depends on specific protein-protein or other interactions — is emphasized by the observation that mutants which produce little or deviant ankyrin exhibit a rapid turnover of both α- and β-spectrin.[5-7] The regular progression of differentiation must involve the well-controlled triggering of different proteolytic processes as well as their cessation. Neither type nor regulation of proteolytic systems has been explored as yet.

The transition from reticulocyte to erythrocyte involves further proteolytic processes that must be instrumental in the disappearance of mitochondria and ribosomes, in the loss of membrane receptors and transport systems, and in the decline of a variety of cytosolic components, mostly enzymes.

Again one is led to expect a variety of specific well-regulated proteolytic systems acting in a strict temporal order and directed towards different targets. Our knowledge with regard to their nature is at this time rudimentary. The only system delineated thus far is that directed toward the degradation of mitochondria, which involves attack by lipoxygenase and the operation of ubiquitin- and ATP-dependent proteolytic systems, as will be discussed later.

A multitude of endopeptidases and exopeptidases occur in reticulocytes, many of which are still present in erythrocytes. Some continue to decline during the aging of the erythrocytes and may be related to the termination of their lifespan.

There are a number of difficulties in assessing the role of proteases, including (1) their capacity, i.e., activity measured under optimal conditions of pH and substrate concentration, is vastly in excess of what can be reasonably assumed to occur in the cell; (2) pH optima of several enzymes are far outside the expected pH in the cytosol or in/at the cell membrane; (3) their natural substrates are not known. The enzymes usually are tested with synthetic low molecular substrates, selected polypeptides such as insulin-chains or glucagon, which do not occur in red cells, or with denatured proteins which occur only exceptionally; (4) protease inhibitors are present in the cells; the interplay between these and the respective proteases under in vivo conditions has not been clarified; (5) there appears to be extensive masking of proteolytic activities, particularly in the cell membrane, which in part may be due to sequestration.

The quantitative importance of hydrolytic proteolytic processes in the reticulocyte is minor as indicated by the fact that more than 90% of proteolysis demonstrable in intact rabbit reticulocytes is ATP-dependent, provided that reticulocytosis has been induced by bleeding and not by phenylhydrazine.[8] On the other hand there are definite indications that at least aminopeptidases take part in the pathway of ATP-dependent proteolysis.[9-11] An analogous role has been suggested for a cytosolic metal-dependent endoprotease, which preferentially attacks large peptides.[12] A likely role of aminopeptidases is connected with protein synthesis and consists of the removal of initiation methionyl residues.[13]

The possibility that acid proteases may take part in the breakdown of proteases at neutral pH is indicated by the observations that β-hemoglobin chains entrapped in human erythrocytes are broken down by cooperative action of cytosolic neutral and membrane-bound acid proteases, and that acid enzymes can degrade endogenous newly synthesized membrane proteins in human reticulocytes.[14-15] These observations are difficult to explain. The authors proposed that membrane binding is a prerequisite for the degradation by membrane-bound proteases and that this binding causes unspecified conformational changes which shift the pH of enzyme action. Because of its preponderance and relatively clear function the discussion of ATP-dependent proteolysis will precede discussion of hydrolytic enzymes.

II. ATP-DEPENDENT PROTEOLYSIS: THE UBIQUITIN-DEPENDENT SYSTEM

For years it was believed that proteolysis is by necessity a simple hydrolytic process depending only on the interaction of proteases with susceptible substrates. Proteolysis was considered to be one of the main characteristics of autolysis which begins with cell death. It was therefore a great surprise when evidence was obtained for the existence of ATP-dependent proteolysis in liver[16,17] and reticulocytes.[18,19] These studies were not followed up until about 20 years later. Meanwhile indications for the existence of powerful proteolytic systems in a variety of both prokaryotes and eukaryotes were obtained (for a review see Goldberg and St. John[1]). It turned out that abnormal or incomplete proteins are readily degraded. The same is true for red blood cells. Normal hemoglobin is resistant to proteolytic attack but abnormal globin molecules containing amino acid analogues,[20-22] puromycyl peptides,[23,24] or mutant globin chains[25,26] — even excess chains produced in thalassemic cells[27-31] — are rapidly broken down.

The breakthrough was achieved when it was shown[32] that the degradation of proteins containing amino acid analogues in lysates from reticulocytes is stimulated by ATP. It was independently demonstrated[34-37] with denatured globin as substrate that the proteolytic system of reticulocytes may be separated into two fractions, one a heat-stable polypeptide and the other containing a variety of components of the proteolytic system. The heat-stable polypeptide was later identified as ubiquitin, consisting of 76 amino acid residues with a COOH-terminal gly-gly grouping[33,34] essential for its activity. The ubiquitin had been previously isolated; it derives its name from its universal distribution. It is found in many types of cells covalently bound to H2-type histones.[38,39]

For ubiquitin- and ATP-dependent proteolysis, the signal event is the conjugation of ubiquitin to substrate proteins by covalent isopeptide bonds in which the COOH-terminal glycine of ubiquitin and the free amino groups of the lysines of the substrate proteins take part.[40-42] In this manner multiple ubiquitins are bound to one substrate protein. The operation of this mechanism as well as the turnover of ubiquitin was shown by immunochemical methods in intact cells.[43] Ubiquitin may be regenerated either by succeeding steps in proteolysis of the conjugates or by an isopeptidase, an enzyme, the function of which is still unclear, that apparently competes with proteolysis.[44-46] Ubiquitin has to be activated to conjugate. The enzyme catalyzing ubiquitin activation was separated and highly purified.[47] It is a dimer, with subunits of 100 kD molecular mass, and catalyzes the overall reaction[38,48,49]:

$$2\ \text{ATP} + 2\ \text{Ub} + \text{E}_{SH} \xrightarrow{\text{Mg}^{2+}} 2\text{PP}_i + \text{AMP} + \text{E}_{S\text{-Ub}}^{\text{AMP-Ub}}$$

in which both an enzyme-bound COOH-terminal ubiquitin thiol ester and a COOH-terminal ubiquitin adenylate are formed. The ubiquitin thiol ester is the proximal donor for subsequent reactions (see Figure 40).

Detailed studies on the mechanism of ubiquitin activation established a reaction cycle which proceeds in a strictly ordered sequence;[49] initially one ATP and one ubiquitin react with the activated enzyme forming a noncovalently enzyme-bound ubiquitin adenylate. This reaction is followed by the conversion of this complex to the enzyme thiol ubiquitin ester. In the third step the first reaction is repeated with the formation of a ternary complex containing two ubiquitins.

In steady-state proteolysis it may be assumed that only one ubiquitin is transferred to an acceptor outside the cycle from the thiol ester position and that only steps two and three of the initial sequence are required. Thus only one ATP is needed for the formation of a substrate protein-ubiquitin conjugate.

It was later established that two further enzymes are interposed between the initial activation of ubiquitin and the formation of substrate-protein-ubiquitin conjugates.[50] One of these, E_2, transmits thiol-bound ubiquitin from the activating enzyme, E_1, to a third enzyme, E_3, which in an as yet unclarified manner mediates the formation of the conjugates. The authors point out that a similar type of enzyme relay system has been found to operate in the oxidation of fatty acids.

Recently it has developed that E_2, which may be considered a carrier protein, constitutes a family of proteins with different specificities with respect to the acceptors, to which they transfer ubiquitin.[50a] Only one of them appears to be functional in the E_3-dependent protein breakdown. This carrier protein has a subunit molecular mass of 14 kD. The complexity of the system is indicated by the observation that polyubiquitin structures, i.e., products of the conjugation among ubiquitin molecules, are found in the ubiquitin-protein conjugates.[50b] It is unclear whether they are artifacts or have a so-far-undetermined role in the proteolytic process.

A new component of the ubiquitin and ATP-dependent proteolytic system appears to be a certain type of tRNA.[50c] It seems to be essential for the breakdown of some proteins only. The possible role of tRNA in proteolysis is at present obscure. Several proteins such as bovine serum albumin are degraded only if they are modified. The susceptibility varies with the type and extent of modification.[50d] Furthermore, differences were found with respect to their dependence on ATP concentration. Neither charge nor denaturation nor aggregation correlated with the degradation rate. Thus, it is obvious that it is an unsolved central problem, which features of the protein structure are recognized by the ubiquitin activating system.

A central problem appears — which features of the protein structure are recognized by the ubiquitin activating system?

The original assumption based on the demonstration of ubiquitin binding to α-NH$_2$ groups of lysine residues meets with the difficulty that in most proteins nearly all lysine residues are exposed due to their strong charge.

Recent work helps to overcome this difficulty, but indicates that the process of ubiquitin-dependent selective proteolysis is complicated, and far from clarified. It was shown that the selective modification of N-terminal α-NH$_2$ groups of proteins prevent their degradation by the ubiquitin-dependent proteolytic system of reticulocytes.[43a] Furthermore, naturally occurring N-α-acetylated proteins are not degraded. On the other hand, blockage of α-NH$_2$ groups of lysine does not prevent any degree of degradation. The conclusion was drawn that the conjugation of ubiquitin with the α-NH$_2$ terminus of a substrate protein precedes and triggers the subsequent conjugation of ubiquitin to lysine. In line with these results is the report that proteins with their lysines modified by guanidination are degraded by ATP-dependent proteolysis in reticulocyte lysates.[43b]

A puzzling feature is the existence of isopeptidase activity by which ubiquitin may be released from its conjugates.[44-46] Ubiquitin activation and isopeptidase activity constitute a kind of ATP-wasting futile cycle. It is conceivable that isopeptidase activity may be modulated so that it may regulate the rate of proteolysis in an inversely proportional manner.

A characteristic of ATP- and ubiquitin-dependent proteolysis is its suppression by hemin,[51,52] which inhibits it by one half in a 25-μM concentration. The site of its action is beyond the reactions of ubiquitin activation.

Vanadate has also been suggested as a useful inhibitor of ATP-dependent proteolysis.[53] At low concentrations, e.g., 0.1 mM, it inhibits this process significantly in a reversible manner, but is without effect on various ATP-independent proteases. Lysosomal enzymes such as cathepsins B, H, and L require vanadate concentrations two orders of magnitude higher for their inhibition. Vanadate does not inhibit the conjugation of ubiquitin with substrate proteins, which indicates that its site of action must be located a step beyond conjugation. Its use on intact cells, however, is compromised by various other effects including decrease in ATP, tendency to hemolysis, and inhibition of phosphate transferases and of protein synthesis.

The entire system of reactions involving ubiquitin is depicted in the accompanying scheme (Figure 40).

In an important study it was demonstrated that the conjugation of ubiquitin to denatured hemoglobin was proportion to the rate of hemoglobin degradation in HeLa cells.[54] In these experiments radio-iodinated ubiquitin was introduced together with hemoglobin in HeLa cells by erythrocyte fusion. Denaturation of hemoglobin was achieved by treatment with low doses of phenylhydrazine. Under these conditions the concentration of globin-ubiquitin conjugates was greatly increased proportionally to the rate of hemoglobin denaturation.

A particle-associated ATP-dependent proteolytic activity was found both in induced and non-induced erythroleukemia cells of mouse and human origin.[54a] It was speculated that this activity is a precursor of the cytosolic system of reticulocytes or a distinct transient system, which is lost before the reticulocyte stage.

III. THE PHYSIOLOGICAL BREAKDOWN OF MITOCHONDRIA DURING MATURATION

As mentioned in the introduction to this chapter, early observations indicated the occurrence of ATP-dependent proteolysis in reticulocytes.[18,19] Later a method was developed to quantify the breakdown of proteins in intact cells, which is based on a combination of isotope dilution and concentration changes in lysine.[55] ATP-dependent proteolysis is an extensive and time-limited process. It is strongly influenced by pH and temperature, being suppressed at low temperature and by low pH values. The extent of proteolysis is proportional to

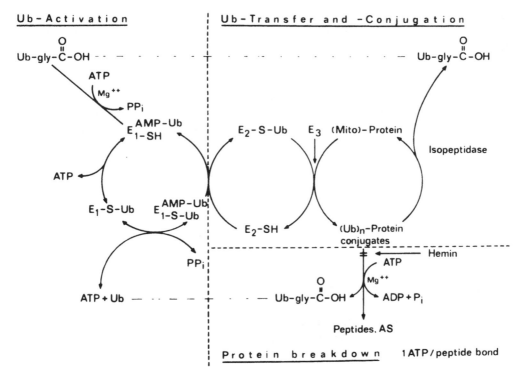

FIGURE 40.   Survey of ubiquitin activation, transfer, and conjugation during protein breakdown. Ub = ubiquitin.

reticulocyte count. The ATP-dependent portion accounts for more than 90% of total proteolysis. Mitochondria are its main substrate, while the plasma membrane, ribosomes, and cytosol, contribute little, if anything.[56] In later work[57] it was established that the degradation of mitochondria is entirely ubiquitin-dependent. Other reports in which ubiquitin-independent proteolysis was found[53,58,59] are explicable by the fact that phenylhydrazine was used to produce reticulocytosis. This problem will be discussed later.

ATP-dependent proteolysis is preceded and triggered by the attack of lipoxygenase on the mitochondria as part of a sequence of events beginning with the unmasking of lipoxygenase mRNA (see Chapter 12). The degradation of mitochondria is determined primarily by their state of susceptibility, which depends on the maturity of reticulocytes.

Native mitochondria are a much better substrate for the ATP-ubiquitin-dependent proteolytic system than are heat-denatured mitochondria.[57] Correspondingly, less than one fourth conjugation of ^{125}I-ubiquitin with heat-denatured reticulocyte stroma was found as compared with native stroma.[57a] It is conceivable, in view of the essential permissive role of α-NH$_2$ groups that the decrease in acceptance of ubiquitin is caused by reactions which block these groups in mitochondrial proteins. We could not observe such a difference in ubiquitin conjugation between native and denatured human serum albumin. Therefore it would appear that heat denaturation causes reactions between proteolysis-susceptible proteins of the stroma and other components.

In other respects ubiquitin conjugation with and release from reticulocyte mitochondria correspond closely to observations on single proteins. Half-maximal ^{125}I-ubiquitin conjugation occurred after about 2 min, similar to the value reported far a single protein.[41] A rapid exchange of ^{125}I-ubiquitin in prelabeled stroma with unlabeled ubiquitin was observed. High ubiquitin concentrations of about 2 μM, inhibit conjugation as found for a single substrate.[49] Hemin blocks proteolysis completely, but does not affect the conjugation of ^{125}I-ubiquitin, which is in agreement with the report by Haas and Rose.[52] However, the maximal extent of conjugation per milligram stroma protein amounts to only one third of that observed with human serum albumin. From the liberation of ^{125}I-ubiquitin under conditions which prevent further conjugation, the presence of an isopeptidase can be deduced, which corresponds to the data of Ciechanover et al.[41] and the recent observations on ubiquitin-lysozyme conjugates by Hershko et al.[43a]

We found that stroma obtained from reticulocytes that had been depleted of ATP and presumably of ubiquitin still exhibited residual ATP-dependent proteolysis, even if no ubiquitin was added.[57] It was interesting to learn if the stroma still had retained some conjugated ubiquitin which might be responsible for the residual proteolysis. Isotopic dilution experiments yielded a value of about 0.3 μM stroma-bound ubiquitin per ℓ cells, which in view of the K_m values of 0.58 μM reported for ubiquitin conjugation[49] accounts fully for residual ATP-dependent proteolytic activity.[57a]

Proteolysis stops after 2 to 3 hr in suspensions of reticulocytes in a salt or minimal Eagle culture medium. The termination of the process is caused by the exhaustion of available substrate, i.e., lipoxygenase-susceptible mitochondria. According to Borsook et al.,[60] in a medium optimal for protein synthesis that is enriched with iron and amino acids, all mitochondria become susceptible and proteolysis proceeds toward completion. The proteolytic system is fully maintained in the optimal medium while its activity declines considerably during in vitro incubation in the minimal medium. According to the original scheme outlined above, only one ATP per ubiquitin ligated to substrate proteins in the steady state should be required. Thus one would expect an ATP consumption for proteolysis that would not exceed the number of lysine residues capable of combining with ubiquitin. In the work on proteolysis of reticulocyte mitochondria, a much higher ratio amounting to about 20 ATP per lysine was observed, indicating that 1 ATP might be required per peptide bond-split. In recent experiments[57] this observation was confirmed under conditions where ATP hydrolysis not connected to proteolysis was much reduced. Furthermore, it was found that the difference in ATP hydrolysis between control assays and those in which proteolysis was inhibited by hemin amounted to about one ATP per peptide bond. In a recent communication[61] it was reported that the proteolytic breakdown of ubiquitin-substrate-protein conjugates produced with denatured proteins is also an ATP-requiring process. The detailed mechanism of the reactions beyond ubiquitin activation that require ATP remains to be clarified.

The ATP-dependent proteolysis in reticulocytes is a strongly maturation-dependent process.[62-64] In fact, proteolytic activity further declines to negligible levels as red cells age. It was assumed[63] that the decline might be due to loss or inactivation of ubiquitin, which appears at variance with the fact that ubiquitin has been found both in the reticulocytes of rabbits and in the erythrocytes of man and rats in high concentrations (>12 μM).[40] On the other hand a decline in high molecular components in proteolytic systems takes place.[65]

At this juncture one may raise the question as to whether lysosomes are involved in proteolysis of reticulocytes. As mentioned previously (Chapter 3) there are few, if any, lysosomes demonstrable in reticulocytes. Nevertheless doubts exist as to whether these organelles are really absent and if they contribute to proteolytic processes. This suspicion is strengthened by the occasional finding of mitochondria in structures which may represent autophagic vacuoles. Several types of experiments[58] on failure of lysosomotropic agents and selective inhibitors of lysosomal proteases to affect protein breakdown serve to reinforce the conclusion that lysosomes play no significant part in the degradation of proteins in reticulocytes. Are ubiquitin and ATP obligatory for protein degradation in reticulocytes?

At first sight there are highly conflicting answers to this question. I hope to be able to reconcile these apparent contradictions in the following account. As a guide line to the literature it is advisable to clearly distinguish different types of substrate, which apparently correlate with distinct types of proteolytic mechanisms: (1) the first type of substrate is the native mitochondria degraded during maturation. This process depends both on ATP and ubiquitin, as discussed previously; (2) a second type of substrate is modified or denatured proteins. With them an ATP requirement for proteolysis, independent of ubiquitin, is demonstrable, as has been stressed by Tanaka and co-workers.[53,59] In line with these reports is the observation that only the breakdown of polylysine of high molecular weight is stimulated by ATP.[66] Thus, high molecular weight appears to be a prerequisite for this ATP-stimulated system. There is also ATP stimulation of the breakdown of proteins, which had incorporated amino acid analogues.[67,68] It is perhaps significant that this proteolytic system always shows basic proteolytic activity and that ATP stimulation is about two- to fivefold. Thus it may not have an absolute ATP requirement; (3) the third type of substrate is low molecular weight polypeptides. Definite evidence has been adduced that globin peptides produced by cyanogen bromide fragmentation are degraded in cell-free extracts of rabbit reticulocytes in an ATP-independent manner.[69] The same holds true for the proteolysis of puromycin peptides, i.e., those that arise from incomplete elongation during translation.[68] Also, the breakdown of polylysines having a molecular mass of 3 kD is not stimulated by ATP;[66] (4) a fourth group is the small peptides. As discussed elsewhere, a variety of peptidases have been found in red cells. For one type, amino peptidases, a clear function within the proteolysis system has been demonstrated. In the presence of bestatin, a specific inhibitor of these enzymes, there is an accumulation of di- and tripeptides during ATP-dependent proteolysis which otherwise does not occur.[9,10] One may therefore conclude that the primary ATP-dependent proteolytic attack is followed by the action of amino peptidases.

All types of the proteolytic activities described exhibit strong maturation dependence. From the foregoing considerations one may assume that three different systems of proteolysis coexist, the share of which under given circumstances may vary according to the substrates occurring in the reticulocyte. The degradation of abnormal proteins including mutants may well proceed both by pathways requiring and independent of ubiquitin. Mutants leading to shortened peptide chains might also be subject to ATP-independent proteolysis. The effects of phenylhydrazine are such that one can expect both normal and damaged mitochondria and denatured and oxidized proteins. Thus it is not surprising that all three types of proteolysis overlap.[8]

Maturation dependence may explain low[70] or absent ATP-dependent proteolysis in human reticulocytes which are commonly much more mature than those experimentally produced in animals.[9,70] In keeping with this explanation is the circumstance that human bone marrow and earlier erythroid precursors exhibit high proteolytic activity.

Finally, two questions will be dealt with: first, whether the mechanism of degradation described for mitochondria is also operative with respect to other components of erythroid cells; and second, whether ubiquitin- and ATP-dependent proteolysis occurs in other types of cells.

As for the first point, one must concede that clear-cut evidence is not available. The processes by which endoplasmic reticulum components of the cell membrane and lysosomes are broken down have not been studied as yet. They must mainly occur before the reticulocyte stage. The difficulties one is faced with are perhaps best exemplified in the case of the methemoglobin reductase system. It consists of two components, one a form of cyt b_5, the other a flavine enzyme, cyt b_5 reductase. Both occur in soluble form in the cytosol of reticulocytes and erythrocytes. In this respect they differ from the corresponding proteins in the liver cell which are located in the endoplasmic reticulum and thus are membrane-bound.

Analysis of amino acid sequences as well as immunologic studies have demonstrated that the soluble enzyme in erythrocytes corresponds to the hydrophilic domain of the liver enzyme with the same functions. It therefore seems cogent to assume that soluble red cell proteins originate by selective posttranslational proteolytic processing by which the hydrophobic C-terminal tail is split off and the membrane binding properties are lost. In line with this postulated course of events, it was found that in Friend erythroleukemia cells, a prototype of an immature erythroid cell, both cyt b_5 and its reductase are present only in membrane-bound form.

Direct evidence for this mechanism could not be obtained thus far. Therefore, model experiments were resorted to. The studies from two groups[71,72] were performed in an analogous manner, namely, by following the liberation of cytochrome b_5 and its reductase from either microsomal cyt b_5[71] or from liver microsomes.[72] The results appear to be contradictory. In the first study it was demonstrated that a membrane-bound protease with an acidic pH optimum of 5.7, which is inhibited by pepstatin, can generate a soluble heme protein. Furthermore, it was shown that this enzyme is one order of magnitude more active in reticulocytes than in erythrocytes. It is likely that this enzyme is identical with proteases previously described. In the second study it was demonstrated that the solubilization from microsomes was strictly dependent on the ubiquitin- and ATP-dependent system. Thus one may choose two possible mechanisms. The one suggested by Schafer and Hultquist[71] presupposes a low pH, since at pH 7 no activity at all was found. Such a condition is inconceivable in the bulk of the cytosol or in the cell membrane and requires a compartmentation of some kind. Indeed, low pH values are found in lysosomes in which many sorts of hydrolases with acid pH optimum are assembled. It therefore appears to be a reasonable assumption that at least some of the acid proteases of red cell membranes are of lysosomal origin and may have become located in the cell membrane by fusion or sequestration of lysosomes. One may conclude that proteolysis by acid proteases may be operative provided that it is located in still intact lysosomes. From this precondition it follows that the liberation of cyt b_5 and its reductase would have to occur at a stage of erythroid differentiation preceding the reticulocyte in which there is an abundance of lysosomes. In accordance with this proposal the bulk of cyt b_5 and its reductase are found in soluble form in the reticulocyte.

What about the other mechanism? Its function cannot be excluded. Perhaps it serves in the degradation of the endoplasmic reticulum, which as a whole is unlikely to be an accessible substrate for lysosomal action. Thus it would represent the first stage of breakdown to be completed by lysosomal and other proteases. The other possibility is that it may continue to function even after lysosomes have ceased to do so.

ATP dependent proteolysis may be responsible for the maturational decline of hexokinase.[73] In these studies it was found that the cytosolic isoenzyme Ib was rapidly degraded by a ubiquitin- and ATP-dependent process, whereas the mitochondria-bound isoenzyme Ia was resistant. The decay of the bound hexokinase corresponded to the disappearance of the mitochondria.[74]

Let me turn to the following question: to what extent are the results on reticulocytes applicable to other types of cells? The general distribution of ATP-dependent proteolysis has been referred to in the introduction of this chapter. Another matter is the role of ubiquitin. The presence of a ubiquitin-dependent protein degradation in HeLa cells[54] has been mentioned. A high degree of ATP dependence was also found for the degradation of both short- and long-lived proteins in growing fibroblasts.[74a] Major steps in the elucidation of the role of ubiquitin were made recently.[3,4] A mutant was studied, derived from a spontaneous mouse mammary carcinoma, which is characterized by a temperature-sensitive cell cycle. At nonpermissive temperatures the generation of new mitotic cells was completely inhibited, the cells remained in an early G2 phase, and the ubiquitin-histone A2 conjugates disappeared. It was shown that it is the specific thermolability of the ubiquitin-activating enzyme (E_1) which is responsible for the effects. Furthermore it was demonstrated that the degradation of short-lived proteins is also temperature-dependent and it was concluded that the contribution of ubiquitin-independent pathways is probably less that 10%. It was suggested that proteolysis may play a role in the regulatory system which is operative in G2 progression during the cell cycle. Furthermore it was hypothesized that there are interrelations between the ubiquitin system and heat shock response.

IV. PROTEASES

The first report on the existence of several proteolytic activities in the membranes of erythrocytes[75] was followed by many studies in which various proteases with different properties were described. They differ with respect to pH optimum, substrates, and sensitivity to inhibitors and may be classified as acid, serine, sulfhydryl, and metal proteases. Proteases have been found both in the stroma, i.e., the cell membrane, and in the cytosol. Inhibitors of proteases occur particularly in the cytosol, which is the main reason for the late discovery of cytosolic enzymes. Proteases often can only be measured after previous separation of the inhibitors. A further problem has been the masking of several kinds of proteolytic activity. Thus, the acid protease that attacks microsomal cyt b_5[53] strictly requires Triton X-100 for its activity to be demonstrable.[71]

In the first studies in this field, membrane-bound proteolytic activity with a pH optimum of about 3.5 was reported to be much higher in reticulocytes of rabbits and men than in the corresponding erythrocytes.[76,77] Later, purification was achieved.[78] Enzyme activity appeared to be due to a single enzyme species. However, later work[79] showed the presence of three acidic proteinases in the membranes of human erythrocytes with molecular masses of 80, 40, and 30 kD, respectively. The enzymes differed from each other in their pH optimum, which ranged from 2.8 to 3.9, and the affinity to their substrate, denatured globin, as well as with respect to their sensitivity to inhibitors. While Hg^{2+} and Cu^{2+} inhibited all three enzyme species, pepstatin only did so to enzymes II and III; hemin only inhibited enzyme III. In a study of differences and similarities between the three acid endoproteases (which were purified to homogeneity), distinct structural differences were found so that the possibility of diversity arising by way of processing was excluded.[80] Despite these differences the enzymes had identical specificities. An enzyme activity which in all respects is identical was reported[81] and found to be localized on the inner side of the cell membrane.

In a comprehensive study of proteolytic activities in the cytosol of mature human erythrocytes the same three acid endopeptidases apparently were found.[82] Furthermore, the decay of these activities was studied in age-fractionated human erythrocytes as well as in reticulocytes and erythrocytes of rabbits.[83] In rabbits only a single acid protease was found which declined to one fourth during maturation. The acid proteases also decreased during aging of human erythrocytes. A remarkable feature was that the activity without added detergents declined much more strongly than that which became unmasked by the detergent treatment. As mentioned previously, acid protease from rabbit reticulocytes, which can generate a soluble heme peptide from liver microsomal cyt b_5,[71] is identical to the enzyme described.[80,82]

Neutral endopeptidases mainly appear to be present in the cytosol,[82] even though some indications of their presence in membranes have been reported.[84] Two neutral proteinase activities were also demonstrated in a 1 M NaCl extract from human and bovine red cell membranes.[85]

Among the cytosolic proteinases at least one is a metal enzyme as indicated by the inhibitory effect of phenanthroline.[86] It was highly purified by Kirschner and Goldberg and characterized extensively.[12] The enzyme has a high molecular mass of 300 kD and is an oligomer. It has preference for large peptides such as insulin, glucagon, and calcitonin and does not attack large proteins such as albumin. The protease, in addition to phenanthroline, is inhibited by EDTA and may be reactivated by Zn^{2+}, Co^{2+}, and Mn^{2+} ions. The pH optimum is 8.5. It occurs both in human and rabbit red cells as well as in other tissues. Two additional endoproteases were found to degrade insulin, one of which is a serine protease. An insulin degrading enzyme purified from human cytosol[87] with a molecular mass of 150 kD appears to be identical to the enzyme isolated by Kirschner and Goldberg.[12] Another high molecular weight proteinase with a molecular mass of 600 kD was partially purified;[88] it attacks both chymotrypsin and trypsin substrates. Its activity was about four times higher in reticulocytes than in erythrocytes. To what extent chymotrypsin-like activity in reticulocyte lysates,[89] which is greatly inhibited by low levels of chymostatin and elastase-like activity, which is strongly suppressed by the specific inhibitor elastatin, are related to the enzymes thus far discussed remains to be clarified. The differences in molecular mass reported among enzymes with similar specificities may well be due to oligomer formation.

Another neutral endopeptidase is characterized by its dependency on Ca^{2+} ions.[90,91-93] It also is located in the cytosol and appears to be a thiol proteinase. This type of enzyme has been called calpain I. It is distinguished by sensitivity to Ca^{2+} ions which cause a half-maximal activation in a concentration of about 10 μM, in contrast to the class of calpain II enzymes which require millimolar Ca^{2+} ion concentrations. The enzyme appears to have a molecular mass of 110 kD with a pH optimum of 7.5. It is a heterodimer with one subunit with an 80-kD mass and the other with 30 kD; apparently activity resides in the larger subunit. Enzymes of both calpain I and calpain II types are widely distributed.[94] There are great similarities between the 30-kD subunits as judged by their immunological cross reactivity,[95] whereas the 80-kD subunits, although immunologically related, are distinct. Small differences in molecular mass among the large subunits of different origin were observed. Recently it was demonstrated that the enzyme is initially present as an inactive proenzyme which is activated at very low Ca^{2+} ion concentrations in the presence of a digestible substrate such as denatured globin.[93] The effect of the substrate requires intact globin chains.[96] Activation involves the dissociation of the enzyme in its subunits and consists in the splitting off of a peptide of about 5-kD mass splitting off. There is also a substrate-conditioned autocatalytic inactivation of the enzyme.[93,96] A noticeable feature of calpain I in red cells is its property to digest bands III and IVa of the cell membrane without attacking spectrin. In other cells it has been shown that enzyme activity of this type is preferentially directed against microtubule-associated proteins, intermediate filaments, and other components of the cytoskeleton.[97-106]

A dual role for the calpain I of red cells has been proposed in the degradation of hemoglobin.[106a] It is suggested that this protease catalyzes a limited proteolysis of hemoglobin and also modifies the inner surface of the erythrocyte membrane. Both these actions are obligatory for the binding of the hemoglobin chains to the cell membrane with the subsequent further breakdown by the membrane-bound proteases.

Side by side with this enzyme occurs a specific large inhibitor protein named calpastatin, with a molecular mass of 240 to 280 kD and appearing to be a tetramer.[88-91] It is also widely distributed and appears to exhibit great molecular diversity.[107] The inhibitor dissociates in its constituent subunits in order to exert its inhibitory effect. The interaction of the enzyme with its inhibitor is induced by Ca^{2+} ions. In human red cells the inhibitor protein is present in large excess so that the enzyme activity is masked.

In the study referred to on the dependency of enzyme activities on cell age,[83] a strong decline in neutral endopeptidase by more than 90% with maturation was found in the red cells of rabbits. There was also an exponential decline during the aging of human erythrocytes with a half-life of about 70 days.

A multitude of exopeptidase activities were found in red cells. In early work indications for the presence of glycyl-glycine dipeptidase, glycyl-leucine dipeptidase, imino- and imidodipeptidase, aminotripeptidase, and leucine-aminopeptidase were found with activities at least one order of magnitude higher than those of the endopeptidases. They were all localized in the cytosol.[108-110] An aminotripeptidase[111,112] was found even earlier in erythrocytes of man and horse. In further studies two aminopeptidases, one of which was classified as aminopeptidase B on account of its dependence on chloride ions and its substrate specificity, were purified from human erythrocytes and characterized.[113] Their molecular mass was in the range of 50 to 80 kD. In the comprehensive studies mentioned earlier[82,83] on human erythrocytes, two dipeptidyl-aminopeptidases classified as II and III with a pH optimum of

Table 17
CYTOSOLIC EXOPEPTIDASE AND ENDOPEPTIDASE
ACTIVITIES IN RETICULOCYTES AND
ERYTHROCYTES OF RABBITS

Enzyme	Proteolytic action (IU/mg Hb)		
	Reticulocyte	Erythrocyte	Decay (%)
Neutral endopeptidase	25	2	92
Acid endopeptidase	29	7	75
Carboxypeptidase	70	5	93
Aminopeptidase A	30	12	60
Aminopeptidase B	55	39	29
Leu-aminopeptidase	78	50	36
Dipeptidyl aminopeptidase II	50	45	10
Dipeptidyl aminopeptidase III	120	117	2

From Melloni, E., Salamino, F., Sparatore, B., Michetti, M., Morelli, A., Benatti, V., De Flora, A., and Pontremoli, S. P., *Biochim. Biophys. Acta*, 675, 110, 1981.

7.0 and two aminopeptidases were found, one with broad specificity and the other of type B. In rabbits, a third aminopeptidase, type A, and a carboxypeptidase were present.

Exopeptidases exhibit variable maturation dependence. In rabbit red cells carboxypeptidase exhibits the strongest decline by more than 90% with maturation, followed by aminopeptidase A, which drops by 60%. At the other end of the spectrum are the dipeptidyl aminopeptidases II and III which change by less than 10%. Only moderate changes were found with respect to aminopeptidase B and leucine-aminopeptidase. These observations are in agreement with earlier reports.[108-110] There may well be a species difference since it has been reported that leucine-aminopeptidase decreases steeply with maturation in red cells of man.[109] The activities of exopeptidases splitting gly-gly, gly-leu, and iminopeptides apparently decline as much as the marker enzyme aspartate aminotransferase. Corresponding results were obtained in a later study[11] in which six dipeptides, containing either leu or gly as one component, and the tripeptide Ala_3 were used as substrate.

A comparison between the activities of cytosolic endopeptidases and exopeptidases in reticulocytes and erythrocytes is presented in Table 17.

REFERENCES

1. **Goldberg, A. L. and St. John, A. C.,** Intracellular protein degradation in mammalian and bacterial cells, *Annu. Rev. Biochem.,* 45, 747, 1976.
2. **Hershko, A. and Ciechanover, A.,** Mechanisms of intracellular protein breakdown, *Annu. Rev. Biochem.,* 51, 335, 1982.
3. **Finley, D., Ciechanover, A., and Varshavsky, A.,** Thermolability of ubiquitin-activating enzyme from the mammalian cell cycle mutant ts85, *Cell,* 37, 43, 1984.
4. **Ciechanover, A., Finley, D., and Varshavsky, A.,** Ubiquitin dependence of selective protein degradation demonstrated in the mammalian cell cycle mutant ts85, *Cell,* 37, 57, 1984.
5. **Moon, R. T. and Lazarides, E.,** Biogenesis of the avian erythroid membrane skeleton: receptor-mediated assembly and stabilization of ankyrin (globin) and spectrin, *J. Cell Biol.,* 98, 1899, 1984.
6. **Blikstad, I. and Lazarides, E.,** Synthesis of spectrin in avian erythroid cells: association of nascent polypeptide chains with the cytoskeleton, *Proc. Natl. Acad. Sci. U.S.A.,* 80, 2637, 1983.

7. **Bodine, D. M., Birkenmeier, C. S., and Barker, J. E.,** Spectrin deficient inherited hemolytic anemias in the mouse: characterization by spectrin synthesis and mRNA activity in reticulocytes, *Cell*, 37, 721, 1984.

8. **Rapoport, S. M. and Dubiel, W.,** The effect of phenylhydrazine on protein breakdown in rabbit reticulocytes, *Biomed. Biochem. Acta*, 43, 23, 1984.

9. **Botbol, V. and Scornik, O. A.,** Degradation of abnormal proteins in intact mouse reticulocytes: accumulation of intermediates in the presence of bestatin, *Proc. Natl. Acad. Sci. U.S.A.*, 76, 710, 1979.

10. **Botbol, V. and Scornik, O. A.,** Intermediates in the degradation of abnormal globin, *J. Biol. Chem.*, 254, 11254, 1979.

11. **McKay, M. J., Atkinson, E. M., Worthington, V. C., and Hipkiss, A. R.,** Changes in proteinase and peptidase activities during reticulocyte maturation, *Biochim. Biophys. Acta*, 759, 42, 1983.

12. **Kirschner, R. S. and Goldberg, A. L.,** A high molecular weight metalloprotease from the cytosol of mammalian cells, *J. Biol. Chem.*, 258, 967, 1983.

13. **Housman, D., Jacobs-Lorena, M., Rajbhandary, U. L., and Lodish, H. F.,** Initiation of haemoglobin synthesis by methionyl-tRNA, *Nature (London)*, 227, 913, 1970.

14. **Melloni, E., Salamino, F., Sparatore, B., Michetti, M., and Pontremoli, S.,** Cooperation between soluble and membrane-bound proteinases in the degradation of β-hemoglobin chains in intact erythrocytes, *Arch. Biochem. Biophys.*, 216, 495, 1982.

15. **Melloni, E., Sparatore, B., Salamino, F., Michetti, M., and Pontremoli, S.,** Proteolysis of human reticulocyte membrane proteins: evidence for a physiological role of the acid endopeptidases, *Arch. Biochem. Biophys.*, 218, 579, 1982.

16. **Simpson, M. V.,** The release of labeled amino acids from the proteins of rat liver slices, *J. Biol. Chem.*, 201, 143, 1953.

17. **Steinberg, D. and Vaughan, M.,** Observations on intracellular protein catabolism studied in vitro, *Arch. Biochem. Biophys.*, 65, 93, 1956.

18. **Schweiger, H. G., Rapoport, S., and Schölzel, E.,** Role of nonprotein nitrogen in the synthesis of haemoglobin in the reticulocyte in vitro, *Nature (London)*, 178, 141, 1956.

19. **Schweiger, H. G. and Rapoport, S.,** Der N-Stoffwechsel bei der Erythrocyten-Reifung: Reststickstoffbildung und Hämoglobinsynthese, *Hoppe-Seyler's Z. Physiol. Chem.*, 306, 33, 1956.

20. **Rabinovitz, M. and Fisher, J. M.,** Formation of a ribosomal lesion in rabbit reticulocytes by the lysine antagonist, S-(β-aminoethyl) cysteine, *Biochem. Biophys. Res. Commun.*, 6, 449, 1962.

21. **Rabinovitz, M. and Fisher, M.,** Characteristics of the inhibition of hemoglobin synthesis in rabbit reticulocytes by threo-α-amino-β-chlorobutyric acid, *Biochim. Biophys. Acta*, 91, 313, 1964.

22. **Asakawa, S., Fisher, S. M., and Rabinovitz, M.,** Proteolytic susceptibility of hemoglobin synthesized in the presence of amino acid analogs, *Mol. Pharmacol.*, 20, 227, 1981.

23. **McIlhinney, A. and Hogan, B. L. M.,** Rapid degradation of puromycyl peptides in hepatoma cells and reticulocytes, *FEBS Lett.*, 40, 297, 1974.

24. **Vissers, M. C. M., Winterbourn, C. C., and Carrel, R. W.,** Rapid proteolysis of unstable globins in human bone marrow, *Br. J. Haematol.*, 53, 417, 1983.

25. **Huehns, E. R.,** The unstable haemoglobins, *Bull. Soc. Chim. Biol.*, 52, 1131, 1970.

26. **Rieder, R. F., Wolf, D. J., Clegg, J. B., and Lee, S. L.,** Rapid post-synthetic destruction of unstable haemoglobin Bushwick, *Nature (London)*, 254, 725, 1975.

27. **Bank, A. and O'Donnell, J. V.,** Intracellular loss of free α-chains in β-thalassemia, *Nature (London)*, 222, 295, 1969.

28. **Clegg, J. B. and Weatherall, D. J.,** Haemoglobin synthesis during erythroid maturation in β-thalassaemia, *Nature (London)*, 240, 190, 1972.

29. **Shaeffer, J. R., Kleve, L., and De Simone, J.,** βˢ Chain turnover in reticulocytes of sickle trait individuals with high or low concentrations of haemoglobin S, *Br. J. Haematol.*, 32, 365, 1976.

30. **Braverman, A. S. and Lester, D.,** Evidence for increased proteolysis in intact β-thalassemia erythroid cells, *Hemoglobin*, 5, 564, 1981.

31. **Shaeffer, J. R.,** Turnover of excess hemoglobin α-chains in β-thalassemic cells is ATP dependent, *J. Biol. Chem.*, 258, 13172, 1983.

32. **Etlinger, J. D. and Goldberg, A. L.,** A soluble ATP-dependent proteolytic system responsible for the degradation of abnormal proteins in reticulocytes, *Proc. Natl. Acad. Sci. U.S.A.*, 74, 54, 1977.

33. **Wilkinson, K. D., Urban, M. K., and Haas, A. L.,** Ubiquitin is the ATP-dependent proteolysis factor I of rabbit reticulocytes, *J. Biol. Chem.*, 255, 7529, 1980.

34. **Wilkinson, D. K. and Audhya, T. K.,** Stimulation of ATP-dependent proteolysis requires ubiquitin with the COOH-terminal sequence arg-gly-gly, *J. Biol. Chem.*, 256, 9235, 1981.

35. **Ciechanover, A., Hod, Y., and Hershko, A.,** A heat-stable polypeptide component of an ATP-dependent proteolytic system from reticulocytes, *Biochem. Biophys. Res. Commun.*, 81, 1100, 1978.

36. **Hershko, A., Ciechanover, A., and Rose, I. A.,** Resolution of the ATP dependent proteolytic system from reticulocytes: a component that interacts with ATP, *Proc. Natl. Acad. Sci. U.S.A.,* 76, 3107, 1979.

37. **Ciechanover, A., Elias, S., Heller, H., Ferber, S., and Hershko, A.,** Characterization of the heat stable polypeptide of the ATP-dependent proteolytic system from reticulocytes, *J. Biol. Chem.,* 255, 7525, 1980.

38. **Goldknopf, I. L. and Busch, H.,** in *The Cell Nucleus,* Vol. 6, Busch, H., Ed., Academic Press, New York, 1978, 149.

39. **Busch, H. and Goldknopf, I. L.,** Ubiquitin-protein conjugates, *Mol. Cell. Biochem.,* 40, 173, 1981.

40. **Ciechanover, A., Heller, H., Katz-Etzion, R., and Hershko, A.,** Activation of the heat-stable polypeptide of the proteolytic system, *Proc. Natl. Acad. Sci. U.S.A.,* 78, 761, 1981.

41. **Chiechanover, A., Heller, H., Elias, S., Haas, L. A., and Hershko, A.,** ATP-dependent conjugation of reticulocyte proteins with the polypeptide required for protein degradation, *Proc. Natl. Acad. Sci. U.S.A.,* 77, 1365, 1980.

42. **Hershko, A., Ciechanover, A., Heller, H., Haas, A. L., and Rose, I. A.,** Proposed role of ATP in protein breakdown: conjugation of proteins with multiple chains of the polypeptide is ATP-dependent, *Proc. Natl. Acad. Sci. U.S.A.,* 77, 1783, 1980.

43. **Hershko, A., Eytan, E., Ciechanover, A., and Haas, A. L.,** Immunochemical analysis of the turnover of ubiquitin-protein conjugates in intact cells, *J. Biol. Chem.,* 257, 13964, 1982.

43a. **Hershko, A., Heller, H., Eytan, E., Kaklij, G., and Rose, I. A.,** Role of the α-amino group of protein in ubiquitin-mediated protein breakdown, *Proc. Natl. Acad. Sci. U.S.A.,* 81, 7021, 1984.

43b. **Hough, R. and Rechsteiner, M.,** Effect of temperature on the degradation of proteins in rabbit reticulocyte lysates and after injection into HeLa cells, *Proc. Natl. Acad. Sci. U.S.A.,* 81, 90, 1984.

44. **Andersen, M. W., Goldknopf, I. L., and Busch, H.,** Protein A24 lyase is an isopeptidase, *FEBS Lett.,* 132, 210, 1981.

45. **Matsui, S.-I., Sandberg, A. A., Negoro, S., Seon, B. K., and Goldstein, G.,** Isopeptidase: a novel eukaryotic enzyme that cleaves isopeptide bonds, *Proc. Natl. Acad. Sci. U.S.A.,* 79, 1535, 1982.

46. **Rose, I. A. and Warms, I. V. B.,** An enzyme with ubiquitin carboxy-terminal esterase activity from reticulocytes, *Biochemistry,* 22, 4234, 1983.

47. **Ciechanover, A., Elias, S., Heller, H., and Hershko, A.,** Covalent affinity purification of ubiquitin-activating enzyme, *J. Biol. Chem.,* 257, 2537, 1982.

48. **Haas, A. L., Warms, J. V. B., Hershko, A., and Rose, I. A.,** Ubiquitin-activating enzyme, *J. Biol. Chem.,* 257, 2543, 1982.

49. **Haas, L. and Rose, I. A.,** The mechanism of ubiquitin activating enzyme, *J. Biol. Chem.,* 257, 10329, 1982.

50. **Hershko, A., Heller, H., Elias, S., and Ciechanover, A.,** Components of ubiquitin-protein ligase system. Resolution, affinity purification, and role in protein breakdown, *J. Biol. Chem.,* 258, 8206, 1983.

50a. **Pickart, M. and Rose, I. A.,** Ubiquitin carboxyl-terminal hydrolase acts on ubiquitin carboxyl-terminal amides, *J. Biol. Chem.,* 260, 7903, 1985.

50b. **Hershko, A. and Heller, H.,** Occurrence of a polyubiquitin structure in ubiquitin-protein conjugates, *Biochem. Biophys. Res. Commun.,* 128, 1079, 1985.

50c. **Ciechanover, A., Wolin, S. L., Steitz, J. A., and Lodish, H. F.,** Transfer RNA is an essential component of the ubiquitin- and ATP-dependent proteolytic system, *Proc. Natl. Acad. Sci. U.S.A.,* 82, 1341, 1985.

50d. **Evans, A. C. and Wilkinson, K. D.,** Ubiquitin-dependent proteolysis of native and alkylated bovine serum albumin: effects of protein structure and ATP concentration on selectivity, *Biochemistry,* 24, 2915, 1985.

51. **Etlinger, J. D. and Goldberg, A. L.,** Control of protein degradation in reticulocytes and reticulocyte extracts by hemin, *J. Biol. Chem.,* 255, 4563, 1980.

52. **Haas, A. and Rose, I. A.,** Hemin inhibits ATP-dependent proteolysis: role of hemin in regulating ubiquitin conjugate degradation, *Proc. Natl. Acad. Sci. U.S.A.,* 78, 6845, 1981.

53. **Tanaka, K., Waxman, L., and Goldberg, A. L.,** Vanadate inhibits the ATP-dependent degradation of proteins in reticulocytes without affecting ubiquitin conjugation, *J. Biol. Chem.,* 259, 2803, 1984.

54. **Chin, D. T., Kuehl, L., and Rechsteiner, M.,** Conjugation of ubiquitin to denatured hemoglobin is proportional to the rate of hemoglobin degradation in HeLa cells, *Proc. Natl. Acad. Sci. U.S.A.,* 79, 5857, 1982.

54a. **Rieder, R., Ibrahim, A., and Etlinger, J. D.,** A particle-associated ATP-dependent proteolytic activity in erythroleukemia cells, *J. Biol. Chem.,* 260, 2015, 1985.

55. **Müller, M., Dubiel, W., Rathmann, J., and Rapoport, S.,** Determination and characteristics of energy-dependent proteolysis in rabbit reticulocytes, *Eur. J. Biochem.,* 109, 405, 1980.

56. **Dubiel, W., Müller, M., Rathmann, J., Hiebsch, C., and Rapoport, S. M.,** Determination and characterization of energy-dependent proteolysis in rabbit reticulocytes, *Acta Biol. Med. Ger.,* 40, 625, 1981.

57. **Rapoport, S., Dubiel, W., and Müller, M.,** Proteolysis of mitochondria in reticulocytes during maturation is ubiquitin-dependent and is accompanied by a high rate of ATP hydrolysis, *FEBS Lett.,* 180, 249, 1985.

57a. **Dubiel, W., Müller, M., and Rapoport, S.,** Kinetics of ^{125}J-ubiquitin conjugation with and liberation from rabbit reticulocyte stroma, *FEBS Lett.,* in press.

58. **Boches, F. S. and Goldberg, A. L.,** Role for the adenosine triphosphate-dependent proteolytic pathway in reticulocyte maturation, *Science,* 215, 978, 1982.

59. **Tanaka, K., Waxman, L., and Goldberg, A. L.,** ATP serves two distinct roles in protein degradation in reticulocytes, one requiring and one independent of ubiquitin, *J. Cell Biol.,* 96, 1580, 1983.

60. **Borsook, H., Jiggins, S., and Wilson, R. T.,** Two inducers of rapid erythroblast multiplication in vitro, *Nature (London),* 230, 328, 1971.

61. **Hershko, A., Leshinsky, E., Ganoth, D., and Heller, H.,** ATP-dependent degradation of ubiquitin-protein conjugates, *Proc. Natl. Acad. Sci. U.S.A.,* 81, 1619, 1984.

62. **McKay, M. J., Daniels, R. S., and Hipkiss, A. R.,** Breakdown of aberrant protein in rabbit reticulocytes decreases with cell age, *Biochem. J.,* 188, 279, 1980.

63. **Speiser, S. and Etlinger, J. D.,** Loss of ATP-dependent proteolysis with maturation of reticulocytes and erythrocytes, *J. Biol. Chem.,* 257, 14122, 1982.

64. **Daniels, R. S., McKay, M. J., Atkinson, E. M., and Hipkiss, A. R.,** Subcellular distribution of abnormal proteins in rabbit reticulocytes: effects of cellular maturation, phenylhydrazine, and inhibitors of ATP synthesis, *FEBS Lett.,* 156, 145, 1983.

65. **Hershko, A., Heller, H., Ganoth, D., and Ciechanover, A.,** Mode of degradation of abnormal globin chains in rabbit reticulocytes, in *Protein Turnover and Lysosome Function,* Segal, H. L. and Doyle, D. J., Eds., Academic Press, New York, 1978, 149.

66. **Worthington, V. C. and Hipkiss, A. R.,** ATP-stimulated breakdown of polylysine in rabbit reticulocyte cell-free extracts: differential effects according to substrate molecular weight, *Biochem. Soc. Trans.,* 10, 96, 1982.

67. **Klemes, J., Etlinger, J. D., and Goldberg, A. L.,** Properties of abnormal proteins degraded rapidly in reticulocytes, *J. Biol. Chem.,* 256, 8436, 1981.

68. **Daniels, R. S., McKay, M. J., Worthington, V. C., and Hipkiss, A. R.,** Effects of ATP and cell development on the metabolism of high molecular weight aggregates of abnormal proteins in rabbit reticulocytes and cell-free extracts, *Biochim. Biophys. Acta,* 717, 220, 1982.

69. **McKay, M. J. and Hipkiss, A. R.,** ATP-independent proteolysis of globin cyanogen bromide peptides in rabbit reticulocyte cell-free extracts, *Eur. J. Biochem.,* 125, 567, 1982.

70. **Hanash, S. M. and Rucknagel, D. L.,** Proteolytic activity in erythrocyte precursors, *Proc. Natl. Acad. Sci. U.S.A.,* 75, 3427, 1978.

71. **Schafer, D. A. and Hultquist, D. E.,** Isolation of an acid protease from rabbit reticulocytes and evidence for its role in processing redox proteins during erythroid maturation, *Biochem. Biophys. Res. Commun.,* 100, 1555, 1981.

72. **Raw, I. and DiFini, F.,** The possible role of ATP-dependent proteolysis on the solubilization of methemoglobin reductase during reticulocyte maturation, *Biochem. Biophys. Res. Commun.,* 116, 357, 1983.

73. **Magnani, M., Stocchi, V., Dacha, M., and Fornaini, G.,** Rabbit red blood cell hexokinase. Evidences for an ATP-dependent decay during cell maturation, *Mol. Cell. Biochem.,* 61, 83, 1984.

74. **Magnani, M., Stocchi, V., Dachà, M., and Fornaini, G.,** Rabbit blood cell hexokinase: intracellular distribution during reticulocytes maturation, *Mol. Cell. Biochem.,* 63, 59, 1984.

74a. **Gronostajski, R. M., Pardee, A. B., and Goldberg, A. L.,** The ATP dependence of the degradation of short- and long-lived proteins in growing fibroblasts, *J. Biol. Chem.,* 260, 3344, 1985.

75. **Morrison, W. L. and Neurath, H.,** Proteolytic enzymes of the formed elements of human blood, *J. Biol. Chem.,* 200, 39, 1953.

76. **Goetze, E., Lindigkeit, R., and Rapoport, S.,** Vergleichende Untersuchungen über die Atmung, den Nucleinsäuregehalt und die Kathepsinwirking der Erythrocyten bei der Entblutungsanämiedes Kaninchens, *Biochem. Z.,* 326, 48, 1954.

77. **Goetze, E. and Rapoport, S.,** Das Kathepsin der Kaninchenerythrocyten und seine Veränderungen bei der Zellreifung, *Biochem. Z.,* 326, 53, 1954.

78. **Reichelt, D., Jacobsohn, E., and Haschen, R.,** Purification and properties of cathepsin D from human erythrocytes, *Biochim. Biophys. Acta,* 341, 15, 1974.

79. **Pontremoli, S., Salamino, F., Sparatore, B., Melloni, E., Morelli, A., Benatti, U., and De Flora, A.,** Isolation and partial characterization of three acidic proteinases in erythrocyte membranes, *Biochem. J.,* 181, 559, 1979.

80. **Pontremoli, S., Sparatore, B., Melloni, E., Salamino, F., Michetti, M., Morelli, A., Benatti, U., and De Flora, A.,** Differences and similarities among three acidic endoproteases associated with human erythrocyte membranes, *Biochim. Biophys. Acta,* 630, 313, 1980.

81. **Murakami, T., Suzuki, Y., and Murachi, T.,** An acidic protease in human erythrocytes and its localization in the inner membrane, *Eur. J. Biochem.,* 96, 221, 1979.

82. **Pontremoli, S., Melloni, E., Salamino, F., Sparatore, B., Michetti, M., Benatti, V., Morelli, A., and De Flora, A.,** Identification of proteolytic activities in the cytosolic compartment of mature human erythrocytes, *Eur. J. Biochem.,* 110, 421, 1980.

83. **Melloni, E., Salamino, F., Sparatore, B., Michetti, M., Morelli, A., Benatti, V., De Flora, A., and Pontremoli, S. P.,** Decay of proteinase and peptidase activities of human and rabbit erythrocytes during cellular aging, *Biochim. Biophys. Acta,* 675, 110, 1981.

84. **Tökes, Z. A. and Chambers, S. M.,** Proteolytic activity associated with the human erythrocyte membranes. Self digestion of isolated human erythrocyte membranes, *Biochim. Biophys. Acta,* 389, 325, 1975.

85. **Scott, G. K. and Kee, T. B.,** Neutral proteases from human and ovine erythrocyte membranes, *Int. J. Biochem.,* 10, 1039, 1979.

86. **Hipkiss, A. R., McKay, M. J., Daniels, R. S., Worthington, V. C., and Atkinson, E. M.,** Selective proteolysis of abnormal proteins of shortened chain length in rabbit reticulocytes, *Acta Biol. Med. Ger.,* 40, 1265, 1981.

87. **Kolb, H. and Standl, E.,** Purification to homogeneity of an insulin-degrading enzyme from human erythrocytes, *Hoppe-Seyler's Z. Physiol. Chem.,* 361, 1029, 1980.

88. **Edmunds, T. and Pennington, R. T.,** A high-molecular weight peptide hydrolase in erythrocytes, *Int. J. Biochem.,* 14, 701, 1982.

89. **Mumford, R. A., Pickett, C. B., Zimmerman, M., and Strauss, A. W.,** Protease activities present in wheat germ and rabbit reticulocyte lysates, *Biochem. Biophys. Res. Commun.,* 103, 565, 1981.

90. **Murakami, T., Hatanaka, M., and Murachi, T.,** The cytosol of human erythrocytes contains a highly Ca^{2+}-sensitive thiol protease (Calpain I) and its specific inhibitor protein (Calpastatin), *J. Biochem.,* 90, 1809, 1981.

91. **Melloni, E., Sparatore, B., Salamino, F., Michetti, M., and Pontremoli, S.,** Cytosolic calcium-dependent proteinase of human erythrocytes: formation of an enzyme-natural inhibitor complex induced by Ca^{2+} ions, *Biochem. Biophys. Res. Commun.,* 106, 731, 1982.

92. **Melloni, E., Sparatore, B., Salamino, F., Michetti, M., and Pontremoli, S.,** Cytosolic calcium dependent neutral proteinase of human erythrocytes: the role of calcium ions on the molecular and catalytic properties of the enzyme, *Biochem. Biophys. Res. Commun.,* 107, 1053, 1982.

93. **Melloni, E., Salamino, F., Sparatore, B., Michetti, M., and Pontremoli, S.,** Ca^{2+}-dependent neutral proteinase from human erythrocytes: activation by Ca^{2+}-ions and substrate and regulation by the endogenous inhibitor, *Biochem. Int.,* 8, 477, 1984.

94. **Yoshimura, N., Hatanaka, M., Kitahara, A., Kawaguchi, N., and Murachi, T.,** Intracellular localization of two distinct Ca^{2+}-proteases (Calpain I and Calpain II) as demonstrated by using discriminative antibodies, *J. Biol. Chem.,* 259, 9847, 1984.

95. **Sasaki, T., Yoshimura, N., Kikuchi, T., Hatanaka, M., Kitahara, A., Sakihama, T., and Murachi, T.,** Similarity and dissimilarity in subunit structures of calpains I and II from various sources as demonstrated by immunological cross-reactivity, *J. Biochem.,* 94, 2055, 1983.

96. **Pontremoli, S., Sparatore, B., Melloni, E., Michetti, M., and Horecker, B. L.,** Activation by hemoglobin of the Ca^{2+}-requiring neutral proteinase of human erythrocytes: structural requirements, *Biochem. Biophys. Res. Commun.,* 123, 331, 1984.

97. **Sandoval, I. V. and Weber, K.,** Calcium-induced inactivation of microtubule formation in brain extracts, *Eur. J. Biochem.,* 92, 463, 1978.

98. **Klein, I., Lehotay, D., and Gondek, M.,** Characterization of a calcium-activated protease that hydrolyzes a microtubule-associated protein, *Arch. Biochem. Biophys.,* 208, 520, 1981.

99. **Davies, P. J. A., Wallach, D., Willingham, M. C., Pastan, I., Yamaguchi, M., and Robson, R. M.,** Filamin-actin interaction, *J. Biol. Chem.,* 253, 4036, 1978.

100. **Phillips, D. R. and Jakábová, M.,** Ca^{2+}-dependent protease in human platelets, *J. Biol. Chem.,* 252, 5602, 1977.

101. **Truglia, J. A. and Stracher, A.,** Purification and characterization of a calcium dependent sulfhydryl protease from human platelets, *Biochem. Biophys. Res. Commun.,* 100, 814, 1981.

102. **Collier, N. C. and Wang, K.,** Purification and properties of human platelet P235, *J. Biol. Chem.,* 257, 6937, 1982.

103. **Gilbert, D. S. and Newby, B. J.,** Neurofilament disguise, destruction and discipline, *Nature (London),* 256, 586, 1975.

104. **Zimmerman, U.-J. P. and Schlaepfer, W. W.,** Characterization of a brain calcium-activated protease that degrades neurofilament proteins, *Biochemistry,* 21, 3977, 1982.

105. **Nelson, W. J. and Traub, P.,** Purification and further characterization of the Ca^{2+}-activated proteinase specific for the intermediate filament proteins vimentin and desmin, *J. Biol. Chem.,* 257, 5544, 1982.

106. **Lazarides, E.,** Intermediate filaments as mechanical integrators of cellular space, *Nature (London),* 283, 249, 1980.

106a. **Pontremoli, S., Melloni, E., Sparatore, B., Michetti, M., and Horecker, B. L.,** A dual role for the Ca^{2+}-requiring proteinase in the degradation of hemoglobin by erythrocyte membrane proteinases, *Proc. Natl. Acad. Sci. U.S.A.,* 81, 6714, 1984.

107. **Takano, E., Yamoto, N., Kannage, R., and Murachi, T.,** Molecular diversity of calpastatin in mammalian organs, *Biochem. Biophys. Res. Commun.,* 122, 912, 1984.

108. **Haschen, R. J.,** Exopeptidasen und Glutamat-Oxalacetat-Transaminase der menschlichen Erythrozyten bei Anämien, *Acta Biol. Med. Ger.,* 9, 15, 1962.

109. **Haschen, R. J., Farr, W., and Groh, F.,** Proteolytische Enzyme un Eiweißstoffwechsel der Kaninchen-Erythrozyten während der Zellreifung in vivo, *Acta Biol. Med. Ger.,* 14, 205, 1965.

110. **Reichelt, D., Jacobasch, G., and Haschen, R. J.,** Erythrozytenalterung und Proteolyse, in *6th Internationales Symposium über Struktur und Funktion der Erythrozyten,* Rapoport, S. and Jung, F., Eds., Akademie-Verlag, Berlin, 1972, 505.

111. **Adams, E., Davis, N. C., and Smith, E. M.,** Peptidases of erythrocytes. III. Tripeptidase, *J. Biol. Chem.,* 199, 845, 1952.

112. **Tsuboi, K. K., Penefsky, Z. J., and Hudson, P. B.,** Enzymes of the human erythrocyte. III. Tripeptidase, purification and specific properties, *Arch. Biochem. Biophys.,* 68, 54, 1957.

113. **Mäkinen, K. K. and Mäkinen, P.-L.,** Purification and characterization of two human erythrocyte arylamidases preferentially hydrolysing N-terminal arginine or lysine residues, *Biochem. J.,* 175, 1051, 1978.

Chapter 14

BREAKDOWN OF PROTEIN SYNTHESIS

I. INTRODUCTION

Protein synthesis ceases during maturation of the reticulocyte. The traditional characteristic of the reticulocyte, substantia reticulofilamentosa, is but an indicator of the number of ribosomes; its disappearance during maturation is a visible sign of the breakdown of ribosomes, which is an essential element of protein synthesis machinery. In the following section the changes that occur during maturation as well as the possible mechanisms of degradation for the four components of this system will be discussed: (1) polysomes and ribosomes; (2) mRNA; (3) the tRNAs; and (4) translation factors.

II. POLYSOMES AND RIBOSOMES

The reticulocyte contains 30,000 to 100,000 ribosomes per cell, more than half of which take the form of polysomes.[1-7] The number of ribosomes decreases with a preferential decline in polysomes so that monosomes progressively predominate.[8-10] The polysome decline may be related both to lower efficiency of initiation by mRNA — as will be discussed in the next section — as well as by changes in ribosomal structure and function.[11]

Accordingly, free cytoplasmic mRNA constitutes a higher portion of the total in mature reticulocytes than in immature ones.[12] The degradation of ribosomes appears to be energy-dependent in some manner.[13] Only recently has this circumstance been the subject of further investigation.[14]

The occurrence of a variety of RNases in reticulocytes by itself is not proof of their involvement in the actual processes of ribosomal RNA breakdown no more than it was possible to relate the multitude of proteases to specific processes in proteolysis. Studies on changes occurring in vivo during experimental anemia likewise are not informative because of variations in cell populations which largely overshadow the changes occurring in a given cell. Therefore, a study was performed on the degradation of ribosomes during an in vitro maturation of reticulocyte-rich suspensions of rabbit red cells in Borsook medium.[15] Ribosomal RNA was isolated from ribosomal subunits and analyzed on denaturing and nondenaturing sucrose gradients.

The following results were obtained: polysomes showed a slow progressive change with the normally dominating pentasomes decreasing from about 60 to 25% during a 12-hr period, while monosomes increased from 15 to 40%. These data are in general agreement with earlier studies in vitro.[2,5] Under nondenaturing conditions dramatic changes in 28S rRNA of the large subunit were not observed even after 24 hr of incubation in vitro. Matters were quite different if denaturing conditions were used. The 28S rRNA obtained — even from nonincubated reticulocytes of peripheral blood — exhibited hidden breaks which were also found in preparations obtained from bone marrow. This result is in keeping with observations in other types of cells including rat liver, HeLa cells, and avian brain, liver, and red cells.[16-20] In such systems it was found that the first nicks are introduced shortly after formation of functional ribosomes.[21,22]

During in vitro incubation of rabbit reticulocytes a high proportion of 28S rRNA was fragmented with two distinct large components of 21S and 18S predominating. After 12 hr the 18S fragment was prominent. Reticulocytes incubated for a further period of time failed to exhibit either intact 28S rRNA or the 21S and 18S intermediate fragments.

These data permit two conclusions: (1) the 28S rRNA of intact reticulocytes contain hidden breaks, presumably held together by double-stranded domains which do not interfere with the function of the ribosomes; and (2) initial cleavage during maturation takes place at distinct sites identical with the preexisting breaks without effect on the three-dimensional structure of the ribosome, which is primarily determined by noncovalent bonds. The specificity of the splitting may be caused by the exposure of distinct regions of the RNA molecule on the surface of the ribosome. Furthermore, the primary cleavage sites may be influenced by the secondary structure of the RNA and the ribosomal proteins. Thus local constellations of ribosomal structure and its interactions with ribosomal proteins may differ in various domains and thus cause differential accessibility of the phosphodiester bonds of RNA to RNases.

The 18S RNA of the small ribosomal subunit behaved rather differently than the 28S rRNA, although under nondenaturing conditions it likewise failed to exhibit significant changes even after an incubation period of 24 hr. Under denaturing conditions initial hidden breaks were obvious but no large fragment could be discerned. After incubation a rapid breakdown ensued with further increase of the heterogeneous breakdown products. Again it would appear that the original hidden breaks did not affect the function of the ribosome and that the preexisting RNA fragments were held together by noncovalent bonds. The heterogeneous distribution of fragments suggests a greater number of sites accessible to degradation by RNases.

The alterations suffered by ribosomal RNA must be viewed in context with changes in ribosomal proteins which apparently are slow as compared to the rapid breakdown of mitochondria. One may speculate that after a certain number of nicks in the RNA have occurred, the ribosomes might present a loosened structure more susceptible to proteolytic attack. Such an attack would in turn facilitate further breakdown of the RNAs. During these processes the function of the ribosomes must deteriorate. In fact, the number of ribosomes not only is decreased in mature reticulocytes, but there is also indication for impairment of their function.[23,24]

A specific membrane-bound RNase attacking 28S rRNA and thereby inhibiting protein synthesis has been described by Wreschner et al.[25-28] and others (for a review see Lengyel[30]).

The breakdown of ribonucleic acids continues to produce as final products hypoxanthine from adenine and guanine and uracil and cytidine, while ribose is utilized as described previously[31] (see Chapter 6). Specific pyrimidine-5'-nucleotidases presumably play a role in the final stages of degradation.[32,33,34]

Because of the occurrence of sizable deoxyribonucleotidase activities in mature erythrocytes, the hypothesis was proposed that nuclear pycnosis might be accompanied by sufficient karyolytic degradation of DNA so as to require a mechanism for dephosphorylation of deoxyribonucleotides.[34]

III. THE mRNAs

Early and later work indicate a half-life for globin RNA of 10 to 20 hr.[35,36] Considerable evidence supports the assumption that the decrease in globin mRNA is related to its loss of stability due to a shortening of the poly-A tail (for a review see Littauer and Soreq[37]). It was demonstrated by a variety of experimental approaches that the presence of a tail consisting of more than 30 adenosine residues confers stability to the mRNA.[38] Definite proof that breakdown of poly-A occurs during maturation of reticulocytes and is related to its stability was provided by Maniatis et al.[39] They found that 10S mRNA obtained from reticulocyte-rich human blood could be separated in fractions of differing poly-A content. When injected into *Xenopus oocytes* the poly-A-rich mRNA was much more stable than the poly-A-poor one, which presumably represented the more mature fraction. In an analogous study on fractionation of purified rabbit globin mRNA according to the length of the poly-A tail, its shortening during maturation was demonstrated.[40] The average length of about 25 nucleotides in the poly-A sequence of globin mRNA in reticulocytes is in accordance with the assumption about the critical role of the poly-A tail for the biological stability of mRNA.[41] Thus, one may conclude that the disappearance of mRNA is caused by the progressive shortening of the poly-A tail. A critical point may be reached if its length is reduced to 30 nucleotides or less, which may represent a threshold of stability.

It remains an open question as to what extent specific poly-A binding proteins play a role in the protection of poly-A and what their fate might be during maturation.[42-44]

Other alterations in mRNA are suggested by the observation that mRNA with a shortened poly-A tail are considerably less able to direct translation.[40] A decreased efficiency in initiation of translation had been found previously in mature reticulocytes.[11] Since other work has clearly shown that the removal of poly-A by itself does not interfere with initiation, another site(s) is probably affected.

In contradiction to these results is a recent report, in which the decay rates of mRNA in mouse erythroleukemia cells were studied. The induction of erythroid differentiation by dimethyl sulfoxide led to considerable increases in the rate of decay of several mRNA species without correlation between the tendency to lose poly(A) and the rate of mRNA decay.[29] Furthermore, a large accumulation of poly(A)-deficient mRNAs was found, a finding that would indicate that the stability of mRNA is not determined solely by the presence of poly(A).

IV. THE tRNAs

The tRNAs are much more stable during maturation than any other part of the protein synthesis system.[45]

Their initial number, 6,000 to 60,000 per reticulocyte, with $tRNA_{Ala}$ and $tRNA_{val}$ predominating, decreases with a half-life of 50 to 60 hr. Even in mature erythrocytes about one sixth remain. The tRNAs appear to remain intact with only a small portion being without the C-terminal pCpCp A-terminus, even though $(2'-5')$ A_n specific phosphodiesterase, which occurs in reticulocytes, is capable of degrading it.[46] Thus it would appear that this group of components cannot be a limiting factor in the decline of protein synthesis.

It is overall difficult to draw a clear picture of the sequence of events and of the mechanism causing the breakdown of protein synthesis despite a multitude of observations, since all of them deal with a single aspect in the complex process. Based on the decline in the proportion of polysomes during maturation, one would surmise that the major changes affect mRNA by a retarded initiation of protein synthesis. It is of course possible that the efficiency of initiation is affected by alterations in quality or quantity of translation factors. The possibility of mRNA nicking is not excluded. The incomplete proteins produced would be rapidly degraded (see Chapter 13).

REFERENCES

1. **Warner, J. R., Knopf, P. M., and Rich, A.,** A multiple ribosomal structure in protein synthesis, *Biochemistry,* 49, 122, 1963.
2. **Marks, P. A., Rifkind, R. A., and Danon, D.,** Polyribosomes and protein synthesis during reticulocyte maturation in vitro, *Proc. Natl. Acad. Sci. U.S.A.,* 50, 336, 1963.
3. **Rifkind, R. A., Luzzatto, L., and Marks, P. A.,** Size of polyribosomes in intact reticulocytes, *Proc. Natl. Acad. Sci. U.S.A.,* 52, 1227, 1964.
4. **Mathias, A. P., Williamson, R., Huxley, H. E., and Page, S.,** Occurrence and function of polysomes in rabbit reticulocytes, *J. Mol. Biol.,* 9, 154, 1964.

5. **Danon, D., Zehavi-Willner, T., and Bevman, C. R.,** Alterations in polyribosomes of reticulocytes maturing in vivo, *Proc. Natl. Acad. Sci. U.S.A.,* 54, 873, 1965.

6. **Miller, A. and Maunsbach, A. B.,** Electron microscopic audioradiography of rabbit reticulocytes active and inactive in protein synthesis, *Science,* 151, 1000, 1966.

7. **Smith, D. W. E. and McNamara, A. L.,** The distribution of transfer ribonucleic acid in rabbit reticulocytes, *J. Biol. Chem.,* 249, 1330, 1974.

8. **Rifkind, R. A., Danon, D., and Marks, P. A.,** Alterations in polysomes during erythroid cell maturation, *J. Cell Biol.,* 22, 599, 1964.

9. **Knopf, P. M. and Lamfrom, H.,** Changes in the ribosome distribution during incubation of rabbit reticulocytes in vitro, *Biochim. Biophys. Acta,* 95, 398, 1965.

10. **Marbaix, G., Burny, A., Huez, G., Lebleu, B., and Temmerman, J.,** Evolution of the polyribosome distribution during in vivo reticulocyte maturation, *Eur. J. Biochem.,* 13, 322, 1970.

11. **Herzberg, M., Revel, M., and Danon, D.,** The influence of ribosomal factors during the maturation of reticulocytes, *Eur. J. Biochem.,* 11, 148, 1969.

12. **Marbaix, G., Huez, C., Nokin, P., and Cleuter, Y.,** Free cytoplasmic α-globin messenger RNA appears during the maturation of rabbit reticulocytes, *FEBS Lett.,* 66, 269, 1976.

13. **Schweiger, H. G. and Rapoport, S. M.,** Der N-Stoffwechsel bei der Erythrocytenreifung: die N-Bilanz unter endogenen Bedingungen, *Hoppe-Seyler's Z. Physiol. Chem.,* 313, 97, 1958.

14. **Park, E. A. and Morgan, H. E.,** Energy dependence of RNA degradation in rabbit reticulocytes, *Am. J. Physiol.,* 247, C390, 1984.

15. **Andree, H. H., Bretschneider, K., Thiele, B., and Rapoport, S. M.,** Breakdown of ribosomal RNA in rabbit reticulocytes, *Acta Biol. Med. Ger.,* 39, 995, 1980.

16. **Fujisawa, T., Abe, S., Satake, M., and Ogata, K.,** Conversion of rat liver nucleolar 29.5-S RNA to 28-S RNA in vitro, *Biochim. Biophys. Acta,* 324, 241, 1973.

17. **Awata, S. and Natori, Y.,** Turnover of rat liver 28 S ribosomal RNA: nicking as the initial step of degradation, *Biochim. Biophys. Acta,* 478, 486, 1977.

18. **Nair, C. N. and Knight, E., Jr.,** Turnover of HeLa ribosomal RNA. The characterization of a·class of RNA in HeLa cytoplasm derived from 28 S RNA, *J. Cell Biol.,* 50, 787, 1971.

19. **Judes, C., Fuchs, J. P., and Jacob, M.,** Les RNA cytoplamiques mineurs: mise en evidence et caracterisation d'un RNA 21 S du poulet, *Biochimie,* 54, 1031, 1972.

20. **Sanchez De Jiminez, E. and Lotina, B.,** Degradation of ribosomal RNA during red cell differentiation, *Biochem. Biophys. Res. Commun.,* 48, 1323, 1972.

21. **Tsurugi, K., Morita, T., and Ogata, K.,** Mode of degradation of ribosomes in regenerating rat liver in vivo, *Eur. J. Biochem.,* 45, 119, 1974.

22. **Ishikawa, H. and Newburgh, R. W.,** Studies of the thermal conversion of a 28 S RNA of galleria mellonella (L.) to an 18 S product, *J. Mol. Biol.,* 64, 135, 1972.

23. **Rowley, P. T. and Morris, J. A.,** Protein synthesis in the maturing reticulocyte, *J. Biol. Chem.,* 242, 1533, 1967.

24. **Rowley, P. T., Midthun, R. A., and Adams, M. H.,** Solubilization of a reticulocyte ribosomal fraction responsible for the decline in ribosomal activity with cell maturation, *Arch. Biochem. Biophys.,* 145, 6, 1971.

25. **Wreschner, D., Melloul, D., and Herzberg, M.,** Specific degradation of ribosomal RNA in rabbit reticulocyte membrane-bound ribosomes, *FEBS Lett.,* 77, 83, 1977.

26. **Wreschner, D., Melloul, D., and Herzberg, M.,** Interaction between membrane functions and protein synthesis in reticulocytes: specific cleavage of 28-S ribosomal RNA by a membrane constituent, *Eur. J. Biochem.,* 85, 233, 1978.

27. **Wreschner, D., Melloul, D., and Herzberg, M.,** Interaction between membrane functions and protein synthesis in reticulocytes. Isolation of RNase M, a membrane component inhibiting protein synthesis through specific endonucleoytic activity, *Eur. J. Biochem.,* 89, 341, 1978.

28. **Wreschner, D. U., Silverman, R. H., James, T. C., Gilbert, C. S., and Kerr, I. M.,** Affinity labeling and characterization of the 5'triphospho oligo ([2'-5']adenyl) adenosine-dependent endoRNase from different mammalian sources, *Eur. J. Biochem.,* 124, 261, 1982.

29. **Krowzynska, A., Yenofsky, R., and Brawerman, G.,** Regulation of messenger RNA stability in mouse erythroleukemia cells, *J. Mol. Biol.,* 181, 231, 1985.

30. **Lengyel, P.,** Biochemistry of interferons and their actions, *Annu. Rev. Biochem.,* 51, 251, 1982.

31. **Bertles, J. F. and Beck, W. S.,** Biochemical aspects of reticulocyte maturation. I. Fate of the ribonucleic acid, *J. Biol. Chem.,* 237, 3770, 1962.

32. **Valentine, W. N., Fink, K., Paglia, D. E., Harris, S. R., and Adams, W. S.,** Hereditary hemolytic anemia with human erythrocyte pyrimidine 5'-nucleotidase deficiency, *J. Clin. Invest.,* 54, 866, 1974.

33. **Paglia, D. E., Valentine, W. N,. Keitt, A. S., Brockway, R. A., and Nakatani, M.,** Pyrimidine nucleotidase deficiency with active dephosphorylation of dTMP: evidence for existence of thymidine nucleotidase in human erythrocytes, *Blood,* 62, 1147, 1983.

34. **Paglia, D. E., Valentine, W. N., and Brockway, R. A.,** Identification of thymidine nucleotidase and deoxyribonucleotidase activities among normal isoenzymes of 5'-nucleotidase in human erythrocytes, *Proc. Natl. Acad. Sci. U.S.A.,* 81, 588, 1984.
35. **Burka, E. R.,** Characteristics of RNA degradation in the erythroid cell, *J. Clin. Invest.,* 48, 1266, 1969.
36. **Lodish, H. F. and Small, B.,** Different lifetimes of reticulocyte messenger RNA, *Cell,* 7, 59, 1976.
37. **Littauer, U. Z. and Soreq, H.,** The regulatory function of poly(A) and adjacent 3' sequences in translated RNA, *Prog. Nucleic Acid Res. Mol. Biol.,* 27, 53, 1982.
38. **Nudel, U., Soreq, H., Littauer, U. Z., Marbaix, G., Huez, G., Leclercq, M., Hubert, E., and Chantrenne, H.,** Globin mRNA species containing poly (A) segments of different lengths, *Eur. J. Biochem.,* 64, 115, 1976.
39. **Maniatis, G. M., Ramirez, F., Cann, A., Marks, P. A., and Bank, A.,** Translation and stability of human globin mRNA in Xenopus oocytes, *J. Clin. Invest.,* 58, 1419, 1976.
40. **Nokin, P., Huez, G., Marbaix, G., Burny, A., and Chantrenne, H.,** Molecular modifications associated with ageing of globin messenger RNA in vivo, *Eur. J. Biochem.,* 62, 509, 1976.
41. **Favre, A., Morel, C., and Scherrer, K.,** The secondary structure and poly (A) content of globin messenger RNA as a pure RNA and in polyribosome-derived ribonucleoprotein complexes, *Eur. J. Biochem.,* 57, 147, 1975.
42. **Blobel, G.,** Protein tightly bound to globin mRNA, *Biochem. Biophys. Res. Commun.,* 47, 88, 1972.
43. **Blobel, G.,** A protein of molecular weight 78,000 bound to the polyadenylate region of eukaryotic messenger RNAs, *Proc. Natl. Acad. Sci. U.S.A.,* 70, 924, 1973.
44. **Standart, N., Vincent, A., and Scherrer, K.,** The polyribosomal poly (A) binding protein is highly conserved in vertetrate species. Comparison in duck, mouse and rabbit, *FEBS Lett.,* 135, 56, 1981.
45. **Smith, D. W. and Weinberg, W. C.,** Transfer RNA in reticulocyte maturation, *Biochim. Biophys. Acta,* 655, 195, 1981.
46. **Schmidt, A., Chernajovsky, Y., Shulman, L., Federman, P., Berissi, H., and Revel, M.,** An interferon-induced phosphodiesterase degrading (2'-5')oligoisoadenylate and the C-C-A terminus of tRNA, *Proc. Natl. Acad. Sci. U.S.A.,* 76, 4788, 1979.

Chapter 15

FINAL REMARKS

In this final chapter it seems appropriate to discuss some general problems related to the maturation of the reticulocyte. At the onset one may question the purpose of studies comparing reticulocytes with erythrocytes. These studies are of course justified by the interest of clinicians and physiologists in assessing the age of red blood cells in the peripheral blood. On that basis the status of erythropoiesis may be assessed. For the investigator of differentiation the reticulocyte is a valuable object of reference. Both of these purposes are clearly phenomenological. Does it therefore make sense to look for specific underlying mechanisms which govern the processes of maturation?

The answer, I hope, has emerged from the preceding chapters. A specific maturational program was discovered which is instrumental in the degradation of mitochondria and constitutes a cascade of interlocking steps as outlined in Chapter 12. On this basis one may ask whether one can discern some further principles which govern maturational events. Their discussion must be speculative by necessity, since there is a dearth of systematic studies. A likely possibility appears to be that ATP-dependent proteolysis may extend beyond the degradation of mitochondria, which must still be considered to be the major process. In favor of additional ATP-dependent proteolyses one may cite the demonstration that the degradation of hexokinase during maturation is catalyzed by the ATP-ubiquitin-dependent system,[1] and the indications that the breakdown of ribosomes is at least partly ATP-dependent.[2,3]

There is some evidence for another regularity, namely, that the enzymes exhibiting drastic declines in their activity are characterized by the presence of SH-groups which are essential for their structure or function. This feature of course is not highly discriminating since it applies to a large variety of enzymes. It is likely that the cooxidative property of lipoxygenase may be significant for the inactivation of the enzymes suceptible to oxidative attack. Such a mechanism would provide a functional and temporal coordination between the degradation of mitochondria on the one hand and alterations in the cytosol including the cytosolic side of the cell membrane.

Another principle may prevail for the maturational changes in the cell membrane, particularly concerning the loss of receptors and transporters. It involves the shedding of vesicles of specific composition as exemplified by the work on transferrin receptors.[4] Similar mechanisms may be postulated for the loss of other functions of the cell membrane. Such processes presumably contribute significantly to the characteristic restructuring of the membrane and the solidification of the membrane skeleton, which occurs during maturation of the reticulocyte (see Chapter 4). It is a remarkable circumstance that the shedding of the transferrin receptor as well as the loss in membrane transport, are, in an as yet unclarified manner, ATP-dependent.[4,5] One may therefore assume that ATP-dependent proteolysis and the events at the cell membrane are in some way connected.

At this point one may ask whether all differences between reticulocytes and erythrocytes are part of a general maturational program effected by specific mechanisms. Such an assumption would neglect the element of biological chance which might express itself in innate differences in the stability of proteins. Such variations have a phylogenetic structural basis and would result in different rates of decay in proteins, which would become obvious as their new formation ceased. It would be an erroneous conception to assume that all changes are essential and must have a biological meaning. The great variation in the enzymatic equipment among erythrocytes of various species argues strongly against such a point of view. One must therefore conclude that the reticulocyte contains many relics of metabolic pathways which have lost their function. Their fate is probably unessential for the maturation process. These considerations lead one to discriminate between essential and nonessential processes for the mechanisms of maturation; accordingly, it must be a goal of research to define and elucidate these essential processes.

Finally, one may consider the question of the relations between the maturation of reticulocytes and aging of erythrocytes. It is remarkable that many changes occurring during maturation continue during the further existence of the erythrocyte, including the decline of a variety of cytosolic enzymes and of some low molecular compounds, e.g., creatine and the polyamines, as well as dimensional changes. None of these alterations are by themselves responsible for the elimination of the aged erythrocyte. Therefore it may be suggested that they are part of a spectrum of changes that are determined by the innate properties of the ensemble of constituents in the erythrocyte under in vivo conditions. Based on this point of view, their study may help to determine the character of the events during maturation of the reticulocyte by retrograde extrapolation.

REFERENCES

1. **Magnani, M., Stocchi, V., Dacha, M., and Fornaini, G.,** Rabbit red blood cell hexokinase. Evidence for an ATP-dependent decay during cell maturation, *Mol. Cell. Biochem.,* 61, 83, 1984.
2. **Schweiger, H. C. and Rapoport, S.,** Der N-Stoffwechsel bei der Erythrozytenreifung: die N-bilanz unter endogenen Bedingungen, *Hoppe-Seyler's Z. Physiol. Chem.,* 313, 97, 1958.
3. **Park, E. A. and Morgan, H. E.,** Energy dependence of RNA degradation in rabbit reticulocytes, *Am. J. Physiol.,* 244, C390, 1984.
4. **Pan, B. T. and Johnstone, R. M.,** Fate of the transferrin receptor during maturation of sheep reticulocytes in vitro; selective externalization of the receptor, *Cell,* 33, 967, 1983.
5. **Weigensberg, A. M. and Blostein, R.,** Energy depletion retards the loss of membrane transport during reticulocyte maturation, *Proc. Natl. Acad. Sci. U.S.A.,* 80, 4978, 1983.

Index

INDEX

A

B